Springer **M**onographs in **M**athematics

Springer

New York
Berlin
Heidelberg
Hong Kong
London
Milan
Paris
Tokyo

Alfred Auslender   Marc Teboulle

Asymptotic Cones and Functions in Optimization and Variational Inequalities

 Springer

Alfred Auslender
UFR de Mathématiques
Université Lyon I
Villeurbanne 69622
France
auslender@wanadoo.fr

Marc Teboulle
School of Mathematical Sciences
Tel Aviv University
Tel Aviv 69978
Israel
teboulle@post.tau.ac.il

Mathematics Subject Classification (2000): 49-02, 49J40, 90C30

Library of Congress Cataloging-in-Publication Data
Auslender, A. (Alfred)
    Asymptotic cones and functions in optimization and variational inequalities / Alfred
Auslender, Marc Teboulle.
      p. cm. — (Springer monographs in mathematics)
    Includes bibliographical references and index.
    ISBN 0-387-95520-8 (alk. paper)
      1. Convex functions. 2. Convex programming. 3. Mathematical optimization.
    4. Variational inequalities (Mathematics) I. Teboulle, M. II. Title. III. Series.
    QA331.5 .A88 2002
    515'.8—dc21   2002070798

ISBN 0-387-95520-8   Printed on acid-free paper.

Printed in the United States of America.

9 8 7 6 5 4 3 2 1   SPIN 10881929

www.springer-ny.com

Photocomposed copy prepared from the authors' LaTeX files.

Springer-Verlag   New York Berlin Heidelberg
A member of BertelsmannSpringer Science+Business Media GmbH

This book is dedicated to

Martine, François, and Jérôme
Rachel, Yoav, Yael, and Keren

Preface

Nonlinear applied analysis and in particular the related fields of continuous optimization and variational inequality problems have gone through major developments over the last three decades and have reached maturity. A pivotal role in these developments has been played by convex analysis, a rich area covering a broad range of problems in mathematical sciences and its applications. Separation of convex sets and the Legendre–Fenchel conjugate transforms are fundamental notions that have laid the ground for these fruitful developments. Two other fundamental notions that have contributed to making convex analysis a powerful analytical tool and that have often been hidden in these developments are the notions of asymptotic sets and functions.

The purpose of this book is to provide a systematic and comprehensive account of asymptotic sets and functions, from which a broad and useful theory emerges in the areas of optimization and variational inequalities. There is a variety of motivations that led mathematicians to study questions revolving around attainment of the infimum in a minimization problem and its stability, duality and minmax theorems, convexification of sets and functions, and maximal monotone maps. In all these topics we are faced with the central problem of handling unbounded situations. This is particularly true when standard compactness hypotheses are not present. The appropriate concepts and tools needed to study such kinds of problems are vital not only in theory but also within the development of numerical methods. For the latter, we need not only to prove that a sequence generated by a given algorithm is well defined, namely an existence result, but also to establish that the produced sequence remains bounded. One can seldom directly apply theorems of classical analysis to answer to such questions. The notions of asymptotic cones and associated asymptotic functions provide a natural and unifying framework to resolve these types of problems. These notions have been used mostly and traditionally in convex analysis, with many results scattered in the literature. Yet these concepts also have a prominent and independent role to play in both convex and nonconvex analysis. This book presents the material reflecting this last point with many parts, including new results and covering convex and nonconvex problems. In particular, our aim is to demonstrate not only the interplay between classical convex-analytic results and the asymptotic machinery, but also the wide potential of the latter in analyzing variational problems.

We expect that this book will be useful to graduate students at an advanced level as well as to researchers and practitioners in the fields of optimization theory, nonlinear programming, and applied mathematical sciences. We decided to use a style with detailed and often transparent proofs. This might sometimes bore the more advanced reader, but should at least make the reading of the book easier and hopefully even enjoyable. The material is presented within the finite-dimensional setting. Our motivation for this choice was to eliminate the obvious complications that would have emerged within a more general topological setting and would have obscured the stream of the main ideas and results. For the more advanced reader, it is noteworthy to realize that most of the notions and properties developed here can be easily extended to reflexive Banach Spaces, assuming a supplementary condition with respect to weak convergence. The extension to more general arbitrary topological spaces is certainly not obvious, but the finite-dimensional setting is rich enough to motivate the interested reader toward the development of corresponding results needed in areas such as partial differential equations and probability analysis.

Structure of the Book

In Chapter 1 we recall the basic mathematical background: elementary convex analysis and set-valued maps. The results are presented without proofs. This material is classical and can be skipped by anyone who has had a standard course in convex analysis. None of this chapter's results rely on any asymptotic notions. Chapter 2 is the heart of the book and gives the fundamental results on asymptotic cones and functions. The interplay between geometry and analysis is emphasized and will be followed consistently in the remaining chapters. Building on the concept of asymptotic cone of the epigraph of a function, the notion of asymptotic function emerges, and calculus at infinity can be developed. The role of asymptotic functions in formulating general optimization problems is described. Chapter 3 studies the existence of optimal solutions for general optimization problems and related stability results, and also demonstrates the power of the asymptotic results developed in Chapter 2. Standard results under coercivity and weak coercivity assumptions imply that the solution set is a nonempty compact set and the sum of a compact set with a linear space, respectively. Here we develop many new properties for the noncoercive and weakly coercive cases through the use of asymptotic sets to derive more general existence results with applications leading to some new theorems "à la Helly" and for the convex feasibility problems. In Chapter 4 we study the subject of minimizing stationary sequences and error bounds. Both topics are central in the study of numerical methods. The concept of well-behaved asymptotic functions and the properties of such functions, which in turn is linked to the problems of error bounds associated with a given subset of a Euclidean space, are introduced. A general framework is developed around these two themes to characterize asymptotic optimality and error bounds for convex inequality systems. Duality theory plays a fundamental role in optimization and is developed in Chapter 5. The abstract perturbational scheme, valid for any optimization problem, is the starting point of the analysis. Under a minimal set of assumptions and thanks to asymptotic calculus, we derive key duality results, which are then applied to cover the classical Lagrange and Fenchel duality as well as minimax theorems, in a simple and unified way. Chapter 6 provides a self-contained introduction to maximal monotone maps and variational inequalities. Solving a convex optimization problem is reduced to solving a generalized equation associated with the subdifferential map. In many areas of applied mathematics, game theory, and equilibrium problems in economy, generalized equations arise and are described in terms of more general maps, in particular maximal monotone maps. The chapter covers the classical material together with some more recent results, streamlining the role of asymptotic functions.

Each chapter ends with some bibliographical notes and references. We did not attempt to give a complete bibliography on the covered topics, which is rather large, and we apologize in advance for any omission in the cited references. Yet, we have tried to cite all the sources that have been used in this book as well as some significant original historical developments, together with more recent references in the field that should help to guide researchers for further reading.

The book can be used as a complementary text to graduate courses in applied analysis and optimization theory. It can also serve as a text for a topics course at the graduate level, based, for example, on Chapters 2, 3, and 5, or as an introduction to variational inequality problems through Chapter 6, which is essentially self-contained.

Alfred Auslender   Lyon, France
Marc Teboulle   Tel-Aviv, Israel

Contents

1
Convex Analysis and Set-Valued Maps: A Review

Convexity plays a fundamental role in the analysis of optimization problems. In this very brief chapter we review some basic concepts on convex sets and convex functions. The material is presented quickly with no proofs and should be familiar to a reader who has had a standard convex analysis course. Some of the topics that might be less familiar and particularly relevant to this monograph will be given in greater details. Similarly, we review the basic definitions and properties of set-valued maps.

1.1 Convex Sets

Throughout this book we will consider only finite-dimensional vector spaces. The n-dimensional real Euclidean vector space will be denoted by $\mathbb{R}^n$. For vectors $x = (x_1, \ldots, x_n) \in \mathbb{R}^n, y \in \mathbb{R}^n$, the inner product between x and y is defined by $\langle x, y \rangle := \sum_{i=1}^{n} x_i y_i$. The norm arising from the inner product will be denoted by $\|x\| := \langle x, x \rangle^{1/2}$.

Definition 1.1.1 *A subset C of $\mathbb{R}^n$ is convex if*

$$tx + (1 - t)y \in C, \ \forall x, y \in C, \ \ \forall t \in [0, 1].$$

Using the algebraic sum of two convex sets,

$$\alpha C_1 + \beta C_2 := \{x \mid x = \alpha c_1 + \beta c_2, c_1 \in C_1, c_2 \in C_2\}, \ \forall \alpha, \beta \in \mathbb{R},$$

the above definition can also be written in the useful and compact notation

$$C \subset \mathbb{R}^n \text{ is convex} \iff tC + (1-t)C \subset C, \ \forall t \in [0,1].$$

The unit simplex of $\mathbb{R}^n$ is the convex set defined by

$$\Delta_n := \left\{ t \in \mathbb{R}^n \mid \sum_{i=1}^n t_i = 1, \ t_i \geq 0, \ i = 1, \dots, n \right\}.$$

Hyperplanes, Convex and Affine Hulls

A hyperplane is a set $H := \{ x \in \mathbb{R}^n \mid \langle x, a \rangle = \alpha \}$, where $a \in \mathbb{R}^n, a \neq 0$ and $\alpha \in \mathbb{R}$. A hyperplane divides the space $\mathbb{R}^n$ into two closed half-spaces $H^+ = \{ x \in \mathbb{R}^n \mid \langle x, a \rangle \geq \alpha \}$ and $H^- = \{ x \in \mathbb{R}^n \mid \langle x, a \rangle \leq \alpha \}$. Clearly, a hyperplane and its associated half-spaces are convex sets.

A set C is an affine manifold (or affine subspace) if

$$x, y \in C \iff tx + (1-t)y \in C, \ \forall t \in \mathbb{R}.$$

The space $\mathbb{R}^n$ itself, points, lines, and hyperplanes in $\mathbb{R}^n$ are affine manifolds. An affine manifold is closed and convex.

Proposition 1.1.1 *For a nonempty set $C \subset \mathbb{R}^n$ the following properties are equivalent:*
(a) C is an affine manifold;
(b) $C = x + M = \{ y \mid y - x \in M \}$, where M is a subspace called the subspace parallel to M;
(c) $C = \{ x \mid Ax = b \}$ for some $A \in \mathbb{R}^{m \times n}, b \in \mathbb{R}^m$.

For a set $C \subset \mathbb{R}^n$, the affine hull of C, denoted by $\operatorname{aff} C$ is the intersection of all affine manifolds containing C; the convex hull of C, denoted by $\operatorname{conv} C$, is the intersection of all convex sets containing C.

Proposition 1.1.2 *For a set $C \subset \mathbb{R}^n$, the following properties hold:*
(a) $\operatorname{conv} C$ is the set of all convex combinations of elements of C, i.e.,

$$\operatorname{conv} C = \left\{ \sum_{i=1}^m t_i x_i \mid x_i \in C, t_i \geq 0, \sum_{i=1}^m t_i = 1 \right\}.$$

(b) $\operatorname{aff} C$ is itself a linear manifold, and $\operatorname{conv} C \subset \operatorname{aff} C$.
(c) $\operatorname{aff} C = \operatorname{aff}(\operatorname{conv} C)$.

Some elementary operations that preserve the convexity of a set are collected in the following proposition.

Proposition 1.1.3 *For any collection $\{ C_i \mid i \in I \}$ of convex sets $C_i \subset \mathbb{R}^{n_i}$ we have:*
(a) $C_1 \times \cdots \times C_m$ is convex in $\mathbb{R}^{n_1} \times \dots \times \mathbb{R}^{n_m}$.
(b) $\cap_{i \in I} C_i$ is convex, here with $n_i = n$ for all i.

(c) The finite sum $\sum_{i=1}^{m} C_i$ is convex, with $n_i = n$ for all i.
(d) The image of a convex set under a linear mapping is convex.

A fundamental characterization of convex sets is provided by Carathéodory's theorem.

Theorem 1.1.1 *For any $C \subset \mathbb{R}^n$, any element of $\operatorname{conv} C$ can be represented as a convex combination of no more than $(n+1)$ elements of C.*

Topological Properties of Convex Sets

We now recall some basic topological concepts associated with convex sets. The closed unit ball in the n dimensional Euclidean space $\mathbb{R}^n$ will be denoted by:
$$\mathbb{B} = \{x \in \mathbb{R}^n \mid \|x\| \leq 1\}.$$
The ball with center x_0 and radius δ can thus be written as $\mathbb{B}(x_0, \delta) := x_0 + \delta\mathbb{B}$. Let $C \subset \mathbb{R}^n$ be a convex set. The interior and closure of C are also convex sets defined respectively by

$$
\begin{aligned}
\operatorname{int} C &:= \{x \in \mathbb{R}^n \mid \exists \varepsilon > 0 \text{ such that } x + \varepsilon\mathbb{B} \subset C\}, \\
\operatorname{cl} C &:= \bigcap_{\varepsilon > 0} (C + \varepsilon\mathbb{B}).
\end{aligned}
$$

The boundary of a set $C \subset \mathbb{R}^n$ is $\operatorname{bd} C := \operatorname{cl} C \setminus \operatorname{int} C$. A point z belongs to the boundary of C if and only if for any $\varepsilon > 0$ the ball $\mathbb{B}(z, \varepsilon)$ contains a point of C as well as a point that is not in C.

For a nonempty convex set of $\mathbb{R}^n$ with $\operatorname{int} C \neq \emptyset$ we have $\operatorname{aff} C = \mathbb{R}^n$. The interior of a convex set C relative to its affine hull is called the relative interior of C and is defined by

$$\operatorname{ri} C := \{x \in \operatorname{aff} C \mid \exists \varepsilon > 0 \text{ such that } (x + \varepsilon\mathbb{B}) \cap \operatorname{aff} C \subset C\}.$$

The relative interior clearly coincides with the interior when the affine hull is the whole space $\mathbb{R}^n$. However, while for a nonempty convex set C we may have $\operatorname{int} C = \emptyset$, in contrast, the relative interior $\operatorname{ri} C$ is not equal to $\emptyset$. This is an important property of the relative interior, which thus should be used in place of the interior when $\operatorname{int} C = \emptyset$.

The difference set $\operatorname{cl} C \setminus \operatorname{ri} C$ is called the relative boundary of C and is denoted by $\operatorname{rbd} C$.

The most important properties of the relative interior, including the line segment principle, are summarized in the following results.

Proposition 1.1.4 *For $C \subset \mathbb{R}^n$ nonempty and convex, the set $\operatorname{ri} C$ is nonempty and has the same affine hull as C. Moreover, when $x \in \operatorname{ri} C$, $y \in \operatorname{cl} C$, one has*
$$tx + (1-t)y \in \operatorname{ri} C, \ \forall t \in (0, 1],$$

and thus $\operatorname{ri} C$ *is convex. Furthermore,*

$$\operatorname{cl} C = \operatorname{cl}(\operatorname{ri} C), \quad \operatorname{ri} C = \operatorname{ri}(\operatorname{cl} C).$$

Proposition 1.1.5 *Let* $C \subset \mathbb{R}^n$ *be nonempty and convex. Then* $x \in \operatorname{ri} C$ *if and only if for every* $y \in C$ *there exists* $\varepsilon > 0$ *such that* $x + \varepsilon(y - x) \in C$.

A useful consequence of this characterization is the following.

Corollary 1.1.1 *Let* $C \subset \mathbb{R}^n$ *be convex. Then* $x \in \operatorname{int} C$ *if and only if for every* $d \in \mathbb{R}^n$ *there exists* $\varepsilon > 0$ *such that* $x + \varepsilon d \in C$.

Proposition 1.1.6 *For a nonempty convex set* $C \subset \mathbb{R}^n$ *one has*
(a) $\operatorname{ri} C \subset C \subset \operatorname{cl} C$,
(b) $\operatorname{cl}\operatorname{cl} C = \operatorname{cl} C;\ \operatorname{ri}(\operatorname{ri} C)) = \operatorname{ri} C$,
(c) $A(\operatorname{cl} C) \subset \operatorname{cl} A(C)$ *and* $\operatorname{ri} A(C) = A(\operatorname{ri} C)$, *where* $A : \mathbb{R}^n \to \mathbb{R}^m$ *is a linear mapping. Moreover, let* $A^{-1}(S) := \{x \in \mathbb{R}^n \mid A(x) \in S\}$ *be the inverse image of* A *for any set* $S \subset \mathbb{R}^n$. *Then, if* $A^{-1}(\operatorname{ri} C) \neq \emptyset$, *one has*

$$\operatorname{ri}(A^{-1}C) = A^{-1}(\operatorname{ri} C), \quad \operatorname{cl}(A^{-1}C) = A^{-1}(\operatorname{cl} C).$$

Proposition 1.1.7 *For a family of convex sets* $C_i \subset \mathbb{R}^n$ *indexed by* $i \in I$ *and such that* $\cap_{i \in I} \operatorname{ri} C_i \neq \emptyset$, *one has*

$$\operatorname{cl} \bigcap_{i \in I} C_i = \bigcap_{i \in I} cl C_i.$$

Furthermore, if I *is a finite index set, then*

$$\operatorname{ri} \bigcap_{i \in I} C_i = \bigcap_{i \in I} ri C_i.$$

As applications of these results one can obtain several other important rules involving relative interiors. The next one is frequently used.

Proposition 1.1.8 *For two convex sets* $C, D \subset \mathbb{R}^n$ *and for any scalars* $\alpha, \beta \in \mathbb{R}$, *one has*

$$\operatorname{ri}(\alpha C + \beta D) = \alpha \operatorname{ri} C + \beta \operatorname{ri} D.$$

Thus, in particular with $\alpha = -\beta = 1$, *we have*

$$0 \in \operatorname{ri}(C - D) \iff \operatorname{ri} C \cap \operatorname{ri} D \neq \emptyset.$$

Proposition 1.1.9 *Let* C *be a convex set in* $\mathbb{R}^{m+p}$. *For each* $y \in \mathbb{R}^m$, *let* $C_y := \{z \in \mathbb{R}^p \mid (y, z) \in C\}$ *and let* $D := \{y \mid C_y \neq \emptyset\}$. *Then*

$$(y, z) \in \operatorname{ri} C \iff y \in \operatorname{ri} D \text{ and } z \in \operatorname{ri} C_y.$$

Separation of Convex Sets

Two fundamental results based on properties of closed convex sets and their interiors are the supporting hyperplane theorem and the separation principle. A hyperplane $H := \{x \in \mathbb{R}^n \mid \langle x, a \rangle = \alpha\}$, with $a \in \mathbb{R}^n, a \neq 0, \alpha \in \mathbb{R}$, is said to separate two sets C_1, C_2 in $\mathbb{R}^n$ if C_1 is contained in the closed half-space $H^+ = \{x \in \mathbb{R}^n \mid \langle x, a \rangle \geq \alpha\}$, while C_2 is contained in the other. The separation is called proper if the hyperplane itself does not actually include both C_1 and C_2. When the sets C_1, C_2 lie in different half-spaces $\{x \in \mathbb{R}^n \mid \langle x, a \rangle \geq \alpha_2\}$, $\{x \in \mathbb{R}^n \mid \langle x, a \rangle \leq \alpha_1\}$, $\alpha_2 > \alpha_1$, then one says that one has strong separation between C_1 and C_2.

Proposition 1.1.10 *Let C_1 and C_2 be nonempty sets in $\mathbb{R}^n$. Then there exists a hyperplane properly separating C_1 and C_2 if and only if there exists a nonzero vector $a \in \mathbb{R}^n$ such that*

$$\inf\{\langle a, x \rangle \mid x \in C_1\} \geq \sup\{\langle a, x \rangle \mid x \in C_2\}$$

and

$$\sup\{\langle a, x \rangle \mid x \in C_1\} > \inf\{\langle a, x \rangle \mid x \in C_2\}.$$

The next two theorems provide the main conditions that guarantee separation between two convex sets.

Proposition 1.1.11 *(Proper separation)*
Let C_1 and C_2 be nonempty convex sets in $\mathbb{R}^n$. Then there exists a hyperplane that separates them properly if and only if $\operatorname{ri} C_1$ and $\operatorname{ri} C_2$ have no point in common.

Proposition 1.1.12 *(Strong separation)*
For two nonempty convex sets C_1 and C_2 in $\mathbb{R}^n$ such that $C_1 \cap C_2 = \emptyset$, with C_1 closed and C_2 compact, there exists a hyperplane that strongly separates them; i.e., there exist a vector $0 \neq a \in \mathbb{R}^n$ and a scalar $\alpha \in \mathbb{R}$ such that

$$\langle a, x_1 \rangle \leq \alpha < \langle a, x_2 \rangle \ \ \forall x_1 \in C_1, x_2 \in C_2.$$

Cones and Polyhedral Sets

Cones are fundamental geometric objects associated with sets. They play a key role in several aspects of mathematics, and will be used extensively throughout this book. Here we recall some elementary properties, well known results, and examples. Further properties and examples will be given in Section 2.7, dealing with real symmetric matrices and related semidefinite optimization problems.

Definition 1.1.2 *A set $K \subset \mathbb{R}^n$ is called a cone if $tx \in K$ for all $x \in K$ and for all $t \geq 0$.*

Examples of convex cones include linear subspaces of $\mathbb{R}^n$ and the nonnegative orthant $\mathbb{R}^n_+ := \{x \mid x_i \geq 0, \ i = 1, \ldots, n\}$. Other cones playing an important role in convex optimization problems are the cone of symmetric real positive semidefinite matrices of order n and the cone of Lorentz. More details are given in Chapter 2, Section 2.7.

Proposition 1.1.13 *For $K \subset \mathbb{R}^n$ the following are equivalent:*
(a) K is a convex cone.
(b) K is a cone such that $K + K \subset K$.

Definition 1.1.3 *A cone $K \subset \mathbb{R}^n$ is pointed if the equation $x_1 + \cdots + x_p = 0$ has no solution with $x_i \in K$ unless $x_i = 0$ for all i.*

Pointedness of convex cones can be checked via the following test. Let $K \subset \mathbb{R}^n$ be a convex cone. Then

$$K \text{ is pointed} \iff K \cap (-K) = \{0\}.$$

Given a cone $K \subset \mathbb{R}^n$, the polar of K is the cone defined by

$$K^* := \{y \in \mathbb{R}^n \mid \langle y, x \rangle \leq 0, \ \forall x \in K\}.$$

Orthogonality of subspaces is a special case of polarity of cones. If M is a subspace of $\mathbb{R}^n$, one has

$$M^* = M^\perp = \{y \in \mathbb{R}^n \mid \langle y, x \rangle = 0, \ \forall x \in M\}.$$

The bipolar is the cone $K^{**} := (K^*)^*$.

Proposition 1.1.14 *For a cone $K \subset \mathbb{R}^n$, the polar cone is closed and convex, and $K^{**} = \mathrm{cl}(\mathrm{conv}\, K)$. If K is also closed and convex, one then has $K^{**} = K$.*

Proposition 1.1.15 *Let $K \subset \mathbb{R}^n$ be a convex cone. Then*

$$\mathrm{int}\, K \neq \emptyset \iff K^* \text{ is pointed.}$$

In fact,
$$x \in \mathrm{int}\, K \iff \langle x, y \rangle < 0, \ \forall \ 0 \neq y \in K^*.$$

An important and useful object in variational problems is the normal cone to a given convex set.

Definition 1.1.4 *Let $C \subset \mathbb{R}^n$ be a nonempty convex set. The normal cone $N_C(\bar{x})$ to C at $\bar{x} \in C$ is defined by*

$$
\begin{aligned}
N_C(\bar{x}) \ &:= \ \{v \in \mathbb{R}^n \mid \langle v, x - \bar{x} \rangle \leq 0 \ \forall x \in C\} \\
&= \ \{v \in \mathbb{R}^n \mid \langle v, \bar{x} \rangle = \sup\{\langle v, x \rangle \mid x \in C\}\}.
\end{aligned}
$$

For all $x \notin C$ one has $N_C(x) := \emptyset$ and

$$x \in \operatorname{int} C \iff N_C(x) = \{0\}.$$

Moreover, $N_C(\bar{x})$ is pointed if and only if $\operatorname{int} C \neq \emptyset$.
Furthermore, for a closed set $C \subset \mathbb{R}^n$, a boundary point of the set C can be characterized through the normal cone of C. In fact, one has

$$x \in \operatorname{bd} C \iff \exists v \neq 0, \;\; v \in N_C(x).$$

Another geometrical object associated with a convex set $C \subset \mathbb{R}^n$, and closely related to the normal cone of C at $\bar{x}$, is the tangent cone of C at $\bar{x}$, given by

$$T_C(\bar{x}) = \operatorname{cl}\{d \in \mathbb{R}^n \mid \exists t > 0 \text{ with } \bar{x} + td \in C\}.$$

For any $\bar{x} \in C$, the cones $N_C(\bar{x})$ and $T_C(\bar{x})$ are polar to each other.
Some useful operations on polar cones are summarized below.

Proposition 1.1.16 *For cones K_i of $\mathbb{R}^n, i = 1, 2$, one has:*
(a) $K_1 \subset K_2 \Longrightarrow K_2^* \subset K_1^*$ and $K_1^{**} \subset K_2^{**}$.
(b) $K = K_1 + K_2 \Longrightarrow K^* = K_1^* \cap K_2^*$.
(c) $K = K_1 \cap K_2$ with K_i closed $\Longrightarrow K^* = \operatorname{cl}(K_1^* + K_2^*)$.
The closure operation can be removed if $0 \in \operatorname{int}(K_1 - K_2)$.
(d) For a family of cones $\{K_i \mid i \in I\}$ in $\mathbb{R}^n$,
$K = \cup_{i \in I} K_i \Longrightarrow K^* = \cap_{i \in I} K_i^*$.
(e) Let $A : \mathbb{R}^n \to \mathbb{R}^m$ be a linear mapping and K a closed convex cone of $\mathbb{R}^m$. Then $\{x \mid Ax \in K\}^* = \operatorname{cl}\{A^T y \mid y \in K\}$. The closure operation can be removed when $0 \in \operatorname{int}(K - \operatorname{rge} A)$, where $\operatorname{rge} A := \{Ax \mid x \in \mathbb{R}^n\}$.

A cone $K \subset \mathbb{R}^n$ is said to be finitely generated if it can be written as

$$K = \left\{ \sum_{i=1}^{p} t_p a_p \mid a_i \in \mathbb{R}^n, \; t_i \geq 0, i = 1, \ldots, p \right\}.$$

For a given nonempty set $C \subset \mathbb{R}^n$, the smallest cone containing the set C is called the positive hull (or conical hull) of C. It is the smallest cone containing the set C and is given by

$$\operatorname{pos} C := \{\lambda x \mid x \in C, \lambda > 0\} \cup \{0\}.$$

The positive hull, $\operatorname{pos} C$, is also said to be the cone generated by C.

Clearly C is itself a cone if and only if $\operatorname{pos} C = C$.
A set $P \subset \mathbb{R}^n$ is called polyhedral if it has the form

$$P = \{x \in \mathbb{R}^n \mid \langle a_i, x \rangle \leq b_i, \; i = 1, \ldots, p\},$$

where $a_i \in \mathbb{R}^n, b_i \in \mathbb{R}, \; i = 1, \ldots, p$. When $b_i = 0, \; \forall i$, then P is called a polyhedral cone.
The next result is a useful separation theorem involving the mixture of a polyhedral set with a convex set.

Proposition 1.1.17 *Let C_1 and C_2 be nonempty convex sets in $\mathbb{R}^n$, with C_1 being a polyhedral set. Then $C_1 \cap \mathrm{ri}\, C_2 \neq \emptyset$ if and only there exists a hyperplane separating C_1 and C_2 properly and not containing C_2.*

Theorem 1.1.2 *(Minkowski–Weyl Theorem)*
A cone K is polyhedral if and only if it is finitely generated.

An important implication of this theorem is the following decomposition formula expressing a polyhedral set.

Proposition 1.1.18 *A set P is polyhedral if and only if there exist a nonempty and finite collection of points $\{a_1, \ldots, a_p\}$ and a finitely generated cone K such that*

$$P = \left\{ x \mid x = y + \sum_{i=1}^{p} t_i a_i, \ y \in K, \ t \in \Delta_p \right\},$$

where Δ_p is the unit simplex of $\mathbb{R}^p$.

This kind of representation by convex hulls of minimal sets can be extended to general closed convex sets but requires the notion of extreme point and extreme ray.

For any two points x and y, the closed (open) line segment joining x and y defined for any $t \in [0,1]$ ($t \in (0,1)$) by $tx + (1-t)y$ is denoted by $[x,y]$ ($]x,y[$).
Let $C \subset \mathbb{R}^n$ be nonempty and convex. A nonempty subset $F \subset C$ is said to be a face of C if

$$\forall x, y \in C \ \text{ with } \ F \cap \,]x,y[\, \neq \emptyset \Longrightarrow [x,y] \subset F.$$

A point $z \in C$ is an extreme point of C if $\{z\}$ is a face, namely, z cannot be written in the form $z = \lambda x + (1-\lambda)y$ with $x, y \in C, x \neq y$, and $\lambda \in (0,1)$. The set of extreme points of C is denoted by $\mathrm{ext}\, C$. An extreme ray of C is the direction of a half-line that is a face of C. We denote by $\mathrm{extray}\, C$ the union of extreme rays of C.

Theorem 1.1.3 *(Krein–Milman) Let C be a nonempty closed convex set containing no lines. Then $C = \mathrm{conv}(\mathrm{ext}\, C \cup \mathrm{extray}\, C)$.*

When C is a compact set, one has $\mathrm{ext}\ C \neq \emptyset$, and the Krein–Milman theorem implies the following,

Theorem 1.1.4 *(Minkowski) Let C be a nonempty compact convex set in $\mathbb{R}^n$. Then $C = \mathrm{conv}\,\mathrm{ext}\, C$.*

1.2   Convex Functions

We denote by $\overline{\mathbb{R}} := [-\infty, +\infty]$ the whole extended real line. It is most convenient, in particular in the context of optimization problems, to work with extended real-valued functions, i.e., functions that take values in $\mathbb{R} \cup \{+\infty\} = (-\infty, +\infty]$, instead of just finite-valued functions, i.e., those taking values in $\mathbb{R} = (-\infty, +\infty)$. Rules of arithmetic are thus extended to include

$$\infty + \infty = \infty, \quad \alpha \cdot \infty = \infty, \ \forall \alpha \geq 0, \quad \inf \emptyset = \infty, \quad \sup \emptyset = -\infty.$$

Let $f : \mathbb{R}^n \to \overline{\mathbb{R}}$. The effective domain of f is the set

$$\operatorname{dom} f := \{x \in \mathbb{R}^n \,|\, f(x) < +\infty\}.$$

A function is called proper if $f(x) < \infty$ for at least one $x \in \mathbb{R}^n$ and $f(x) > -\infty$, $\forall x \in \mathbb{R}^n$ otherwise, the function is called improper. Two important and useful geometrical objects associated with a function f are the epigraph and level set of f, defined, respectively by

$$\begin{aligned}
\operatorname{epi} f &:= \{(x, \alpha) \in \mathbb{R}^n \times \mathbb{R} \,|\, \alpha \geq f(x)\}, \\
\operatorname{lev}(f, \alpha) &:= \{x \in \mathbb{R}^n \,|\, f(x) \leq \alpha\}.
\end{aligned}$$

The epigraph is thus a subset of $\mathbb{R}^{n+1}$ that consists of all points of $\mathbb{R}^{n+1}$ lying on or above the graph of f. From the above definitions one has

$$(x, \alpha) \in \operatorname{epi} f \iff x \in \operatorname{lev}(f, \alpha).$$

For $f : \mathbb{R}^n \to \overline{\mathbb{R}}$, we write

$$\begin{aligned}
\inf f &:= \inf\{f(x) \mid x \in \mathbb{R}^n\}, \\
\arg\min f &= \arg\min\{f(x) \mid x \in \mathbb{R}^n\} := \{x \in \mathbb{R}^n \mid f(x) = \inf f\}.
\end{aligned}$$

An optimization problem can thus be expressed equivalently in terms of its epigraph and level set as

$$\inf f = \inf\{\alpha \mid (x, \alpha) \in \operatorname{epi} f\},$$

while the set of optimal solutions is

$$\arg\min f = \operatorname{lev}(f, \inf f).$$

Some useful operations involving epigraphs and level sets are collected in the next proposition.

Proposition 1.2.1 *Let $f : \mathbb{R} \to \overline{\mathbb{R}}$ and let $L : (x, \alpha) \to x$ be the projection maps mapping from $\mathbb{R}^{n+1}$ to $\mathbb{R}^n$. The following hold:*
(a) $\operatorname{dom} f = L(\operatorname{epi} f)$.
(b) $\operatorname{lev}(f, \alpha) = L(\operatorname{epi} f \cap \{(x, \alpha) \, | x \in \mathbb{R}^n\})$.
(c) For a collection of extended real-valued functions $\{f_i \mid i \in I\}$, with I an arbitrary index sets one has

$$\operatorname{epi}\left(\inf_{i \in I} f_i\right) = \bigcup_{i \in I} \operatorname{epi} f_i, \quad \operatorname{epi}\left(\sup_{i \in I} f_i\right) = \bigcap_{i \in I} \operatorname{epi} f_i.$$

Recall that

$$\liminf_{x \to y} f(x) := \sup_{r > 0} \inf_{x \in \mathbb{B}(y, r)} f(x),$$

and lower limits are characterized via

$$\liminf_{x \to y} f(x) = \min\{\alpha \in \overline{\mathbb{R}} \mid \exists x_n \to y \text{ with } f(x_n) \to \alpha\}.$$

Note that one always has $\liminf_{x \to y} f(x) \le f(y)$.

Definition 1.2.1 *The function $f : \mathbb{R}^n \to \overline{\mathbb{R}}$ is lower semicontinuous (lsc) at x if*

$$f(x) = \liminf_{y \to x} f(y),$$

and lower semicontinuous on $\mathbb{R}^n$ if this holds for every $x \in \mathbb{R}^n$.

Theorem 1.2.1 *Let $f : \mathbb{R}^n \to \overline{\mathbb{R}}$. The following statements are equivalent:*
(a) f is lsc at x.
(b) $\liminf_{n \to \infty} f(x_n) \ge f(x)$, $\forall x_n \to x$.
(c) For each α such that $f(x) > \alpha$ $\exists \delta > 0$ such that $f(y) > \alpha$, $\forall y \in \mathbb{B}(x, \delta)$.

Lower semicontinuity on $\mathbb{R}^n$ of a function can be characterized through its level set and epigraph.

Theorem 1.2.2 *Let $f : \mathbb{R}^n \to \overline{\mathbb{R}}$. The following statements are equivalent:*
(a) f is lsc on $\mathbb{R}^n$.
(b) The epigraph $\operatorname{epi} f$ is closed on $\mathbb{R}^n \times \mathbb{R}$.
(c) The level sets $\operatorname{lev}(f, \alpha)$ are closed in $\mathbb{R}^n$.

When f is not lower semicontinuous, then its epigraph is not closed, but $\operatorname{cl}(\operatorname{epi} f)$ is closed and leads to the lower closure of f, denoted by $\operatorname{cl} f$, and defined such that $\operatorname{epi}(\operatorname{cl} f) = \operatorname{cl}(\operatorname{epi} f)$. In terms of f one then has

$$(\operatorname{cl} f)(x) := \liminf_{y \to x} f(y),$$

and it holds that $\operatorname{cl} f \le f$. This function is the greatest of all the lsc functions g such that $g \le f$.

Definition 1.2.2 *A function $f : \mathbb{R}^n \to \overline{\mathbb{R}}$ with $f \not\equiv \infty$ is called convex if epi f is a nonempty convex set.*

As a consequence, since the effective domain dom f is the projection of epi f, it follows that dom f is convex when f is convex.

Alternatively, for an extended real-valued function $f : \mathbb{R}^n \to \mathbb{R} \cup \{+\infty\}$, f is convex if and only if

$$f(tx + (1-t)y) \leq tf(x) + (1-t)f(y), \ \forall x, y \in \mathbb{R}^n, \ \forall t \in (0,1).$$

The function is called strictly convex if the above inequality is strict for all $x, y \in \mathbb{R}^n$ with $x \neq y$ and $t \in (0,1)$.

A function f is called concave whenever $-f$ is convex. Convexity of f can also be defined through Jensen's inequality, namely, f is convex if and only if

$$f \left(\sum_{i=1}^{p} t_i x_i \right) \leq \sum_{i=1}^{p} t_i f(x_i), \ \forall x_i \in \mathbb{R}^n, \ i = 1, \ldots, p, \ t \in \Delta_p.$$

If f is convex, then $\mathrm{lev}(f, \alpha)$ is a convex set for all $\alpha \in \mathbb{R}$ (the converse statement does not hold).

The indicator function of a set C of $\mathbb{R}^n$, denoted by δ_C, is defined by

$$\delta_C(x) = \begin{cases} 0 & \text{if } x \in C, \\ +\infty & \text{if } x \notin C. \end{cases}$$

If C is a nonempty closed convex subset of $\mathbb{R}^n$, then δ_C is a proper, lsc, and convex function.

The indicator function allows for recovering the definition of a convex real-valued function defined on a convex set C. Indeed, given g a real-valued function on a convex set C, then with $f := g + \delta_C$, Definition 1.2.2 becomes

$$g(tx + (1-t)y) \leq tg(x) + (1-t)g(y), \ \forall x, y \in C, \ \forall t \in [0,1].$$

The following inf-projection operation of a convex function is particularly useful.

Proposition 1.2.2 *Let $\Phi : \mathbb{R}^n \times \mathbb{R}^m \to \mathbb{R} \cup \{+\infty\}$ be convex. Then, $\varphi(u) = \inf_x \Phi(x, u)$ is convex on $\mathbb{R}^m$, and the set $S(u) = \arg\min_x \Phi(x, u)$ is convex for each u.*

For a proper convex function, closedness is the same as lower semicontinuty. Furthermore, one has the following result.

Proposition 1.2.3 *A convex function $f : \mathbb{R}^n \to [-\infty, +\infty]$ is lsc at a point x where it is finite if and only if $f(x) = (\mathrm{cl}\, f)(x)$, and in this case f is proper.*

Proposition 1.2.4 *Let $f : \mathbb{R}^n \to \mathbb{R} \cup \{+\infty\}$ be an improper convex function. Then $f(x) = -\infty$ for every $x \in \mathrm{ri}(\mathrm{dom}\, f)$.*

Proposition 1.2.5 *Let* $f : \mathbb{R}^n \to \mathbb{R} \cup \{+\infty\}$ *be a proper convex function. Then* $\operatorname{cl} f$ *is a lsc proper convex function, and* $\operatorname{dom}(\operatorname{cl} f)$ *and* $\operatorname{dom} f$ *have the same closure and relative interior, as well as the same dimension. Furthermore,* $f = \operatorname{cl} f$ *on* $\operatorname{ri} \operatorname{dom} f$.

It is thus important to remark that a convex function that is lsc at some point and takes the value $-\infty$ cannot take values other than $+\infty$ and $-\infty$.

Proposition 1.2.6 *Let* $f : \mathbb{R}^n \to \mathbb{R} \cup \{+\infty\}$ *be a proper convex function. Then*

$$(\operatorname{cl} f)(x) = \lim_{t \to 0^+} f(x + t(y - x)), \ \forall x \in \mathbb{R}^n, \ \forall y \in \operatorname{ri}(\operatorname{dom} f).$$

If in addition f *is assumed lsc, then*

$$f(x) = \lim_{t \to 0^+} f(x + t(y - x)), \ \forall x, \ \forall y \in \operatorname{dom} f.$$

Proposition 1.2.7 *Let* $f_i : \mathbb{R}^n \to \mathbb{R} \cup \{+\infty\}$, $i = 1, \ldots, p$, *be a collection of proper convex functions. If* $\cap_{i=1}^p \operatorname{ri} \operatorname{dom} f_i \neq \emptyset$, *then*

$$\operatorname{cl}(f_1 + \cdots + f_p) = \operatorname{cl} f_1 + \cdots + \operatorname{cl} f_p.$$

Proposition 1.2.8 *Let* $f : \mathbb{R}^n \to \mathbb{R} \cup \{+\infty\}$ *be a proper convex function with level sets* $\operatorname{lev}(f, \alpha)$. *Then for any* $\alpha \in \mathbb{R}$ *with* $\alpha > \inf f$, *one has*

$$\operatorname{ri}(\operatorname{lev}(f, \alpha)) = \{x \in \operatorname{ri}(\operatorname{dom} f) \mid f(x) < \alpha\}.$$

Theorem 1.2.3 *Let* $f : \mathbb{R}^n \to \mathbb{R} \cup \{+\infty\}$ *be a convex function. Then* f *is relatively continuous on any relatively open convex set* C *contained in* $\operatorname{dom} f$, *in particular on* $\operatorname{ri}(\operatorname{dom} f)$.

Recall that a function $f : \mathbb{R}^n \to \mathbb{R} \cup \{+\infty\}$ is called Lipschitz with constant L on a subset C of $\operatorname{dom} f$ if there exists $L \geq 0$ such that

$$|f(x) - f(y)| \leq L\|x - y\|, \ \ \forall x, y \in C,$$

and f is called locally Lipschitz on C if for every $c \in C$ there exists a neighborhood V of c such that f is Lipschitz on $V \cap C$.

Theorem 1.2.4 *Let* $f : \mathbb{R}^n \to \mathbb{R} \cup \{+\infty\}$ *be a proper convex function and let* C *be any compact subset of* $\operatorname{ri}(\operatorname{dom} f)$. *Then* f *is Lipschitz relative to* C.

An important and useful class of convex functions is the class of polyhedral convex functions.

Definition 1.2.3 *A polyhedral convex function is a convex function whose epigraph is polyhedral, which holds, if and only if it is finitely generated. Such a function, if proper, is lsc.*

For a nonconvex function f there is a natural procedure to convexify it via the use of the convex hull of the epigraph of f.

Definition 1.2.4 *For a function $f : \mathbb{R}^n \to \mathbb{R} \cup \{+\infty\}$, the convex hull of f is denoted by* conv f *and is defined for each $x \in \mathbb{R}^n$ by*

$$
\begin{aligned}
\text{conv } f(x) \quad &:= \quad \inf \{r \in \mathbb{R} \mid (x, r) \in \text{conv(epi } f)\} \\
&= \quad \inf \left\{ \sum_{i=1}^{n+1} t_i f(x_i) \mid \sum_{i=1}^{n+1} t_i x_i = x, \ x_i \in \text{dom } f, \ t \in \Delta_{n+1} \right\}.
\end{aligned}
$$

Equivalently, it is the greatest convex function majorized by f.

Similarly, for a collection of functions $\{f_i \mid i \in I\}$ on $\mathbb{R}^n$, where I is an arbitrary index set, the convex hull is denoted by conv$\{f_i \mid i \in I\}$ and is the greatest convex function h such that $h(x) \leq f_i(x)$ for every $x \in \mathbb{R}^n$ and every $i \in I$. Whenever the index set $I := \{1, \ldots, p\}$ is finite, the convex hull h of the finite family f_i is then given by

$$
\begin{aligned}
h(x) \quad &:= \quad \inf \{r \in \mathbb{R} \mid (x, r) \in \text{conv}(\cup_{i=1}^{p} \text{ epi } f_i)\} \\
&= \quad \inf \left\{ \sum_{i=1}^{p} t_i f_i(x_i) \mid \sum_{i=1}^{p} t_i x_i = x, \ x_i \in \text{dom } f_i, \ t \in \Delta_p \right\}.
\end{aligned}
$$

Proposition 1.2.9 *Let $f_i : \mathbb{R}^n \to \mathbb{R} \cup \{+\infty\}$, $i \in I$, be an arbitrary collection of proper convex functions. Then for any $x \in \mathbb{R}^n$,*

$$
\text{conv}\{f_i \mid i \in I\}(x) = \inf \left\{ \sum_{i \in I} t_i f_i(x) \mid \sum_{i \in I} t_i x_i = x \right\},
$$

where the infimum is taken over all expressions of x as a convex combination in which at most $(n+1)$ of the t_i are positive.

Legendre–Fenchel Transform and Conjugate Functions

Duality plays a fundamental role in optimization problems, convex and nonconvex ones. A key player in any duality framework is the Legendre–Fenchel transform, also called the conjugate of a given function.

Definition 1.2.5 *For any function $f : \mathbb{R}^n \to \mathbb{R} \cup \{+\infty\}$, the function $f^* : \mathbb{R}^n \to \mathbb{R} \cup \{+\infty\}$ defined by*

$$
f^*(y) := \sup_{x} \{\langle x, y \rangle - f(x)\}
$$

is called the conjugate to f. The biconjugate of f is defined by

$$
f^{**}(x) := \sup_{y} \{\langle x, y \rangle - f^*(y)\}.
$$

Whenever conv f is proper, one always has that both f^* and f^{**} are proper, lsc, and convex, and the following relations hold:

$$f^{**} = \text{cl}(\text{conv } f) \quad \text{and} \quad f^{**} \leq \text{cl } f.$$

Theorem 1.2.5 *(Fenchel–Moreau) Let $f : \mathbb{R}^n \to \mathbb{R} \cup \{+\infty\}$ be convex. The conjugate function f^* is proper, lsc, and convex if and only if f is proper. Moreover, $(\text{cl } f)^* = f^*$ and $f^{**} = \text{cl } f$.*

From the definition of the conjugate function, we immediately obtain Fenchel's inequality:

$$\langle x, y \rangle \leq f(x) + f^*(y), \quad \forall x \in \text{dom } f, \ y \in \text{dom } f^*.$$

Definition 1.2.6 *Let $f, g : \mathbb{R}^n \to \mathbb{R} \cup \{+\infty\}$ be proper functions. The infimal convolution of the function f with g is the function h defined by*

$$h(x) := (f \Box g)(x) = \inf\{f(x_1) + g(x_2) \mid x_1 + x_2 = x\}, \ \forall x \in \mathbb{R}^n.$$

The above definition can be extended as well for a finite collection $\{f_i \mid i = 1, \ldots, p\}$ of proper functions, i.e.,

$$(f_1 \Box \cdots \Box f_p)(x) := \inf\{f_1(x_1) + \cdots + f_p(x_p) \mid x_1 + \cdots x_p = x\}.$$

The next proposition gives some important results relating conjugates and their infimal convolutions.

Proposition 1.2.10 *Let $f_i : \mathbb{R}^n \to \mathbb{R} \cup \{+\infty\}$, $i = 1, \ldots, p$, be a collection of proper and convex functions. Then:*
(a) $(f_1 \Box \cdots \Box f_p)^* = f_1^* + \cdots + f_p^*$,
(b) $(\text{cl } f_1 + \cdots + \text{cl } f_p)^* = \text{cl}(f_1^* \Box \cdots \Box f_p^*)$.

Polyhedrality is preserved under the conjugacy operation.

Proposition 1.2.11 *Let f, g be proper polyhedral convex. Then:*
(a) $f + g$ is polyhedral.
(c) The conjugate of f is polyhedral.

Proposition 1.2.12 *Let $f_i : \mathbb{R}^n \to \mathbb{R} \cup \{+\infty\}$, $i \in I$, be an arbitrary collection of functions with conjugates f_i^*. Then:*
(a) $(\inf\{f_i \mid i \in I\})^* = \sup\{f_i^* \mid i \in I\}$.
(b) $\text{conv}\{f_i \mid i \in I\}^* = \sup\{f_i^* \mid i \in I\}$, and $\sup\{\text{cl} f_i \mid i \in I\})^* = \text{cl}(\text{conv}\{f_i^* \mid i \in I\})$, whenever the functions f_i are proper convex.

Differentiability and Subdifferentiability of Convex Functions
Under differentiability assumptions, one can check the convexity of a function via the following useful tests. For a function f defined on $\mathbb{R}^n$ and sufficiently smooth, the gradient of f at x, denoted by $\nabla f(x)$, is a vector in $\mathbb{R}^n$ whose i-th component is the partial derivative of f with respect to x_i, while the Hessian of f at x, when f is twice differentiable, is a symmetric $n \times n$ matrix, whose (i,j) element is the second-order partial derivative $\partial^2 f(x)/\partial x_i \partial x_j$, and is denoted by $\nabla^2 f(x)$.

Theorem 1.2.6 *Let f be a differentiable function on an open convex subset S of $\mathbb{R}^n$. Each of the following conditions is necessary and sufficient for f to be convex on S:*
(a) $f(x) - f(y) \geq \langle x - y, \nabla f(y) \rangle$, $\forall x, y \in S$.
(b) $\langle \nabla f(x) - \nabla f(y), x - y \rangle \geq 0$, $\forall x, y \in S$.
(c) $\nabla^2 f(x)$ is positive semidefinite for all $x \in S$ whenever f is twice differentiable on S.

For nondifferentiable functions, convexity plays a fundamental role that allows one to define the concept of subgradients and subdifferential maps. A subgradient of an extended function of $\mathbb{R}^n$ at a point x with $f(x)$ finite is any vector $g \in \mathbb{R}^n$ satisfying

$$f(y) \geq f(x) + \langle g, y - x \rangle, \ \forall y.$$

The set of all subgradients of f at $x \in \operatorname{dom} f$ is called the subdifferential of the function f at x and is denoted by $\partial f(x)$; that is,

$$\partial f(x) := \{ g \mid f(y) \geq f(x) + \langle g, y - x \rangle, \ \forall y \in \mathbb{R}^n \}.$$

When $x \notin \operatorname{dom} f$ we set $\partial f(x) = \emptyset$. The subdifferential of f at x can be viewed as a set-valued map ∂f that assigns to each point $x \in \mathbb{R}^n$ a certain subset $\partial f(x)$ of $\mathbb{R}^n$, see Section 1.4 below for more on set-valued maps. In general, $\partial f(x)$ may be empty. When $\partial f(x) \neq \emptyset$, the function f is then called subdifferentiable at x. A remarkable and evident property is that $\partial f(x)$ is a closed convex set. Indeed, from its definition, the subdifferential of f at x is nothing but an infinite intersection of closed half-spaces.
Let $f : \mathbb{R}^n \to \mathbb{R} \cup \{+\infty\}$ be a convex function, and let x be any point where f is finite and $d \in \mathbb{R}^n$. Then the function

$$\tau \to \frac{f(x + \tau d) - f(x)}{\tau}$$

is nondecreasing in $\tau > 0$. As a consequence, the limit

$$f'(x; d) := \lim_{\tau \to 0^+} \frac{f(x + \tau d) - f(x)}{\tau}$$

exists (finite or infinite) for all $d \in \mathbb{R}^n$ and is called the directional derivative of f at x.

Proposition 1.2.13 *Let* $f : \mathbb{R}^n \to \mathbb{R} \cup \{+\infty\}$ *be convex and let* $x \in \operatorname{dom} f$. *Then* g *is a subgradient of* f *at* x *if and only if* $f'(x; d) \geq \langle g, d \rangle,\ \forall d \in \mathbb{R}^n$.

Proposition 1.2.14 *Let* $f : \mathbb{R}^n \to \mathbb{R} \cup \{+\infty\}$ *be a proper convex function and let* $x \in \operatorname{dom} f$ *be such that there exists a nonzero vector of the form* $d := y - x$ *with* $y \in \operatorname{ri}(\operatorname{dom} f)$ *and such that* $f'(x; d) > -\infty$. *Then* $\partial f(x) \neq \emptyset$. *In particular, for any* $x \in \operatorname{ri}(\operatorname{dom} f)$ *one has* $\partial f(x) \neq \emptyset$.

Given subgradients of f one might identify the directional derivative. In fact, the directional derivative and the subdifferential of f are related by the following fundamental max-formula.

Proposition 1.2.15 *Let* $f : \mathbb{R}^n \to \mathbb{R} \cup \{+\infty\}$ *be a proper convex function. Then for any* $x \in \operatorname{ri}(\operatorname{dom} f)$ *and any direction* $d \in \mathbb{R}^n$ *one has* $d \to f'(x; d)$ *is lsc and proper as a function of* d, *and*

$$f'(x; d) = \max\{\langle g, d \rangle \mid g \in \partial f(x)\}.$$

We have already observed that $\partial f(x)$ is a closed convex set. The next result provides a simple criterion to determine when this set is nonempty and bounded, and hence compact.

Proposition 1.2.16 *Let* $f : \mathbb{R}^n \to \mathbb{R} \cup \{+\infty\}$ *be a proper convex function. Then* $\partial f(x)$ *is nonempty and bounded if and only if* $x \in \operatorname{int} \operatorname{dom} f$.

Let $\operatorname{dom} \partial f := \{x \mid \partial f(x) \neq \emptyset\}$ denote the domain of ∂f. This set is not necessarily convex. However, from the above results it follows that for any proper convex function $f : \mathbb{R}^n \to \mathbb{R} \cup \{+\infty\}$, one has the relation

$$\operatorname{ri}(\operatorname{dom} f) \subset \operatorname{dom} \partial f \subset \operatorname{dom} f. \tag{1.1}$$

A proper convex function f is differentiable at $x \in \operatorname{dom} f$ if and only if the subdifferential reduces to a singleton, which is the gradient of f, i.e., one has $\partial f(x) = \{\nabla f(x)\}$. In that case, one has $f'(x; d) = \langle \nabla f(x), d \rangle,\ \forall d \in \mathbb{R}^n$.

There exists an interesting and fundamental interplay between conjugate functions and subdifferentials of convex functions that is quite at the heart of many significant results in convex optimization problems and that will be used all along in this monograph. We begin with a simple but most significant result played by subgradients in convex optimization, which follows immediately from the definition of the subdifferential.

Proposition 1.2.17 *Let* $f : \mathbb{R}^n \to \mathbb{R} \cup \{+\infty\}$ *be a proper convex function. A point* $x \in \mathbb{R}^n$ *is a global minimizer of* f *if and only if* $0 \in \partial f(x)$.

Using the definitions of the subdifferential and of the conjugate function, it is possible to characterize subdifferentiabililty in the following useful and important way.

Proposition 1.2.18 *Let* $f : \mathbb{R}^n \to \mathbb{R} \cup \{+\infty\}$ *be a proper function. Then*

$$f(x) + f^*(y) \geq \langle x, y \rangle, \ \forall x, y,$$

and equality holds if and only if $y \in \partial f(x)$, *or equivalently, if and only if* $x \in \partial f^*(y)$ *whenever* f *is in addition lsc and convex.*

Proposition 1.2.19 *Let* $f : \mathbb{R}^n \to \mathbb{R} \cup \{+\infty\}$ *and* $g : \mathbb{R}^n \to \mathbb{R} \cup \{+\infty\}$ *be proper convex functions on* $\mathbb{R}^n$. *Then for any* $x \in \mathbb{R}^n$,

$$\partial(f + g)(x) \supset \partial f(x) + \partial g(x).$$

Proposition 1.2.20 *Let* $f : \mathbb{R}^n \to \overline{\mathbb{R}}$, *and* x *with* $f(x)$ *finite. Then:*
(a) $\partial f(x) \neq \emptyset \implies f(x) = f^{**}(x)$.
(b) $f(x) = f^{**}(x) \implies \partial f(x) = \partial f^{**}(x)$.
(c) $y \in \partial f(x) \implies x \in \partial f^*(y)$.

Let $C \subset \mathbb{R}^n$ be a convex set with normal cone N_C as given in Definition 1.1.4. Then, using the subgradient inequality, one immediately sees that the normal cone reduces to the subdifferential of the indicator function of the set C; i.e.,

$$N_C(x) = \partial \delta_C(x), \ \forall x \in \mathbb{R}^n. \tag{1.2}$$

A useful representation for the normal cone of a polyhedral set is given by the following proposition.

Proposition 1.2.21 *Let* $(a_i, \alpha_i) \in \mathbb{R}^n \times \mathbb{R}$ *for* $i = 1, \ldots, p$, *with* $a_i \neq 0$, *and* $C_i = \{x \in \mathbb{R}^n \mid \langle a_i, x \rangle \leq \alpha_i\}$, $C = \cap_{i=1}^p C_i$. *Then for any* $x \in C$ *one has* $N_C(x) = \sum_{i=1}^p N_{C_i}(x)$, *with*

$$N_{C_i}(x) = \begin{cases} \mathbb{R}_+ a_i & \text{if } \langle a_i, x \rangle = \alpha_i, \\ \{0\} & \text{otherwise.} \end{cases}$$

The normal cone of the level set of a convex function can be written explicitly:

Proposition 1.2.22 *Let* $f : \mathbb{R}^n \to \mathbb{R} \cup \{+\infty\}$ *be a proper convex function and let* $x \in \mathrm{ri}(\mathrm{dom}\, f)$ *be such that* $f(x)$ *is not the minimum. Then for each* $\alpha \in \mathbb{R}$ *one has*

$$N_{\mathrm{lev}(f,\alpha)}(x) = \begin{cases} \mathbb{R}_+ \partial f(x) & \text{if } f(x) = \alpha, \\ \{0\} & \text{if } f(x) < \alpha. \end{cases}$$

1.3   Support Functions

Support functions play an important role in convex analysis and optimization. In fact they allow a characterization of closed convex sets and of the position of a point relative to a set with powerful analytical tools. This permits one to translate geometrical questions on convex sets into questions in terms of convex functions, which often facilitate the way of handling these sets.

Definition 1.3.1 *Given a nonempty set C of $\mathbb{R}^n$, the function $\sigma_C : \mathbb{R}^n \to \mathbb{R} \cup \{+\infty\}$ defined by*

$$\sigma_C(d) := \sup\{\langle x, d \rangle \mid x \in C\}$$

is called the support function of C. The domain of σ_C defined by

$$\operatorname{dom} \sigma_C := \{x \in \mathbb{R}^n \mid \sigma_C(x) < +\infty\}$$

is called the barrier cone of C and is denoted by $b(C)$.

One thus has

$$\sigma_C(d) \leq \alpha \iff C \subset \{x \mid \langle x, d \rangle \leq \alpha\},$$

characterizing the closed half-space that contains C.

Definition 1.3.2 *A function $\pi : \mathbb{R}^n \to \mathbb{R} \cup \{+\infty\}$ is positively homogeneous if $0 \in \operatorname{dom} \pi$ and $\pi(\lambda x) = \lambda \pi(x)$ for all x and all $\lambda > 0$. It is sublinear if in addition,*

$$\pi(x + y) \leq \pi(x) + \pi(y), \ \forall x, y \in \mathbb{R}^n.$$

Examples of positively homogeneous functions that are also sublinear include norms and linear functions. The indicator function δ_C of a set C is positively homogeneous if and only if C is a cone.

Proposition 1.3.1 *A function $\pi : \mathbb{R}^n \to \mathbb{R} \cup \{+\infty\}$ is convex and positively homogeneous if and only if epi π is a nonempty convex cone in $\mathbb{R}^n \times \mathbb{R}$.*

From the definition of a support function we immediately obtain the following basic properties.

Proposition 1.3.2 *For any nonempty set C of $\mathbb{R}^n$, the support function σ_C satisfies the following:*
(a) σ_C is an lsc convex positively homogeneous function.
(b) The barrier cone $\operatorname{dom} \sigma_C$ is a convex cone (not necessarily closed).
(c) $\sigma_C = \sigma_{\operatorname{cl} C} = \sigma_{\operatorname{conv} C} = \sigma_{\operatorname{cl conv} C}$.
(d) $\sigma_C < +\infty$ if and only if C is bounded.
(e) If C is convex, then $\sigma_C = \sigma_{\operatorname{ri} C}$.

Example 1.3.1 *(Examples of particular support functions)*
(i) $\sigma_{\mathbb{R}^n} = \delta_{\{0\}}$, $\sigma_\emptyset = -\infty$, $\sigma_{\{0\}} \equiv 0$.
(ii) $\sigma_{\operatorname{conv}\{a_1, \ldots, a_p\}}(d) = \max_{1 \leq i \leq p} \langle d, a_i \rangle$.
(iii) $\max_{1 \leq j \leq m} d_j = \sup \left\{ \langle d, x \rangle \mid x \geq 0, \ \sum_{j=1}^m x_j = 1 \right\}$.
(iv) For any cone $K \subset \mathbb{R}^n$, $\sigma_K = \delta_{K^*}$.

The support functions possess a structure allowing the development of calculus rules. We list below some important formulas for support functions and barrier cones associated with arbitrary (not necessarily convex) nonempty sets of $\mathbb{R}^n$.

Proposition 1.3.3 *Let C, D be nonempty sets of $\mathbb{R}^n$.*
(a) $C \subset D \implies \operatorname{dom} \sigma_D \subset \operatorname{dom} \sigma_C$ and $\sigma_C(d) \leq \sigma_D(d)$, $\forall d \in \mathbb{R}^n$.
(b) $\sigma_{C+D}(d) = \sigma_C(d) + \sigma_D(d)$, and $\operatorname{dom} \sigma_{C+D} = \operatorname{dom} \sigma_C \cap \operatorname{dom} \sigma_D$.
If D is a cone, then the formula above reduces to

$$\sigma_{C+D}(d) = \begin{cases} \sigma_C(d) & \text{if } d \in D^*, \\ +\infty & \text{otherwise,} \end{cases}$$

and $\operatorname{dom} \sigma_{C+D} = \operatorname{dom} \sigma_C \cap D^*$.
(c) For any point $y \in \mathbb{R}^n$, $\sigma_{C+y}(d) = \sigma_C(d) + \langle d, y \rangle$.
(d) For a family of nonempty sets $\{C\}_{i \in I}$ with $C := \operatorname{cl} \operatorname{conv}(\cup_{i \in I} C_i)$ one has

$$\sigma_C(d) = \sup_{i \in I} \sigma_{C_i}(d) \quad \text{and} \quad \operatorname{dom} C = \cap_{i \in I} \operatorname{dom} \sigma_{C_i}.$$

(e) Let $A : \mathbb{R}^n \to \mathbb{R}^m$ be a linear mapping. For any nonempty set C of $\mathbb{R}^n$ we have

$$\sigma_{A(C)}(d) = \sigma_{\operatorname{cl} A(C)}(d) = \sigma_C(A^T d), \quad \forall d \in \mathbb{R}^n.$$

(f) For any two closed convex sets C_1, C_2 in $\mathbb{R}^n$ such that $C_1 \cap C_2 \neq \emptyset$ one has

$$\sigma_{C_1 \cap C_2}(d) = \operatorname{cl}(\sigma_{C_1} \square \sigma_{C_2})(d), \quad \forall d.$$

Turning to nonempty closed convex sets, we now state the fundamental correspondence between a closed convex set and its support function, namely the analytic version of the so called Hahn–Banach theorem.

Theorem 1.3.1 *Let $C \subset \mathbb{R}^n$. The closed convex hull of C is given by*

$$\operatorname{cl} \operatorname{conv} C = \{x \in \mathbb{R}^n \mid \langle x, d \rangle \leq \sigma_C(d), \forall d \in \mathbb{R}^n\}.$$

As a result of this theorem a closed convex set is completely determined by its support function. Key properties can then be usefully expressed via support functions.

Theorem 1.3.2 *Let C be a nonempty convex set in $\mathbb{R}^n$ and define*

$$B(d) := \sigma_C(d) + \sigma_C(-d), \ d \in \mathbb{R}^n.$$

Then $B(d) \geq 0$ and:
(a) $x \in \operatorname{aff} C$ if and only if $\langle x, d \rangle = \sigma_C(d)$, $\forall d$ with $B(d) = 0$.
(b) $x \in \operatorname{ri} C$ if and only if $\langle x, d \rangle < \sigma_C(d)$, $\forall d$ with $B(d) > 0$.
(c) $x \in \operatorname{int} C$ if and only if $\langle x, d \rangle < \sigma_C(d)$, $\forall d \neq 0$.
(d) $x \in \operatorname{cl} C$ if and only if $\langle x, d \rangle \leq \sigma_C(d)$, $\forall d$.

Corollary 1.3.1 *For any convex sets C, D in $\mathbb{R}^n$ one has*

$$\operatorname{cl} C \subset \operatorname{cl} D \ \text{ if and only if } \ \sigma_C \leq \sigma_D.$$

By its definition, the support function of a set $C \subset \mathbb{R}^n$ is in fact the conjugate function of the corresponding indicator function of that set, $\sigma_C = \delta_C^*$. Therefore, one has the relation

$$\sigma_C^*(\cdot) = (\delta_C^*(\cdot))^* = \operatorname{cl}(\delta_C(\cdot)) = \delta_{\operatorname{cl} C}(\cdot),$$

and when C is a closed convex set, this reduces to

$$\sigma_C^*(\cdot) = \delta_C^{**}(\cdot) = \delta_C(\cdot).$$

1.4   Set-Valued Maps

Let X and Y be two sets of $\mathbb{R}^n$. A set-valued map S from X to Y is a map that assigns to each point $x \in X$ a subset $S(x)$ of Y. Actually, it is convenient to characterize a set-valued map by its graph. The graph of S is the subset of $X \times Y$ defined by

$$\operatorname{gph} S := \{(x, y) \mid y \in S(x)\}.$$

Conversely, every nonempty subset G of $X \times Y$ is the graph of the set-valued map S uniquely determined by

$$S(x) = \{y \in Y \mid (x, y) \in G\}.$$

We use the notation $S : X \rightrightarrows Y$ to quantify the multivalued aspect of S whenever $S(x)$ is a set containing more than one element. When S is single-valued, we mean that $S(x) = \{y\}$ or simply $S(x) = y$, and in that case we use the notation $S : X \to Y$.

We shall say that a set-valued map S is convex-valued or closed-valued, etc., when $S(x)$ is convex or closed, etc.

The domain and range of $S : X \rightrightarrows Y$ are defined by the sets

$$\begin{aligned}
\operatorname{dom} S &:= \{x \in X \mid S(x) \neq \emptyset\}, \\
\operatorname{rge} S &:= \{y \mid \exists x \text{ such that } y \in S(x)\} = \bigcup_{x \in X} S(x).
\end{aligned}$$

Thus, the domain and range of S are respectively the images of $\operatorname{gph} S$ under the projections $(x, y) \to x$ and $(x, y) \to y$.

The inverse of any set-valued map always exists and is defined by

$$S^{-1}(y) := \{x \in X \mid y \in S(x)\},$$

or equivalently,

$$x \in S^{-1}(y) \Longleftrightarrow (x, y) \in \operatorname{gph} S.$$

The image of a set C is defined by

$$S(C) = \bigcup_{x \in C} S(x) = \{u \mid S^{-1}(u) \cap C \neq \emptyset\}.$$

One has the following relations:

$$\operatorname{dom} S^{-1} = \operatorname{rge} S, \ \ \operatorname{rge} S^{-1} = \operatorname{dom} S, \ \ (S^{-1})^{-1} = S.$$

In this book we will work with X, Y that are subsets of the finite-dimensional spaces $\mathbb{R}^n$ and $\mathbb{R}^m$. An easy way to extend a set-valued map to the whole space $\mathbb{R}^n$ is formally achieved by defining the map

$$\hat{S}(x) = \begin{cases} S(x) & \text{if } x \in X, \\ \emptyset & \text{if } x \notin X. \end{cases}$$

We will consider here set-valued maps $S : \mathbb{R}^n \rightrightarrows \mathbb{R}^m$ with the understanding that $S(x) \neq \emptyset$ for any $x \in \operatorname{dom} S$.
A single-valued map $F : X \to \mathbb{R}^m$ given on $X \subset \mathbb{R}^n$ can be treated in terms of the set-valued map $S : \mathbb{R}^n \to \mathbb{R}^m$ by setting

$$S(x) = \begin{cases} \{F(x)\} & \text{if } x \in X, \\ \emptyset & \text{if } x \notin X, \end{cases}$$

so that $\operatorname{dom} S = X$.
Basic operations with set-valued maps such as Minkowski addition and scalar multiplication and composition can be defined to generate new maps as follows:

$$(S + T)(x) := S(x) + T(x), \ \ (\lambda S)(x) := \lambda S(x), \ \lambda \in \mathbb{R},$$

and the composition of $S : \mathbb{R}^n \to \mathbb{R}^m$ with $T : \mathbb{R}^m \rightrightarrows \mathbb{R}^p$ by

$$(T \circ S)(x) := T(S(x)) = \bigcup_{y \in S(x)} T(y) = \{z \mid S(x) \cap T^{-1}(z) \neq \emptyset\},$$

to get the mapping $T \circ S : \mathbb{R}^n \to \mathbb{R}^p$.

Semicontinuity and Continuity of Set-Valued Maps

Definition 1.4.1 *A set-valued map $S : \mathbb{R}^n \rightrightarrows \mathbb{R}^m$ is upper semicontinuous (usc) at $\bar{x}$ if for each open set N containing $S(\bar{x})$ there exists a neighborhood $U(\bar{x})$ such that*

$$x \in U(\bar{x}) \implies S(x) \subset N.$$

Alternatively, upper semicontinuity of the set-valued map S at $\bar{x}$ can be expressed as

$$\forall \varepsilon > 0, \quad \exists \delta > 0 \text{ such that } S(x) \subset B_\varepsilon(S(\bar{x})), \ \forall x \in \mathbb{B}_\delta(\bar{x}),$$

where we define for any set C, $B_\varepsilon(C) := \{z \mid \text{dist}(z, C) < \varepsilon\}$.

Clearly, when S is a single-valued map on $\mathbb{R}^n$, i.e., a function, then for any $x \in \mathbb{R}^n$, $S(x)$ reduces to a singleton, and thus the above formulation coincides with the definition of continuous functions. The latter formulation also leads to the translation of upper semicontinuity in terms of infinite sequences. Let $\{x_k\} \in \text{dom}\, S$ be a convergent sequence with limit $\bar{x}$. Then, the set-valued map is upper semicontinuous at $\bar{x}$ if

$$\lim_{k \to \infty} \text{dist}(y_k, S(\bar{x})) = 0, \quad \forall \, y_k \in S(x_k).$$

Definition 1.4.2 *A set-valued map $S : \mathbb{R}^n \rightrightarrows \mathbb{R}^m$ is lower semicontinuous (lsc) at $\bar{x}$ if for each open set N with $N \cap S(\bar{x}) \neq \emptyset$, there exists a neighborhood $U(\bar{x})$ such that*

$$x \in U(\bar{x}) \implies S(x) \cap N \neq \emptyset.$$

S is called continuous at $\bar{x}$ if it is both lsc and usc at $\bar{x}$. We say that S is lsc (usc) in $\mathbb{R}^n$ if it is lsc (usc) at every point $\bar{x} \in \mathbb{R}^n$, and continuous in $\mathbb{R}^n$ if S is both lsc and usc in $\mathbb{R}^n$.

Proposition 1.4.1 *A set-valued map $S : \mathbb{R}^n \rightrightarrows \mathbb{R}^m$ is lsc at $\bar{x}$ if and only if for any sequence $\{x_k\} \in \mathbb{R}^n$ converging to $\bar{x}$, and any $\bar{y} \in S(\bar{x})$, there exist $k_0 \in \mathbb{N}$ and a sequence $\{y_k\} \in S(x_k)$ for $k \geq k_0$ that converges to $\bar{y}$.*

Proposition 1.4.2 *Let $\Phi : \mathbb{R}^n \times \mathbb{R}^m \to \mathbb{R} \cup \{+\infty\}$ and let S be a set-valued map $S : \mathbb{R}^n \rightrightarrows \mathbb{R}^m$. Define the function $\varphi(x) = \inf\{\Phi(x, u) \mid u \in S(x)\}$. If S is lsc at $\bar{x}$ and Φ is usc on $\{\bar{x}\} \times S(\bar{x})$, then the function φ is usc at $\bar{x}$.*

Definition 1.4.3 *A set-valued map $S : \mathbb{R}^n \rightrightarrows \mathbb{R}^m$ is closed at $\bar{x}$ if for any sequence $\{x_k\} \in \mathbb{R}^n$ and any sequence $\{y_k\} \in \mathbb{R}^m$ one has*

$$x_k \to \bar{x}, y_k \to \bar{y}, y_k \in S(x_k) \implies \bar{y} \in S(\bar{x}).$$

From this definition one can verify the following useful equivalence: $S : \mathbb{R}^n \rightrightarrows \mathbb{R}^n$ is closed at $\bar{x}$ if $\forall \bar{y} \notin S(\bar{x})$ there exist two neighborhoods $U(\bar{x}), V(\bar{y})$ such that $x \in U(\bar{x}) \implies S(x) \cap V(\bar{y}) = \emptyset$.

Definition 1.4.4 *A set-valued map $S : \mathbb{R}^n \rightrightarrows \mathbb{R}^m$ is locally bounded at $\bar{x}$ if for some neighborhood $N(\bar{x})$ the set $\bigcup_{x \in N(\bar{x})} S(x)$ is bounded. S is called locally bounded on $\mathbb{R}^n$ if this holds for every $\bar{x} \in \mathbb{R}^n$. It is bounded on $\mathbb{R}^n$ if $\text{rge}\, S$ is a bounded subset of $\mathbb{R}^m$.*

From the above definition one clearly has that if S is locally bounded at $\bar{x}$, then the set $S(\bar{x})$ is bounded, and in particular that S is locally bounded at any point $\bar{x} \notin \operatorname{cl} \operatorname{dom} S$. Furthermore, in terms of sequences one has that $S : \mathbb{R}^n \rightrightarrows \mathbb{R}^m$ is locally bounded if

$\{x_k\}$ is a bounded sequence, $u_k \in S(x_k) \implies \{u_k\}$ is a bounded sequence.

The next results give some important relations between upper semicontinuous and locally bounded set-valued maps.

Theorem 1.4.1 *Suppose that the set-valued map $S : \mathbb{R}^n \rightrightarrows \mathbb{R}^m$ is usc at $\bar{x}$ and $S(\bar{x})$ is closed. Then S is closed at $\bar{x}$. Conversely, if S is locally bounded and closed at $\bar{x}$ then S is usc at that point.*

Theorem 1.4.2 *Let S be a set-valued map $S : \mathbb{R}^n \rightrightarrows \mathbb{R}^m$ that is usc, and let C be a compact set in $\mathbb{R}^n$. Then $S(C)$ is compact.*

Corollary 1.4.1 *Let S be a closed set-valued map $S : \mathbb{R}^n \rightarrow \mathbb{R}^m$. Then S is locally bounded at x if and only if $S(x)$ is bounded.*

As already mentioned, the subdifferential map of a convex function can be seen as a set-valued map that enjoys the following important properties.

Proposition 1.4.3 *Let $f : \mathbb{R}^n \rightarrow \mathbb{R} \cup \{+\infty\}$ be a proper convex function. Then ∂f is upper semicontinuous and locally bounded at every point $x \in \operatorname{int}(\operatorname{dom} f)$. Moreover, if f is assumed lsc, then ∂f is closed.*

We end this section by recalling some useful and basic operations preserving lower/upper semicontinuity of set-valued maps.

Proposition 1.4.4 *Let $\{S_i \,|i \in I\}$ be a family of set-valued maps defined in a finite-dimensional vector space with appropriate dimensions. Then the following properties hold:*
(a) The composition map $S_1 \circ S_2$ of lsc (usc) maps S_i, $i = 1, 2$ is lsc (usc).
(b) The union of lsc maps $\cup_{i \in I} S_i$ is an lsc map.
(c) The intersection of usc maps $\cap_{i \in I} S_i$ is a usc map.
(d) The Cartesian product of a finite number of lsc (usc) maps $\Pi_{i=1}^{m} S_i$ is an lsc (usc) map.

1.5   Notes and References

Convex analysis has emerged as the most powerful and elegant theory underlying the development of optimization and variational analysis. The number of references including research papers, books, and monograph dealing with various aspects of convex analysis and its applications is today very large. The results briefly outlined in this chapter are far from reflecting all the convex-analytic machinery. Thus this chapter is not an introduction to convex analysis. Rather, the writing of this chapter is intended to provide to the reader a concise review of the elements of convex analysis that will be needed and used throughout this monograph. For details and proofs of all the results summarized in this chapter, the reader is referred to the following short bibliography. Standard reference on convex analysis include the books and monographs by J. J. Moreau,[103], which study mostly convex functions and their properties on arbitrary topological vector spaces. R. T. Rockafellar [119], is a comprehensive book on all aspects of convex analysis in $\mathbb{R}^n$. Laurent [85], which is a book with fundamental results on the theory and applications of convex analysis to optimization and approximation theory within the framework of abstract spaces. Ekeland and Temam [67], which deals with variational and infinite-dimensional problems. A recent concise and accessible account of convex analysis is given in Borwein and Lewis [38]. Other classics and useful books studying mainly properties of convex sets are H. G. Eggleston [65] and F. A. Valentine [129].

2
Asymptotic Cones and Functions

This chapter provides the fundamental tools used throughout this monograph. For a given subset of $\mathbb{R}^n$ we are interested in studying its behavior at infinity. This leads to the concepts of asymptotic cone and asymptotic function through its epigraph. Using elementary real analysis and geometrical concepts we develop a complete mathematical treatment to handle the asymptotic behavior of sets, functions, and other induced functional operations.

2.1 Definitions of Asymptotic Cones

The set of natural numbers is denoted by $\mathbb{N}$, so that $k \in \mathbb{N}$ means $k = 1, 2, \ldots$. A sequence $\{x_k\}_{k \in \mathbb{N}}$ or simply $\{x_k\}$ in $\mathbb{R}^n$ is said to converge to x if $\|x_k - x\| \to 0$ as $k \to \infty$, and this will be indicated by the notation $x_k \to x$ or $x = \lim_{k \to \infty} x_k$. We say that x is a cluster point of $\{x_k\}$ if some subsequence converges to x. Recall that every bounded sequence in $\mathbb{R}^n$ has at least one cluster point. A sequence in $\mathbb{R}^n$ converges to x if and only if it is bounded and has x as its unique cluster point.

Let $\{x_k\}$ be a sequence in $\mathbb{R}^n$. We are interested in knowing how to handle situations when the sequence $\{x_k\} \subset \mathbb{R}^n$ is unbounded. To derive some convergence properties, we are led to consider directions $d_k := x_k \|x_k\|^{-1}$ with $x_k \neq 0$, $k \in \mathbb{N}$. From classical analysis, the Bolzano–Weierstrass theorem implies that we can extract a convergent subsequence $d = \lim_{k \in K} d_k$, $K \subset \mathbb{N}$, with $d \neq 0$. Now suppose that the sequence $\{x_k\} \subset \mathbb{R}^n$ is such that

$\|x_k\| \to +\infty$. Then

$$\exists\, t_k := \|x_k\|,\ k \in K \subset \mathbb{N},\ \text{such that}\ \lim_{k \in K} t_k = +\infty\ \text{and}\ \lim_{k \in K} \frac{x_k}{t_k} = d.$$

This leads us to introduce the following concepts.

Definition 2.1.1 *A sequence $\{x_k\} \subset \mathbb{R}^n$ is said to converge to a direction $d \in \mathbb{R}^n$ if*

$$\exists\{t_k\},\ \text{with}\ t_k \to +\infty\ \text{such that}\ \lim_{k \to \infty} \frac{x_k}{t_k} = d.$$

Definition 2.1.2 *Let C be a nonempty set in $\mathbb{R}^n$. Then the asymptotic cone of the set C, denoted by C_∞, is the set of vectors $d \in \mathbb{R}^n$ that are limits in direction of the sequences $\{x_k\} \subset C$, namely*

$$C_\infty = \left\{ d \in \mathbb{R}^n \mid \exists t_k \to +\infty, \exists x_k \in C\ \text{with}\ \lim_{k \to \infty} \frac{x_k}{t_k} = d \right\}.$$

From the definition we immediately deduce the following elementary facts.

Proposition 2.1.1 *Let $C \subset \mathbb{R}^n$ be nonempty. Then:*
(a) C_∞ is a closed cone.
(b) $(\mathrm{cl}\, C)_\infty = C_\infty$.
(c) If C is a cone, then $C_\infty = \mathrm{cl}\, C$.

The importance of the asymptotic cone is revealed by the following key property, which is an immediate consequence of its definition.

Proposition 2.1.2 *A set $C \subset \mathbb{R}^n$ is bounded if and only if $C_\infty = \{0\}$.*

Proof. It is clear that C_∞ cannot contain any nonzero direction if C is bounded. Conversely, if C is unbounded, then there exists a sequence $\{x_k\} \subset C$ with $x_k \neq 0$, $\forall k \in \mathbb{N}$, such that $t_k := \|x_k\| \to \infty$ and thus the vectors $d_k = t_k^{-1} x_k \in \{d : \|d\| = 1\}$. Therefore, we can extract a subsequence of $\{d_k\}$ such that $\lim_{k \in K} d_k = d$, $K \subset \mathbb{N}$, and with $\|d\| = 1$. This nonzero vector d is an element of C_∞ by Definition 2.1.2, a contradiction. $\square$

Associated with the asymptotic cone C_∞ is the following related concept, which will help us in simplifying the definition of C_∞ in the particular case where $C \subset \mathbb{R}^n$ is assumed convex.

Definition 2.1.3 *Let $C \subset \mathbb{R}^n$ be nonempty and define*

$$C_\infty^1 := \left\{ d \in \mathbb{R}^n \mid \forall t_k \to +\infty, \exists x_k \in C\ \text{with}\ \lim_{k \to \infty} \frac{x_k}{t_k} = d \right\}.$$

We say that C is asymptotically regular if $C_\infty = C_\infty^1$.

Proposition 2.1.3 *Let C be a nonempty convex set in $\mathbb{R}^n$. Then C is asymptotically regular.*

Proof. The inclusion $C_\infty^1 \subset C_\infty$ clearly holds from the definitions of C_∞^1 and C_∞, respectively. Let $d \in C_\infty$. Then $\exists \{x_k\} \in C$, $\exists s_k \to \infty$ such that $d = \lim_{k\to\infty} s_k^{-1} x_k$. Let $x \in C$ and define $d_k = s_k^{-1}(x_k - x)$. Then we have

$$d = \lim_{k\to\infty} d_k, \quad x + s_k d_k \in C.$$

Now let $\{t_k\}$ be an arbitrary sequence such that $\lim_{k\to\infty} t_k = +\infty$. For any fixed $m \in N$, there exists $k(m)$ with $\lim_{m\to\infty} k(m) = +\infty$ such that $t_m \leq s_{k(m)}$, and since C is convex, we have $x'_m := x + t_m d_{k(m)} \in C$. Hence, $d = \lim_{m\to\infty} t_m^{-1} x'_m$, showing that $d \in C_\infty^1$. $\qquad\square$

We note that a set can be nonconvex, yet asymptotically regular. Indeed, consider, for example, sets defined by $C := S + K$, with S compact and K a closed convex cone. Then clearly C is not necessarily convex, but it can be easily seen that $C_\infty = C_\infty^1$.

Remark 2.1.1 Note that the definitions of C_∞ and C_∞^1 are related to the theory of set convergence of Painlevé–Kuratowski. Indeed, for a family $\{C_t\}_{t>0}$ of susbsets of $\mathbb{R}^n$, the outer limit as $t \to +\infty$ is the set

$$\limsup_{t\to+\infty} C_t = \left\{ x \mid \liminf_{t\to+\infty} d(x, C_t) = 0 \right\},$$

while the inner limit as $t \to +\infty$ is the set

$$\liminf_{t\to+\infty} C_t = \left\{ x \mid \limsup_{t\to+\infty} d(x, C_t) = 0 \right\}.$$

It can then be verified that the corresponding asymptotic cones can be written as

$$C_\infty = \limsup_{t\to+\infty} t^{-1} C, \quad C_\infty^1 = \liminf_{t\to+\infty} t^{-1} C.$$

Proposition 2.1.4 *Let $C \subset \mathbb{R}^n$ be nonempty and define the normalized set*

$$C_N := \left\{ d \in \mathbb{R}^n \mid \exists \{x_k\} \in C, \|x_k\| \to +\infty \text{ with } d = \lim_{k\to\infty} \frac{x_k}{\|x_k\|} \right\}.$$

Then $C_\infty = \operatorname{pos} C_N$, where for any set C, $\operatorname{pos} C = \{\lambda x \mid x \in C, \lambda \geq 0\}$.

Proof. Clearly, one always has $\operatorname{pos} C_N \subset C_\infty$. Conversely, let $0 \neq d \in C_\infty$. Then there exists $t_k \to \infty$, $x_k \in C$ such that

$$d = \lim_{k\to\infty} t_k^{-1} x_k = \lim_{k\to\infty} t_k^{-1} \|x_k\| \frac{x_k}{\|x_k\|}, \quad \text{with } \|x_k\| \to \infty.$$

Thus the sequence $\{t_k^{-1}\|x_k\|\}$ is a nonnegative bounded sequence, and by the Bolzano–Weierstrass theorem, there exists a subsequence $\{t_k^{-1}\|x_k\|\}_{k\in K}$ with $K \subset \mathbb{N}$ such that $\lim_{k\in K} t_k^{-1}\|x_k\| = \lambda \geq 0$, which means that $d = \lambda d_N$ with $d_N \in C_N$, namely $d \in \text{pos}\, C_N$. $\qquad\square$

We now turn to some useful formulations of the asymptotic cone for convex sets.

Proposition 2.1.5 *Let C be a nonempty convex set in $\mathbb{R}^n$. Then the asymptotic cone C_∞ is a closed convex cone. Moreover, define the following sets:*

$$D(x) := \{d \in \mathbb{R}^n \mid x + td \in \text{cl}\, C,\ \forall t > 0\}\ \forall x \in C,$$
$$E := \{d \in \mathbb{R}^n \mid \exists x \in C \text{ such that } x + td \in \text{cl}\, C,\ \forall t > 0\},$$
$$F := \{d \in \mathbb{R}^n \mid d + \text{cl}\, C \subset \text{cl}\, C\}.$$

Then $D(x)$ is in fact independent of x, which is thus now denoted by D, and $C_\infty = D = E = F$.

Proof. We already know that C_∞ is a closed cone. The convexity property simply follows from Proposition 2.1.3, which ensures that $C_\infty^1 = C_\infty$. We now prove the three equivalent formulations. Let $d \in C_\infty$, $x \in C$, $t > 0$. From Definition 2.1.2,

$$\exists t_k \to +\infty\ \exists d_k \to d, \text{ with } x + t_k d_k \in C.$$

Take k sufficiently large such that $t \leq t_k$, then since C is convex, we have

$$x + td_k = (1 - t_k^{-1}t)x + t_k^{-1}t(x + t_k d_k) \in C.$$

Passing to the limit, we thus have $x + td \in \text{cl}\, C$, $\forall t > 0$, thus proving the inclusion $C_\infty \subset D(x)$, $\forall x \in C$. Clearly, we also have the inclusion $D(x) \subset E$, $\forall x \in C$. We now show that $E \subset C_\infty$. For that, let $d \in E$ and $x \in C$ be such that $x(t) := x + td \in \text{cl}\, C$, $\forall t > 0$. Then, since $d = \lim_{t\to\infty} t^{-1}x(t)$ and $x(t) \in \text{cl}\, C$, we have $d \in (\text{cl}\, C)_\infty = C_\infty$ (by Proposition 2.1.1), which also proves that $D(x)$ is in fact independent of x, and we can write $D(x) = D = C_\infty = E$. Finally, let $d \in C_\infty$. Using the representation $C_\infty = D$ we thus have $x + d \in \text{cl}\, C$, $\forall x \in C$, and hence $d + \text{cl}\, C \subset \text{cl}\, C$, proving the inclusion $C_\infty \subset F$. Now, if $d \in F$, then $d + \text{cl}\, C \subset \text{cl}\, C$, and thus

$$2d + \text{cl}\, C = d + (d + \text{cl}\, C) \subset d + \text{cl}\, C \subset \text{cl}\, C,$$

and by induction $\text{cl}\, C + md \subset \text{cl}\, C, \forall m \in \mathbb{N}$. Therefore, $\forall x \in \text{cl}\, C$, $d_m := x + md \in \text{cl}\, C$ and $d = \lim_{m\to\infty} m^{-1}d_m$, namely $d \in C_\infty$, showing that $F \subset C_\infty$, and hence the proof of the three equivalent formulations for C_∞ is completed. $\qquad\square$

Remark 2.1.2 When C is a closed convex set the asymptotic cone is also called the recession cone. However, we will keep the terminology asymptotic cone throughout this book.

From Proposition 2.1.5, using the set D as a representation of C_∞, an alternative useful representation of the asymptotic cone of a closed convex set is

$$C_\infty = \bigcap_{t>0} t^{-1}(C - x), \quad \forall x \in C.$$

It is important to note that in the definition of the set D given above, the closure assumption on the convex set $C \subset \mathbb{R}^n$ is crucial and cannot be removed. Indeed, consider the convex set

$$C = \{x \in \mathbb{R}^2 : x_1 > 0, x_2 > 0\} \cup \{(0,0)\}.$$

Then from Definition 2.1.2 we have $C_\infty = \mathbb{R}_+^2$, while any of the three formulations given in Proposition 2.1.5 would lead to the wrong result $C_\infty = C$.

Proposition 2.1.6 *For any nonempty closed convex set $C \subset \mathbb{R}^n$, one has $C = C + C_\infty$.*

Proof. The inclusion $C \subset C + C_\infty$ is clear. Let $x \in C + C_\infty$. Then there exists $c \in C$, $d \in C_\infty$ with $x = c + d$. Thus, there exists $d_k \in C$, $t_k \to \infty$ such that $t_k^{-1}d_k \to d$, which implies that for k sufficiently large $(1 - t_k^{-1})c + t_k^{-1}d_k \in C$ and $(1 - t_k^{-1})c + t_k^{-1}d_k \to x \in C$. □

Proposition 2.1.7 *For any nonempty closed convex set $C \subset \mathbb{R}^n$ that contains no lines one has $C = \mathrm{conv}(\mathrm{ext}\, C) + C_\infty$.*

Proof. Thanks to Proposition 2.1.6, we have only to prove that

$$C \subset \mathrm{conv}(\mathrm{ext}\, C) + C_\infty.$$

Let $x \in C$. Then from Theorem 1.1.3, we can write

$$x = \sum_{i=1}^{k} \lambda_i x_i + \sum_{i=k+1}^{m} \lambda_i d_i, \quad \sum_{i=1}^{m} \lambda_i = 1, \quad \lambda_i \geq 0, \quad i = 1, \ldots, m,$$

with $x_i \in \mathrm{ext}\, C$, $d_i \in \mathrm{extray}\, C$, $i = 1, \ldots, m$. Thus $d_i \in \mathrm{extray}\, C$ implies that $d_i = e_i + v_i$, where e_i, the endpoint of the ray, is an extreme point of C and $v_i \in C_\infty$, from which it follows that $x \in \mathrm{conv}(\mathrm{ext}\, C) + C_\infty$. □

We now prove the following useful result, which as we shall see is often used as a technical device in analyzing closedness properties involving operations on closed sets in $\mathbb{R}^n$.

Lemma 2.1.1 *Let C be a nonempty closed set of $\mathbb{R}^n$, and K the cone in $\mathbb{R}^{n+1}$ generated by $\{(1, x) \mid x \in C\}$, i.e., $K = \mathrm{pos}\{(1, x) \mid x \in C\}$. Define $D := \{(0, x) \mid x \in C_\infty\}$. Then $\mathrm{cl}\, K = K \cup D$.*

Proof. The cone $K \subset \mathbb{R}^{n+1}$ generated by $\{(1, x) \mid x \in C\}$ can be written as

$$K = \{\lambda(1, x) \mid \lambda \geq 0, x \in C\} = \{\lambda(1, x) \mid \lambda > 0, x \in C\} \cup \{0\}.$$

Let $y = (t, x) \in \mathrm{cl}\, K$. Then $\exists t_k \geq 0$, $t_k \to t$, $x^k \in C$ such that $t_k(1, x^k) \to y$, $t_k x^k \to x$. If $t = 0$, $x \in C_\infty$, and thus $y \in D$. Otherwise, $t^{-1}x \in C$ and $y = t(1, t^{-1}x) \in K$, showing that $\mathrm{cl}\, K \subset K \cup D$. Conversely, any $y \in D$ can be written as $y = \lim_{k\to\infty} t_k(1, x^k)$, with $t_k \to 0$, $x^k \in C$, so that $D \subset \mathrm{cl}\, K$, and hence since $K \subset \mathrm{cl}\, K$, that $D \cup K \subset \mathrm{cl}\, K$. $\qquad\square$

Some further elementary operations on asymptotic cones of nonempty convex sets are collected below.

Proposition 2.1.8 *Let $C \subset \mathbb{R}^n$ be nonempty and convex. Then:*
(a) $(\mathrm{cl}\, C)_\infty = (\mathrm{ri}\, C)_\infty = C_\infty$.
(b) For any $x \in \mathrm{ri}\, C$, one has $d \in (\mathrm{cl}\, C)_\infty \iff x + td \in \mathrm{ri}\, C, \forall t > 0$.
(c) $C \subset C_\infty \implies \mathrm{ri}\, C_\infty = \mathrm{ri}\,\mathrm{pos}\, C$.
(d) C is closed with $0 \in C \implies C_\infty = \{d : t^{-1}d \in C, , \forall t > 0\}$.

Proof. Since C is nonempty and convex, the relative interior of C is nonempty, and thus applying Proposition 2.1.1(b) to $\mathrm{ri}\, C$ we obtain $(\mathrm{ri}\, C)_\infty = (\mathrm{cl}\,\mathrm{ri}\, C)_\infty = (\mathrm{cl}\, C)_\infty = C_\infty$, where in the second equation we use Proposition 1.1.4, and (a) is proved, while (b) follows from (a) and the line segment principle, cf. Proposition 1.1.4. To prove (c), we first note that $\mathrm{pos}\, C$ is convex. Then under the assumption $C \subset C_\infty$ one has $\mathrm{pos}\, C \subset \mathrm{pos}\, C_\infty = C_\infty$ and hence $\mathrm{cl}\,\mathrm{pos}\, C \subset \mathrm{cl}\, C_\infty = C_\infty$, which together with the fact $C_\infty \subset \mathrm{cl}\,\mathrm{pos}\, C$ gives that $\mathrm{cl}\,\mathrm{pos}\, C = C_\infty = \mathrm{cl}\, C_\infty$, which is equivalent to $\mathrm{ri}\,\mathrm{pos}\, C = \mathrm{ri}\, C_\infty$. In part (d), since $0 \in C$, the representation of C_∞ follows from the representation of C_∞ via the set D given in Proposition 2.1.5. $\qquad\square$

Some useful operations with asymptotic cones are now given for arbitrary sets in $\mathbb{R}^n$.

Proposition 2.1.9 *Let $C_i \subset \mathbb{R}^n$, $i \in I$, be an arbitrary index set. Then:*
(a) $(\cap_{i\in I} C_i)_\infty \subset \cap_{i\in I}(C_i)_\infty$, whenever $\cap_{i\in I} C_i$ is nonempty.
(b) $(\cup_{i\in I} C_i)_\infty \supset \cup_{i\in I}(C_i)_\infty$.
The inclusion in (a) holds as an equation for closed convex sets C_i having a nonempty intersection. The inclusion (b) holds as an equation when I is a finite index set.

Proof. Let d be any point in the set $C := (\cap_{i \in I} C_i)_\infty$, which is closed by definition of the asymptotic cone and satisfies

$$\exists t_k \to \infty, \ \exists x_k \in \cap_{i \in I} C_i \text{ such that } \frac{x_k}{t_k} \to d.$$

Therefore, $\exists t_k \to \infty$, $\exists x_k \in C_i$, $\forall i \in I$ such that $t_k^{-1} x_k \to d$, implying that $d \in (C_i)_\infty$ for all $i \in I$ and proving (a). The relation (b) is proved in a similar way through the definition of the corresponding asymptotic cone. Finally, the special result for the convex case follows by applying the characterization given in Proposition 2.1.5. □

Proposition 2.1.10 *For any sets $C_i \subset \mathbb{R}^n$, $i = 1, \ldots, m$, one has*

$$(C_1 \times \cdots \times C_m)_\infty \subset (C_1)_\infty \times \cdots \times (C_m)_\infty.$$

The inclusion holds as an equation if every C_i is nonempty and convex.

Proof. The inclusion is proved by direct use of the definition of the asymptotic cone, and in the convex case using the characterization provided in Proposition 2.1.5. □

Proposition 2.1.11 *Let $A : \mathbb{R}^n \to \mathbb{R}^m$ be a linear mapping and C a closed convex set in $\mathbb{R}^n$ such that the inverse image of C is nonempty. Then $(A^{-1}(C))_\infty = A^{-1}(C_\infty)$.*

Proof. Since A is continuous and C is closed and convex, $A^{-1}(C)$ is closed and convex. Let $x \in A^{-1}(C)$. Then $d \in (A^{-1}(C))_\infty$ if and only if $A(x + td) = Ax + tAd \subset C$, $\forall t \geq 0$, but the latter means that $Ad \in C_\infty$, namely $d \in A^{-1}(C_\infty)$. □

We close the section by giving explicit formulas of asymptotic cones for some other important sets.

Example 2.1.1 (i) Let C be a cone in $\mathbb{R}^n$. Then $C_\infty = \operatorname{cl} C$.
(ii) Let C be an affine set. Then C_∞ is the linear subspace parallel to C.
(iii) Let C be a polyhedral convex set $C := \{x \in \mathbb{R}^n \mid Ax \leq b\}$, where A is an $m \times n$ matrix and $b \in \mathbb{R}^m$. Then $C_\infty = \{d \in \mathbb{R}^n \mid Ad \leq 0\}$.

2.2 Dual Characterization of Asymptotic Cones

There exists a close connection between the support function of a set and its asymptotic cone. We show below that for a closed convex set $C \subset \mathbb{R}^n$, one can in fact give a *dual* characterization of the asymptotic cone, via the barrier cone of C.

Theorem 2.2.1 *Let $C \subset \mathbb{R}^n$ be nonempty and let C_∞^* denote the polar cone of C_∞. Then the following relations hold:*
(a) $\operatorname{dom} \sigma_C \subset C_\infty^*$.
(b) If $\operatorname{int} C_\infty^* \neq \emptyset$, *then* $\operatorname{int} C_\infty^* \subset \operatorname{dom} \sigma_C$.
(c) If C is convex, then $(\operatorname{dom} \sigma_C)^* = C_\infty$.

Proof. (a) Let $y \notin C_\infty^*$. Then $\exists\, 0 \neq d \in C_\infty$ such that $\langle d, y \rangle > 0$. Since $d \in C_\infty$, $\exists t_k \to \infty$, $\exists x_k \in C$ with $t_k^{-1} x_k \to d$, and with $\langle d, y \rangle > 0$, it follows that $\langle x_k, y \rangle \to +\infty$, proving that $y \notin \operatorname{dom} \sigma_C$.
(b) Let $y \notin \operatorname{dom} \sigma_C$. Then $\exists x_k \in C$ with $\langle x_k, y \rangle \to +\infty$. Passing to subsequences if necessary we can assume without loss of generality that $\|x_k\|^{-1} x_k \to d \neq 0$, with $d \in C_\infty$, and hence $\langle \|x_k\|^{-1} x_k, y \rangle \geq 0$. Therefore, $\forall \varepsilon > 0$ we have $\langle d, y + \varepsilon d \rangle \geq \varepsilon \|d\|^2 > 0$, showing that $y + \varepsilon d \notin C_\infty^*$, i.e., $y \notin \operatorname{int} C_\infty^*$.
(c) Since C is assumed convex, then C_∞ is a closed convex cone and $C_\infty^{**} = C_\infty$. Using (a) we thus have

$$\operatorname{dom} \sigma_C \subset C_\infty^* \implies C_\infty = C_\infty^{**} \subset (\operatorname{dom} \sigma_C)^*.$$

We now prove the reverse inclusion. Let $d \in (\operatorname{dom} \sigma_C)^*$, $t > 0$, and let $\bar{x}$ be any point in C. Since $td \in (\operatorname{dom} \sigma_C)^*$, we have for any $y \in \operatorname{dom} \sigma_C$,

$$\begin{aligned} \langle \bar{x} + td, y \rangle &= \langle \bar{x}, y \rangle + \langle td, y \rangle \leq \langle \bar{x}, y \rangle \\ &\leq \sup_{x \in C} \langle x, y \rangle = \sigma_C(y). \end{aligned}$$

Since for any $y \notin \operatorname{dom} \sigma_C$ we have $\sigma_C(y) = +\infty$, the above inequality remains valid for any $y \in \mathbb{R}^n$. Therefore, invoking Theorem 1.3.1 we obtain for all $t > 0$, $\bar{x} + td \in \operatorname{cl} C$, which by Proposition 2.1.5 proves that $d \in C_\infty$. $\square$

2.3 Closedness Criteria

Given a nonempty closed set of $\mathbb{R}^n$, we are interested in answering the following question: Under what conditions does the image of a closed set under a linear mapping remain closed. This type of result is of fundamental importance and is at the root of several closedness criteria that are particularly useful for analyzing the existence of solutions of extremum problems. Asymptotic cones play a key role in the derivation of such results.
Let $C \subset \mathbb{R}^n$ be a nonempty closed set and let $A : \mathbb{R}^n \to \mathbb{R}^m$ be a linear mapping. Let $\ker A := \{x : Ax = 0\} := A^{-1}(0)$. As we shall see, a basic

sufficient condition that guarantees that $A(C)$ is closed is

$$\ker A \cap C_\infty = \{0\}. \tag{2.1}$$

Moreover, under the condition (2.1) we also have $A(C_\infty) = (A(C))_\infty$. For a convex set, a weaker sufficient condition to guarantee closedness of $A(C)$ is

$$\ker A \cap C_\infty \quad \text{is a linear subspace.} \tag{2.2}$$

Unfortunately, even in the convex case, the above condition can fail, as shown in the following two examples.

Example 2.3.1 Let $C = \{x \in \mathbb{R}^2 \mid x_2 \geq x_1^2\}$ and let A be the linear mapping $A : \mathbb{R}^2 \to \mathbb{R}^2$ given by $A(x_1, x_2) = (x_1, 0)$, i.e., the projection onto the x_1-axis. Then, one has

$$
\begin{aligned}
C_\infty &= \{d \in \mathbb{R}^2 \mid d_1 = 0, d_2 \geq 0\}, \\
A(C) &= \{x \in \mathbb{R}^2 \mid x_2 = 0\} = x_1\text{-axis}, \\
\ker A &= \{x \in \mathbb{R}^2 \mid x_1 = 0\} = x_2\text{-axis},
\end{aligned}
$$

and condition (2.2) fails, yet $A(C)$ is a closed set.

Example 2.3.2 Let $D = \{x \in \mathbb{R}^2_+ \mid x_1 + x_2 \geq 1\}$ and A be the linear mapping as in Example 2.3.1. One easily see that $\ker A \cap D_\infty = \{x \in \mathbb{R}^2 \mid x_1 = 0, x_2 \geq 0\}$, and thus condition 2.2 fails; yet $A(D)$ is a closed set.

As we shall see, the sets C and D are quite different in nature. The set C belongs to a class of sets called *continuous sets*, while the set D is in the class of *asymptotically linear sets*. These two classes of sets are quite general and are very different one from the other and will be introduced later on. We thus need conditions that can handle the situations just described. In fact, the next general result will establish a necessary and sufficient condition for preserving closedness of $A(C)$.

Theorem 2.3.1 *Let C be any nonempty closed set of $\mathbb{R}^n$ and $A : \mathbb{R}^n \to \mathbb{R}^m$ a linear mapping. Let $\{y_k\}$ be any sequence in $A(C)$ converging to y and define the set*

$$S := \left\{ \{x_k\} \mid x_k \in C : \|x_k\| \to \infty, \|Ax_k - y\| \leq \|y - y_k\|, \frac{x_k}{\|x_k\|} \to \bar{x} \right\}. \tag{2.3}$$

Then a necessary and sufficient condition for the set $A(C)$ to be closed is that for each y and each sequence $\{y_k\} \in A(C)$ converging to y, (a) either S is empty or (b) for each sequence $\{x_k\} \in S$ there exists a sequence $\{z_k, \rho_k\} \subset \mathbb{R}^n \times \mathbb{R}_{++}$ such that for k sufficiently large,

$$x_k - \rho_k z_k \in C, \ \|A(x_k - \rho_k z_k) - y\| \leq \|y - y_k\|, \ \rho_k \in (0, \|x_k\|), \tag{2.4}$$

and $z_k \to z$ with $\|z - \bar{x}\| < 1$.

Proof. First, suppose that $y_k \in A(C)$, $y_k \to y$. Then there exists $x_k \in C$ such that $y_k = Ax_k$, which implies $\|Ax_k - y\| = \|y_k - y\|$. Thus, $S = \emptyset$ means that a subsequence of $\{x_k\}$ is bounded and has a cluster point, say $x^* \in C$. Passing to the limit in the latter relation implies that $\|Ax^* - y\| = 0$, showing that $A(C)$ is closed. We now prove the second case, described in (b). Suppose that $S \neq \emptyset$, and (2.4) holds, and let $y_k \to y$ with $y_k = Ax'_k$, $x'_k \in C$. Define the set $S_k := \{x \in C \mid \|Ax - y\| \leq \|y - y_k\|\}$ and consider the optimization problem $\inf\{\|x\| \mid x \in S_k\}$. By (2.3), we clearly have $x'_k \in S_k$, and hence S_k is nonempty. Furthermore, S_k is closed, and then the existence of $x_k \in \operatorname{argmin}\{\|x\| : x \in S_k\}$ is guaranteed. If we can show that the sequence $\{x_k\}$ is bounded, we are done. Indeed, in that case there would exist a subsequence $\{x_k\}_{k \in K}$, $K \subset \mathbb{N}$, such that $\lim_{k \in K} x_k = x \in C$ and with $Ax = y$, proving the closedness of $A(C)$. To prove this, suppose that $\{x_k\}$ is unbounded. Passing to subsequences if necessary, we can suppose without loss of generality that $\|x_k\|^{-1} x_k \to \bar{x} \neq 0$, and then using (2.3) there exists $\{z_k, \rho_k\}$ satisfying (2.4). As a consequence, $x_k - \rho_k z_k \in S_k$, and since by definition of x_k we have $\|x_k\| \leq \|x\|, \forall x \in S_k$, we obtain in particular that $\|x_k\| \leq \|x_k - \rho_k z_k\|$. Now,

$$
\begin{aligned}
\|x_k - \rho_k z_k\| &= \|(1 - \|x_k\|^{-1}\rho_k)x_k + \rho_k(\|x_k\|^{-1}x_k - z_k)\|, \\
&\leq (1 - \|x_k\|^{-1}\rho_k)\|x_k\| + \rho_k\|\|x_k\|^{-1}x_k - z_k\|, \\
&= \|x_k\| + \rho_k(\|\|x_k\|^{-1}x_k - z_k\| - 1),
\end{aligned}
$$

and hence $\|\|x_k\|^{-1}x_k - z_k\| \geq 1$. Passing to the limit we get $\|\bar{x} - z\| \geq 1$, contradicting assumption (2.4). Conversely, suppose that $A(C)$ is closed and let $y_k \to y$ with $y_k \in A(C)$. Then, since $A(C)$ is closed, there exists $x \in C$ with $Ax = y$. Set $z_k = \|x_k\|^{-1}(x_k - x)$, $\rho_k = \|x_k\|$, and then (2.4) is clearly satisfied. $\qquad \square$

In general, it is easier and more convenient to choose in Theorem 2.3.1 $z_k = \bar{x}$ or more generally $z_k = \alpha\bar{x}$, $\alpha \in (0, 2)$, in assumption (2.4). In this case with $\alpha = 1$ we can say more about the image of C as shown in the theorem given below. As an application of that theorem we will also show how situations previously discussed, illustrating the failure of condition (2.2) to characterize closedness, can now be adequately handled.

For any nonempty closed set C of $\mathbb{R}^n$ and any linear map $A : \mathbb{R}^n \to \mathbb{R}^m$, it is immediate, using the definition of the asymptotic cone, to verify the inclusion $A(C_\infty) \subset A(C)_\infty$. However, the reverse inclusion requires a much finer analysis.

Theorem 2.3.2 *Let C be any nonempty closed set of $\mathbb{R}^n$ and $A : \mathbb{R}^n \to \mathbb{R}^m$ a linear mapping. Let $\{y_k\}$ be any sequence in $A(C)$ converging to y and let S be the set defined in (2.3). Suppose that for each y and any sequence $y_k \in A(C)$ converging to y, either S is empty or for any sequence $\{x_k\} \in S$ there exists a sequence $\{\rho_k\} \subset \mathbb{R}_{++}$ such that for k sufficiently* *large, $x_k - \rho_k \bar{x} \in C$, $\rho_k \leq \|x_k\|$. Then $A(C)$ is closed, and we have*

$$A(C_\infty) = (A(C))_\infty. \tag{2.5}$$

Proof. The closedness of $A(C)$ follows as an immediate consequence of the previous theorem. Indeed, if we take $z_k = z = \bar{x}$, it follows from (2.3) that $A\bar{x} = 0$, and (2.4) holds. Now, to prove (2.5) we first note that the inclusion $A(C)_\infty \supset A(C_\infty)$ follows immediately from the definition of the asymptotic cone. Thus it remains to prove the reverse inclusion $(A(C))_\infty \subset A(C_\infty)$. Let $y \in (A(C))_\infty$ and let $u_k \in C$, $t_k \to +\infty$, and $y_k = Au_k$ with $t_k^{-1} y_k \to y$. Define $S_k = \{x \in C \mid Ax = y_k\}$. Then, S_k is a nonempty closed set, and the existence of $x_k \in \arg\min\{\|x\| \mid x \in S_k\}$ is guaranteed. We prove now that we cannot have $\lim_{k\infty} t_k^{-1} \|x_k\| = +\infty$. Indeed, in the contrary case, we would have $\lim_{k\infty} \|x_k\| = +\infty$. Without loss of generality, we can suppose that $\|x_k\|^{-1} x_k \to \bar{x}$ with $\|\bar{x}\| = 1$. We thus obtain

$$A\bar{x} = \lim_{k \to \infty} A \frac{x_k}{\|x_k\|} = \lim_{k \to \infty} \frac{y_k}{\|x_k\|} = \lim_{k \to \infty} \frac{y_k}{t_k} \frac{t_k}{\|x_k\|} = 0.$$

Since (2.3) holds, there exists $\rho_k \in (0, \|x_k\|]$ such that $x_k - \rho_k \bar{x} \in C$ and since $A\bar{x} = 0$, it follows that $x_k - \rho_k \bar{x} \in S_k$. By definition of x_k we have $\|x_k\| \leq \|x_k - \rho_k \bar{x}\|$, and as in the proof of Theorem 2.3.1 it follows that $\|\|x_k\|^{-1} x_k - \bar{x}\| \geq 1$ and passing to the limit, we get a contradiction. Finally, since the sequence $\{t_k^{-1} \|x_k\|\}$ is bounded, passing to subsequences if necessary, we can conclude that $\{t_k^{-1} x_k\}$ converges to some x_∞. Since $x_k \in S_k$, we have $x_\infty \in C_\infty$ and $t_k^{-1} A x_k \to y = A x_\infty$, proving the desired result. $\qquad\square$

We consider now the situation in which Theorem 2.3.2 is satisfied and give further applications.

Corollary 2.3.1 *Let C be a nonempty closed set of $\mathbb{R}^n$ and $A : \mathbb{R}^n \to \mathbb{R}^m$ a linear mapping. Let $L(C) = C_\infty \cap -C_\infty$ and $L = L(C) \cap \ker A$ and suppose that the following two conditions hold:*
(a) For k sufficiently large, $C_k + L \subset C$, with $C_k = \{x \in C : \|x\| \geq k\}$.
(b) $z \in \ker A \cap C_\infty$ implies $z \in -C_\infty$.
Then $A(C)$ is closed and $(A(C))_\infty = A(C_\infty)$.

Proof. Let $\{y_k\} \subset A(C)$ be a sequence converging to y and let $\{x_k\} \in S$ be an unbounded sequence satisfying (2.3). Then clearly, $\bar{x} \in \ker A \cap C_\infty$, and from assumption (b) it follows that $\bar{x} \in -C_\infty$, which together with $\bar{x} \in C_\infty$ implies that $-\bar{x} \in L$. Let $\rho > 0$. Then by assumption (a), for k sufficiently large we have $x_k - \rho \bar{x} \in C$, and thus we can apply Theorem 2.3.2. $\qquad\square$

Remark 2.3.1 It is possible to replace assumption (a) in Corollary 2.3.1 by the somewhat stronger assumption $C + L \subset C$.

Corollary 2.3.2 *Let C be a nonempty closed set of $\mathbb{R}^n$ and $A : \mathbb{R}^n \to \mathbb{R}^m$ a linear mapping. Then, $A(C)$ is closed and $A(C_\infty) = (A(C))_\infty$ under either of the following conditions:*
(a) $\ker A \cap C_\infty = \{0\}$,
(b) C is convex and $\ker A \cap C_\infty$ is a linear subspace.

Proof. Assume (a) holds, and in order to use Theorem 2.3.2 let $\{y_k\} \subset A(C)$ converge to y. Then there cannot exist any sequence $x_k \in C$ satisfying (2.3). Indeed, in the contrary case, we would have

$$\left\| A \frac{x_k}{\|x_k\|} - \frac{y_k}{\|x_k\|} \right\| \leq \frac{\|y - y_k\|}{\|x_k\|},$$

and passing to the limit we obtain $A\bar{x} = 0$. Furthermore, since $x_k \in C$, it follows that $\bar{x} \in C_\infty$ with $\|\bar{x}\| = 1$, in contradiction to the assumption (a), asserting $\ker A \cap C_\infty = \{0\}$. Now assume (b). Then since C is convex, assumption (a) in Corollary 2.3.1 is satisfied by Proposition 2.1.5. Furthermore, since $\ker A \cap C_\infty$ is a linear subspace, the assumption $z \in \ker A \cap C_\infty$ implies that $z \in -C_\infty$, and assumption (b) in Corollary 2.3.1 holds. $\square$

Corollary 2.3.3 *Let $S \subset \mathbb{R}^n$ be closed with $0 \notin S$. Then*

$$\mathrm{cl}\,\mathrm{pos}\, S = \mathrm{pos}\, S \cup S_\infty.$$

Furthermore, if S is bounded, then $\mathrm{pos}\, S$ is closed.

Proof. Let K be the cone in $\mathbb{R}^{n+1}$ generated by $\{(1, x) \mid x \in S\}$, and let $A : (\alpha, x) \to x$. Then $\mathrm{pos}\, S = A(K)$. By Lemma 2.1.1 one has $\mathrm{cl}\, K = K \cup \{(0, x) \mid x \in S_\infty\}$ and $A(\mathrm{cl}\, K) = \mathrm{pos}\, S \cup S_\infty$. Invoking Corollary 2.3.2 (a) with $C := \mathrm{cl}\, K$ and since $0 \notin S$ noting that in that case $\ker A \cap C_\infty = \{0\}$, it follows that $\mathrm{pos}\, S \cup S_\infty$ is closed. Now, since $\mathrm{pos}\, S \subset \mathrm{pos}\, S \cup S_\infty$ and $S_\infty \subset \mathrm{cl}\,\mathrm{pos}\, S$, then

$$\mathrm{cl}\,\mathrm{pos}\, S \subset \mathrm{cl}(\mathrm{pos}\, S \cup S_\infty) = \mathrm{pos}\, S \cup S_\infty \subset \mathrm{pos}\, S \cup \mathrm{cl}\,\mathrm{pos}\, S \subset \mathrm{cl}\,\mathrm{pos}\, S,$$

and hence $\mathrm{pos}\, S \cup S_\infty = \mathrm{cl}\,\mathrm{pos}\, S$. Finally, when S is bounded $S_\infty = \{0\}$, and this proves the last statement. $\square$

Asymptotic Linear Sets

We introduce now the class of asymptotically linear sets. As we shall see, this notion is of great interest. We begin with the basic definition of this concept.

Definition 2.3.1 *Let C be a nonempty closed set of $\mathbb{R}^n$. Then C is said to be an asymptotically linear set if for each $\rho > 0$ and each sequence $\{x_k\}$ satisfying*

$$x_k \in C, \quad \|x_k\| \to +\infty, \quad x_k \|x_k\|^{-1} \to \bar{x}, \tag{2.6}$$

there exists $k_0 \in \mathbb{N}$ such that

$$x_k - \rho \bar{x} \in C \quad \forall k \geq k_0. \tag{2.7}$$

If C is a closed convex set, we also have $x_k + \rho \bar{x} \in C$. This justifies the terminology "asymptotically linear".

As a consequence of Definition 2.3.1 a set that is the intersection of a finite number of sets, each of them being the union of a finite number of asymptotically linear sets, is also an asymptotically linear set.

Definition 2.3.2 *A set $C \subset \mathbb{R}^n$ is said to be a simple asymptotically polyhedral set if there exists a nonnegative integer l for which the set $C_l := C \cap \{x : \|x\| \geq l\}$ admits the decomposition*

$$C_l = K + M,$$

with K compact and M a polyhedral cone.
The set C is said to be asymptotically polyhedral if it is the intersection of a finite number of sets, each of them being the union of a finite number of simple asymptotically polyhedral sets.

When $l = 0$, the simplest set of this type is clearly a polyhedral set.

Recall that when C is convex without lines, the Krein–Milman theorem (cf. Theorem 1.1.3), always permits the decomposition $C_l = K + M$, where K is the convex hull of the extreme points of C, $M = C_\infty$, and $l = 0$. In this case when C_∞ is polyhedral, and when the set of extreme points is bounded (which is the case if C is a polyhedral set), then C is asymptotically polyhedral. It is also easy to note that the extension to sets for which $l > 0$ gives a wider class of sets.

Proposition 2.3.1 *An asymptotically polyhedral set C of $\mathbb{R}^n$ is asymptotically linear.*

Proof. Obviously the proposition holds if it holds for simple asymptotically polyhedral sets. Thus, we suppose that C is a simple asymptotically polyhedral set and let $\{x_k\}$ be a sequence satisfying (2.6), and $\rho > 0$. If (2.7) is not satisfied, there exists a subsequence of $\{x_k - \rho \bar{x}\} \notin C$. Without loss of generality we can suppose that $x_k - \rho \bar{x} \notin C$, and that $\|x_k\| \geq l$ for each k.

Since M is a polyhedral cone, M is finitely generated (cf. Chapter 1) and there exist rays d_i, $i = 1, \ldots, r$, such that

$$M = \left\{ y \mid \exists \lambda_i \geq 0 \; i = 1 \ldots r \text{ such that } y = \sum_{i=1}^{r} \lambda_i d_i \right\}.$$

Then for each k, since $x_k \in C_l$, there exist $y_k \in K$, a subset $I_k \subset \{1, \ldots, m\}$, and $\lambda_i^k \geq 0$ for $i \in I_k$ such that

$$x_k = y_k + z_k, \quad z_k = \sum_{i \in I_k} \lambda_i^k d_i,$$

and we can assume that the vectors $\{d_i\}, i \in I_k$, are linearly independent. Indeed it can be easily seen that M is a finite union of sets of the same type but with the vectors d_i linearly independent.

Since $\{y_k\}$ is bounded, it follows that $\lim_{k \to \infty} \|z_k\| = +\infty$. As a consequence, there exist a subsequence z_{k_m} and a nonempty set of indices I such that the following hold:

$$\lim_{m \to \infty} \lambda_i^{k_m} = +\infty \ , \quad \frac{\lambda_i^{k_m}}{\sum_{i \in I} \lambda_i^{k_m}} \to \mu_i \geq 0 \ \ \forall i \in I, \quad \lambda_i^{k_m} \geq 0 \ \forall i \notin I,$$

the vectors $d_i, i \in I$, are linearly independent, and the sequence $\{\lambda_i^{k_m}\}$ is bounded for $i \notin I$. Now let $w_k = \sum_{i \in I} \lambda_i^k d_i$, $w = \sum_{i \in I} \mu_i d_i$. Then $w \neq 0$ and

$$\lim_{m \to \infty} \frac{w_{k_m}}{\sum_{i \in I} \lambda_i^{k_m}} = w, \quad \bar{x} = \frac{w}{\|w\|},$$

and we thus have

$$x_{k_m} - \rho \bar{x} = y_{k_m} + \sum_{i \in I} \left(\lambda_i^{k_m} - \rho \frac{\mu_i}{\|w\|} \right) d_i + \sum_{i \notin I} \lambda_i^{k_m} d_i.$$

Since for m sufficiently large $\lambda_i^{k_m} - \rho \frac{\mu_i}{\|w\|} \geq 0$ for $i \in I$, it follows that $x_{k_m} - \rho \bar{x} \in C$ for m sufficiently large, which is impossible. □

As we shall see in the next chapter, where we introduce the class of asymptotically stable functions, which contains in particular convex polynomial functions, we can enlarge considerably the class of asymptotically linear sets.

Theorem 2.3.3 *Let C be asymptotically linear. Then $A(C)$ is closed and $A(C_\infty) = (A(C))_\infty$.*

Proof. Let $\{y_k\} \subset A(C)$ converge to y and let $\{x_k\}$ be any sequence satisfying (2.3). Then, since C is asymptotically linear, by definition there exists $\rho_k > 0$ such that for k sufficiently large $x_k - \rho_k \bar{x} \in C$, $\rho_k \leq \|x_k\|$, and we can thus invoke Theorem 2.3.2, which implies the results. □

Closure of the Sum of Closed Convex Sets

The sum of closed convex sets might not be closed even when the sets are themselves closed. It is thus important to find conditions under which closedness is preserved. Obviously, if C_1, C_2 are closed and one of the two sets is also bounded, then $C_1 + C_2$ is also closed. As another application of Theorem 2.3.2, we now derive weaker and general conditions preserving closedness under appropriate assumptions on the sets involved.

Definition 2.3.3 *Let $C_i \subset \mathbb{R}^n$, $i = 1, \ldots, m$, be a collection of nonempty closed sets such that for any $z_i \in (C_i)_\infty, i = 1, \ldots, m$, $\sum_{i=1}^m z_i = 0$. Then:*
(a) If one has $z_i = 0$ for all $i = 1, \ldots, m$, the collection C_i is said to be in general position.
(b) If for any $i = 1, \ldots, m$ one has

(i) $z_i \in -(C_i)_\infty$,

(ii) $z_i + C_i \subset C_i$,

the collection C_i is said to be in relative general position.

Remark 2.3.2 Inclusion (ii) is in fact equivalent to $\mathbb{R}_+ z_i + C_i \subset C_i$.

The condition (a) imposed on the set C_i in the definition is stronger than the second one, asking for relative general position of the sets C_i, but is often easier to use for proving closedness of the sum of sets; see Corollary 2.3.4 below.

Theorem 2.3.4 *Let $C_i \subset \mathbb{R}^n, i = 1, \ldots, m$, be nonempty closed sets, which are supposed to be in relative general position. Then $\sum_{i=1}^m C_i$ is a closed set, and we have*

$$(C_1 + \cdots + C_m)_\infty \subset (C_1)_\infty + \cdots + (C_m)_\infty,$$

where the inclusion holds as an equation if in addition the C_i are all convex.

Proof. Let $C := C_1 \times \cdots \times C_m$ and define the linear map $A : (x_1, \ldots, x_m) \to x_1 + \cdots + x_m$, $x_i \in \mathbb{R}^n$. Then $A(C) = C_1 + \ldots + C_m$. Since by Proposition 2.1.10 one has $C_\infty \subset (C_1)_\infty \times \ldots \times (C_m)_\infty$, then $\forall z \in \ker A \cap C_\infty$ it follows from hypothesis (i) and (ii) of Definition 2.3.3 that $-z + C \subset C$. Therefore, $-2z + C = -z + (-z + C) \subset C$, and by recurrence, $-kz + C \subset C$, $\forall k \in \mathbb{N}_*$. Thus, $\forall x \in C$, $x_k = x - kz \in C$ and $\lim_{k \to \infty} \frac{x_k}{k} = -z$, implying $-z \in C_\infty$. Invoking Corollary 2.3.1 (with hypothesis (a) of that theorem replaced by (ii) of Definition 2.3.3, cf. Remark 2.3.2, the result concerning the inclusion is proved. If the C_i are in addition convex, the reverse inclusion is an immediate consequence of Proposition 2.1.5. □

As another application of Theorem 2.3.4 we obtain the following simple test to check the closedness of the sum of nonempty closed sets.

Corollary 2.3.4 *Let $C_i \subset \mathbb{R}^n, i = 1, \ldots, m$, be nonempty closed sets. Suppose that $z_i \in (C_i)_\infty$ with $\sum_{i=1}^m z_i = 0$ implies that $z_i = 0 \ \forall i = 1, \ldots, m$. Then $\sum_{i=1}^m C_i$ is a closed set*

The next application of Theorem 2.3.2 concerns a closedness criterion of the set of convex combinations of a finite number of nonempty closed sets C_i, $i = 1, \ldots, m$, of $\mathbb{R}^n$. Define the following two sets:

$$S \ := \ \left\{ x \in \mathbb{R}^n \mid x = \sum_{i=1}^m \lambda_i x_i, \lambda \in \Delta_m, \ x_i \in C_i, \ i = 1, \ldots, m \right\} (2.8)$$

$$T \ := \ \left\{ x \in \mathbb{R}^n \mid x = \sum_{i=1}^m \lambda_i * x_i, \lambda \in \Delta_m \right\} \tag{2.9}$$

where

$$\lambda_i * x_i := \begin{cases} \lambda_i x_i, & \text{with } x_i \in C_i \text{ if } \lambda_i > 0, \\ x_i, & \text{with } x_i \in (C_i)_\infty \text{ if } \lambda_i = 0, \end{cases}$$

and Δ_m denotes the simplex in $\mathbb{R}^m$.

Theorem 2.3.5 *For a finite collection of nonempty closed sets $C_i \subset \mathbb{R}^n$, $i = 1, \ldots, m$, that are in relative general position, one has $\operatorname{cl} S = T$.*

Proof. Let K_i be a cone in $\mathbb{R}^{n+1}$ generated by $\{(1, x_i) \mid x_i \in C_i\}$ and let $D_i := \{(0, x_i) \mid x_i \in (C_i)_\infty\}$. Applying Lemma 2.1.1 for each $i = 1, \ldots, m$ we have

$$\operatorname{cl} K_i = K_i \cup D_i. \tag{2.10}$$

Furthermore, since K_i is a cone, we also have

$$\operatorname{cl} K_i = (K_i)_\infty = (\operatorname{cl} K_i)_\infty = K_i \cup D_i. \tag{2.11}$$

We now want to apply Theorem 2.3.4 to the sets $\operatorname{cl} K_i$. Let $z_1, \ldots, z_m$ be vectors such that

$$z_i := (t_i, d_i) \in (\operatorname{cl} K_i)_\infty, \ \forall i \in [1, m], \ \sum_{i=1}^m z_i = 0.$$

By (2.11), $z_i = (t_i, d_i) \in (\operatorname{cl} K_i)_\infty$ implies that $z_i = (t_i, d_i) \in K_i \cup D_i$, which in turns implies $t_i = 0$, $d_i \in (C_i)_\infty$, and $\sum_{i=1}^m d_i = 0$, i.e., $z_i = (0, d_i)$ with $d_i \in (C_i)_\infty$, and therefore, since hypothesis (i) of Definition 2.3.3(b) is satisfied for the sets C_i, it is also satisfied for the sets $\operatorname{cl} K_i$, namely $-z_i \in \operatorname{cl} K_i$, $\forall i = 1, \ldots, m$. Now, we need to verify that hypothesis (ii) of Definition 2.3.3(b) holds for the sets $\operatorname{cl} K_i$, given that it holds for C_i; i.e., we have to prove that

$$z_i + \operatorname{cl} K_i \subset \operatorname{cl} K_i, \ \forall i = 1, \ldots, m. \tag{2.12}$$

Let $z_i = (0, d_i)$, $d_i \in (C_i)_\infty$ and take any $0 \neq y_i \in \operatorname{cl} K_i$. Then by (2.10), $y_i \in K_i$ or $y_i \in D_i$. If $y_i \in K_i$, one has $y_i = \lambda_i(1, \bar{y}_i), \lambda_i > 0$, with $\bar{y}_i \in C_i$ and

$$y_i + z_i = (\lambda_i, \lambda_i \bar{y}_i) + (0, d_i) = \lambda_i \left(1, \bar{y}_i + \frac{d_i}{\lambda_i}\right).$$

Since $\bar{y}_i \in C_i$, $\lambda_i^{-1} d_i \in (C_i)_\infty$, and the sets C_i satisfy hypothesis (ii) of Definition 2.3.3(b), then $u_i := \bar{y}_i + \lambda_i^{-1} d_i \in C_i$, and hence $y_i + z_i = \lambda_i(1, u_i) \in K_i$. Otherwise, if $y_i \in D_i$, one has $y_i = (0, \bar{d}_i)$, $\bar{d}_i \in (C_i)_\infty$, and $y_i + z_i = (0, \bar{d}_i + d_i)$. But since $\bar{d}_i \in (C_i)_\infty$, then for all $i = 1, \ldots, m$,

$$\exists t_i^k > 0, t_i^k \to 0^+, \exists x_i^k \in C_i \text{ such that } \bar{d}_i = \lim_{k \to \infty} t_i^k x_i^k,$$

so that

$$\bar{d}_i + d_i = \lim_{k \to \infty} t_i^k \left(x_i^k + \frac{d_i}{t_i^k}\right), \text{ with } x_i^k + \frac{d_i}{t_i^k} \in C_i,$$

by hypothesis (b)(ii) of Definition 2.3.3 on the sets C_i. This implies that $\bar{d}_i + d_i \in (C_i)_\infty$ and therefore $y_i + z_i \in D_i$, proving (2.12). We can thus apply Theorem 2.3.4 to the sets $\operatorname{cl} K_i$ to conclude that $\operatorname{cl} K_1 + \cdots + \operatorname{cl} K_m$ is closed. But since one always has

$$\operatorname{cl}(K_1 + \cdots + K_m) = \operatorname{cl}(\operatorname{cl} K_1 + \cdots + \operatorname{cl} K_m),$$

it follows that

$$\operatorname{cl}(K_1 + \cdots + K_m) = \operatorname{cl} K_1 + \cdots + \operatorname{cl} K_m. \tag{2.13}$$

Now consider the hyperplane $H := \{(1, x) \,|\, x \in \mathbb{R}^n\}$ and let

$$\begin{aligned} E &:= (\operatorname{cl} K_1 + \cdots + \operatorname{cl} K_m) \cap H, \\ F &:= \operatorname{cl}(K_1 + \cdots + K_m) \cap H. \end{aligned}$$

Since by (2.11) one has $\operatorname{cl} K_i = K_i \cup D_i$, one can verify that in fact $E = (1, T)$, with T defined in (2.9). Now $y := (t, x) \in F$ if and only if $y = \lim_{k \to \infty} y^k$ with

$$y^k = \sum \lambda_i^k (1, x_i^k), \ \lambda_i^k \geq 0, x_i^k \in C_i \text{ and } \lim_{k \to \infty} \sum_{i=1}^m \lambda_i^k = 1.$$

Define $\mu_i^k := \lambda_i^k (\sum_{i=1}^m \lambda_i^k)^{-1}$. Then $y \in F$ if and only if $y = (1, x)$ with $x = \lim_{k \to \infty} \sum_{i=1}^m \mu_i^k x_i^k$ with $x_i^k \in C_i$. In other words, one has $F = (1, \operatorname{cl} S)$. Finally, by the definition of E and F and from (2.13) it follows that $T = \operatorname{cl} S$. □

The convex case is much simpler and follows as an immediate consequence of Theorem 2.3.5.

Corollary 2.3.5 *Let $C_i \subset \mathbb{R}^n, i = 1, \ldots, m$, be nonempty closed convex sets such that for any $z_i \in (C_i)_\infty \; \forall i = 1, \ldots, m$, and $\sum_{i=1}^m z_i = 0$ one has $z_i \in -(C_i)_\infty$ for each $i = 1, \ldots, m$. Then*

$$T = \mathrm{cl}\, S = \mathrm{cl}(\mathrm{conv} \cup_{i=1}^m C_i),$$

where S, T are as defined in (2.8)–(2.9).

Proof. Note that when C_i is convex for each i, then one has $S = \mathrm{conv}(\cup_{i=1}^m C_i)$, and hypothesis (ii) of Definition 2.3.3(b) holds. Applying Theorem 2.3.5, the result follows. $\square$

As another consequence of Theorem 2.3.5 we obtain a decomposition formula for the closed convex hull of an arbitrary closed set in $\mathbb{R}^n$, in terms of its convex hull and the associated asymptotic cone. This leads to the following notions.

Definition 2.3.4 *Let C be a nonempty set of $\mathbb{R}^n$ and suppose that for any vectors $c_1, \ldots, c_{n+1}$ such that $\sum_{i=1}^{n+1} c_i = 0$, $c_i \in C_\infty$, $i = 1, \ldots, n+1$, one has*

$$c_i \in -C_\infty, \quad c_i + C \subset C, \; \forall i = 1, \ldots, n+1.$$

Then C is said to be weakly semibounded. If instead of the above inclusion we have the stronger assumption

$$c_i = 0 \quad \forall i = 1 \ldots m,$$

C is said to be semibounded.

Another way to say that a set C is semibounded is to suppose that C_∞ is a pointed cone; cf. Definition 1.1.3.

Corollary 2.3.6 *Let C be a closed weakly semibounded set. Then*

$$\mathrm{cl}\,\mathrm{conv}\, C = \mathrm{conv}\, C + \mathrm{conv}\, C_\infty.$$

Proof. Apply Theorem 2.3.5 with $m = n + 1$ and $C_i = C \; \forall i$. $\square$

Lemma 2.3.1 *Let K be a closed pointed cone in $\mathbb{R}^n$. Then:*
(a) $\exists \theta > 0$ such that $\|x_i\| \leq \theta \| \sum_{i=1}^{n+1} x_i \|$, $\forall x_i \in K, i = 1, \ldots, n+1$.
(b) $\mathrm{conv}\, K$ is closed pointed cone.

Proof. (a) The proof is by contradiction. Then for all $j \in \mathbb{N}$, there exist for $i = 1, \ldots, n+1$, $u_i^j \in K$ such that $\|u_i^j\| > j \| \sum_{i=1}^{n+1} u_i^j \|$. Since K is pointed, it follows that $\sum_{i=1}^{n+1} u_i^j \neq 0$, and then if we set

$$x_i^j = \left\| \frac{u_i^j}{\sum_{i=1}^{n+1} u_i^j} \right\|, \quad y^j = \sum_{i=1}^{n+1} x_i^j,$$

we have

$$\forall i \quad x_i^j \in K, \ \|y^j\| = 1, \ \lambda^j := \left(\max_{1 \leq i \leq n+1} \|x_i^j)\| \right)^{-1} \to 0.$$

Since $\|\lambda^j x_i^j\| \leq 1$, there exists a subsequence $\{\lambda^{j_l}, x_i^{j_l}, \ i = 1, \ldots, n+1\}$ such that for each i, $\lambda^{j_l} x_i^{j_l} \to z_i$. Now since K is a closed cone, $z_i \in K$ and since $\| \sum_{i=1}^{n+1} \lambda^{j_l} x_i^{j_l} \| = \lambda^{j_l}$, it follows that $\sum_{i=1}^{n+1} z_i = 0$. Furthermore, since one has $1 \leq \sum_{i=1}^{n+1} \lambda^j \|x_i^j\| \leq n+1$, then at least one z_i is nonzero, which contradicts the hypothesis that K is pointed.

(b) The fact that conv K is pointed is obvious. Now let us prove that it is closed, and let $x^j \in \text{conv } K$ with $x^j \to x$. Then by Theorem 1.1.1 there exist $x_i^j \in K$ with $i = 1, \ldots, n+1$ and such that $x^j = \sum_{i=1}^{n+1} x_i^j$. By part (a) the sequences $\{x_i^j\}_{j \in \mathbb{N}}$ are bounded. Passing to subsequences if necessary, it follows that there exist $\{x_i^j\}$ with $x_i^j \to y_i$. Then, $x = \sum_{i=1}^{n+1} y_i \in \text{conv } K$, and we can conclude that conv K is closed. $\qquad \square$

Lemma 2.3.2 *Let C be a nonempty set of $\mathbb{R}^n$. Then:*
(a) $(\text{conv } C)_\infty = \text{conv } C_\infty$.
(b) If in addition C is semibounded then $\text{conv } C$ is semibounded.

Proof. (a) Clearly, one has that $(\text{conv } C)_\infty$ is a convex set. Therefore, the inclusion $C_\infty \subset (\text{conv } C)_\infty$ implies that $\text{conv } C_\infty \subset (\text{conv } C)_\infty$. To prove the reverse inclusion $(\text{conv } C)_\infty \subset \text{conv } C_\infty$, let $d \in (\text{conv } C)_\infty$. Then there exist $x_k \in \text{conv } C$, $t_k \to +\infty$ such that $t_k^{-1} x_k \to d$. Every $x_k \in \text{conv } C$ can be written as

$$x_k = \sum_{i=0}^n \lambda_k^i x_k^i, \text{ with } x_k^i \in C, \ \lambda_k^i \geq 0, \sum_{i=0}^n \lambda_k^i = 1.$$

Since $\{t_k^{-1} \lambda_k^i x_k^i\}$ is a bounded sequence, passing to subsequences if necessary, it follows that $t_k^{-1} \lambda_k^i x_k^i \to d^i \in C_\infty$ and therefore $d = \sum_{i=0}^n d^i$, so that $d \in \text{conv } C_\infty$. To prove (b), note that $C \subset \mathbb{R}^n$ being semi-bounded means that C_∞ is pointed. Invoking Lemma 2.3.1(b), the latter implies that $\text{conv } C_\infty$ is pointed, which in turn implies that $(\text{conv } C)_\infty$ is pointed and hence $\text{conv } C$ is semibounded. $\qquad \square$

Lemma 2.3.3 *Let C be a nonempty convex and semibounded set of $\mathbb{R}^n$. Then*

$$\text{int } C_\infty \neq \emptyset \iff \text{cl dom } \sigma_C \text{ is pointed.}$$

Proof. Applying Theorem 2.2.1 to the closed convex set $\text{cl } C$ one has $C_\infty = (\text{cl } C)_\infty = (\text{cl dom } \sigma_C)^*$, and the desired result follows from Proposition 1.1.15. $\qquad \square$

2.4 Continuous Convex Sets

Convex compact sets in $\mathbb{R}^n$ enjoy many important properties that are not shared by closed convex sets in general. For example, if C and D are compact and convex, then the sets $C + D$ and $\operatorname{conv}(C \cup D)$ enjoy the same property. Moreover, if $C \cap D = \emptyset$, then C and D can be strongly separated by a hyperplane and $\delta(C, D) = \inf\{\|c - d\| \mid c \in C, d \in D\} > 0$. In general, none of these properties remain true for arbitrary closed convex sets in $\mathbb{R}^n$. However, there is a natural class of closed convex sets in $\mathbb{R}^n$ that are not necessarily compact, yet which enjoy these properties. This is the class of *continuous* sets.

Definition 2.4.1 *A nonempty closed convex set of $\mathbb{R}^n$ is called continuous if its support function $x \to \sigma_C(x)$ is continuous for any $x \neq 0$.*

We recall that if $\sigma_C(x) = +\infty$, the continuity of σ_C at x means that there exists a neighborhood V of x such that $\sigma_C(y) = +\infty$ for all $y \in V$.
Continuous convex sets can be characterized in a more geometric sense using the concepts of boundary rays and asymptotes.

Definition 2.4.2 *Let C be a nonempty closed set of $\mathbb{R}^n$. A boundary ray of the set C is a half-line that is contained in the boundary of C. An asymptote of C is a half-line ρ contained in $\mathbb{R}^n \setminus C$ such that*

$$\delta(\rho, C) := \inf\{\|x - y\| \mid x \in \rho, y \in C\} = 0.$$

Proposition 2.4.1 *Let C be a nonempty closed convex set of $\mathbb{R}^n$, $x \in C$, and suppose that the half-line $\{y + t\rho : t \geq 0\}$ with $\rho \in C_\infty$ and $y \in \mathbb{R}^n$ has an empty intersection with C. For $\lambda \in \mathbb{R}$, let $x(\lambda) = (1 - \lambda)x + \lambda y$. Then there exist $\lambda \in [0, 1], s \geq 0$ such that the half-line $D(\lambda, s) = \{x(\lambda) + t\rho \mid t \geq s\}$ is a boundary ray or an asymptote of C.*

Proof. For any $\lambda \in \mathbb{R}$ and any $t \geq 0$, define $G(\lambda, t) = \operatorname{dist}(x(\lambda) + t\rho, C)$ and $g(\lambda) = \inf\{G(\lambda, t) \mid t \geq 0\}$. Then, since C is a closed convex set, the function g is finite and convex on $\mathbb{R}$, and as a consequence, is continuous. Let $E := \{\lambda \in [0, 1] \mid g(\lambda) = 0\}$. Since $x \in C$ and $\rho \in C_\infty$, one has $0 \in E$ and since g is continuous, it follows that E is a nonempty compact set. Therefore, there exists $\lambda^* \in [0, 1]$ that maximizes λ on E. If $\lambda^* = 1$, then by definition $D(1, 0)$ is an asymptote of C; otherwise $\lambda^* < 1$. Now suppose that $D(\lambda^*, 0)$ is not an asymptote. Then $D(\lambda^*, 0) \cap C \neq \emptyset$. Take $u \in D(\lambda^*, 0) \cap C$. Then $D := \{u + t\rho \mid t \geq 0\} \subset C$. Without loss of generality we can suppose that $\operatorname{int} C \neq \emptyset$ (otherwise, D would be a boundary ray), and if D were not a boundary ray, there would exist $s \geq 0$ such that $z := u + s\rho \in \text{int } C$. As a consequence, there would exist $\Delta\lambda > 0$ such that $\{x(\lambda^* + \Delta\lambda) + t\rho \mid t \geq 0\} \cap C \neq \emptyset$, a contradiction with the fact that λ^* is the maximum of λ on E. $\qquad\square$

The next result gives an important characterization of continuous convex sets.

Theorem 2.4.1 *A nonempty closed convex subset C of $\mathbb{R}^n$ is continuous if and only if C has no boundary ray or asymptote.*

Proof. Assume first that C has no boundary ray or asymptote and suppose that C is not continuous at some point $u \neq 0$. Then there exists a sequence $\{u_k\}$ converging to u such that $\sigma_C(u_k)$ does not converge to $\sigma_C(u)$. Since σ_C is lsc, it follows that $\alpha := \limsup_{k\to\infty} \sigma_C(u_k) > \sigma_C(u)$. Furthermore, there exist $\beta \in \mathbb{R}$ with $\sigma_C(u) < \beta \leq \alpha$ and a sequence $\{x_{k_j}\} \subset C$ such that

$$\lim_{j\to\infty} \langle u_{k_j}, x_{k_j} \rangle \geq \beta. \tag{2.14}$$

When $\alpha \in \mathbb{R}$, taking $\beta = \alpha$, the latter inequality follows from the definition of the support functional, and when $\alpha = +\infty$, it follows for the same reason. Let us prove now that the sequence $\{x_{k_j}\}$ cannot be bounded. Assume the contrary. Then with $\{x_{k_j}\}$ bounded, without loss of generality we can suppose that $x_{k_j} \to x \in C$, and taking the limit in (2.14) it follows that $\sigma_C(u) < \beta \leq \langle u, x \rangle$, which is impossible. Thus, without loss of generality we can suppose that $\|x_{k_j}\| \to +\infty$ and $x_{k_j}\|x_{k_j}\|^{-1} \to \bar{x}$. Dividing both members of (2.14) by $\|x_{k_j}\|$ and passing to the limit when $j \to +\infty$, it follows that $\langle u, \bar{x} \rangle \geq 0$. Now let $y \notin C$, with $\langle y, u \rangle > \sigma_C(u)$; as a consequence it follows that $\langle y + \lambda\bar{x}, u \rangle > \sigma_C(u)$, $\forall \lambda > 0$, and then invoking Theorem 1.3.2, this implies that $y + \lambda\bar{x} \notin C$, $\forall \lambda > 0$. Therefore, by Proposition 2.4.1, there exists some ray of direction $\bar{x}$, which is a boundary ray or an asymptote of C, in contradiction to the hypothesis of the theorem. We now prove the reverse statement. Suppose that σ_C is continuous, and there exist some ray y and a direction $\bar{x}$ such that the set $D = \{\bar{x} + \lambda y \mid \lambda \geq 0\}$ is a boundary ray or an asymptote. In both cases, invoking Proposition 1.1.11, it can be separated from C by a hyperplane; i.e., there exist $0 \neq a \in \mathbb{R}^n$, $b \in \mathbb{R}$ such that

$$\inf_{z \in D} \langle a, z \rangle \geq b, \quad \sup_{x \in C} \langle a, x \rangle \leq b. \tag{2.15}$$

Let us prove that σ_C is not continuous at a. Since for $z \in D$ one has $\langle a, z \rangle = \langle a, \bar{x} \rangle + \lambda \langle a, y \rangle$, letting $\lambda \to +\infty$, it follows from the first inequality in (2.15) that $\langle a, y \rangle \geq 0$. Suppose that D is a boundary ray. Then from the second inequality in (2.15) it follows that $\lambda \langle a, y \rangle \leq b - \langle a, \bar{x} \rangle$, and letting $\lambda \to +\infty$, this implies $\langle a, y \rangle \leq 0$, so that we have proved that $\langle a, y \rangle = 0$. The same holds if D is an asymptote. Indeed, by definition there exist

$z_k = \bar{x} + \lambda_k y$, with $\lambda_k \to +\infty$, $x_k \in C$ such that $\|x_k - z_k\| \to 0$. Then using the second inequality in (2.15) we obtain

$$\langle a, x_k \rangle = \langle a, \bar{x} \rangle + \lambda_k \langle a, y \rangle + \langle a, x_k - z_k \rangle \le b,$$

and passing to the limit we obtain $\langle a, y \rangle \le 0$, and hence $\langle a, y \rangle = 0$. Now from the second inequality in (2.15) one has that $\sigma_C(a)$ is finite. Let $\varepsilon > 0$ and let us prove that $\sigma_C(a + \varepsilon y) = +\infty$. This will prove that σ_C is not continuous at a. We consider two cases. If D is a boundary ray, then for all $\lambda > 0$,

$$\sigma_C(a + \varepsilon y) \ge \langle a + \varepsilon y, \bar{x} + \lambda y \rangle = \langle a, \bar{x} \rangle + \varepsilon \langle y, \bar{x} \rangle + \varepsilon \lambda \|y\|^2 + \lambda \langle a, y \rangle,$$

and letting $\lambda \to +\infty$, it follows that $\sigma_C(a + \varepsilon y) = +\infty$. If D is an asymptote,

$$\sigma_C(a + \varepsilon y) \ge \langle a + \varepsilon y, x_k \rangle = \langle a + \varepsilon y, \bar{x} \rangle + \langle a + \varepsilon y, x_k - z_k \rangle + \lambda_k \langle a + \varepsilon y, y \rangle.$$

Since $\langle a + \varepsilon y, x_k - z_k \rangle \to 0$ and since $\lambda_k \langle a + \varepsilon y, y \rangle = \varepsilon \lambda_k \|y\|^2$, it follows that $\sigma_C(a + \varepsilon y) = +\infty$. $\square$

As previously discussed in the beginning of Section 2.3, a key question in convex analysis is to know when the image of a closed convex set under a linear map remains closed. As demonstrated Example 2.3.1, where the set is continuous, the standard sufficient condition (2.2) fails, and in Theorem 2.3.2 we gave a necessary and sufficient condition for preserving closedness of the image of a closed convex set. However, thanks to this theorem, the next result shows that for continuous convex sets, the closedness property is guaranteed.

Proposition 2.4.2 *Let $C \subset \mathbb{R}^n$ be a nonempty closed convex and continuous set and let $A : \mathbb{R}^n \to \mathbb{R}^m$ be a linear map. Then $A(C)$ is closed. Furthermore, if A is a surjective map, then $A(C)$ is a continuous set.*

Proof. In order to prove that $A(C)$ is closed we can suppose that aff $A(C) \ne A(C)$. Otherwise, there is nothing to prove. Then there exists $u \in \mathrm{bd}(\mathrm{cl}\, AC)$ with $u \notin \mathrm{ri}\, A(C)$, and invoking the separation Proposition 1.1.11 there exists $0 \ne a \in \mathbb{R}^n$ such that the hyperplane $H = \{x \in \mathbb{R}^n \mid \langle a, x \rangle = \beta\}$ separates properly $\{u\}$ and $\mathrm{cl}\, A(C)$, namely, one has

$$\langle a, u \rangle = \beta, \ \langle a, z \rangle \le \beta, \ \forall z \in \mathrm{cl}\, A(C), \ \inf_{z \in \mathrm{cl}\, A(C)} \langle a, z \rangle < \beta,$$

so that

$$\langle A^T a, x \rangle \le \beta, \ \forall x \in C.$$

Now take d and x satisfying

$$\langle A^T a, d \rangle = \beta, \ \langle A^T a, x \rangle < \beta,$$

and set $z = 2d - x$. In order to prove that $A(C)$ is closed, let $\{y_k\}$ be a sequence converging to y. Then invoking Theorem 2.3.2 it is sufficient to prove only that there cannot exist any sequence $\{x_k\}$ satisfying (2.3). Suppose the contrary, and let $\{x_k\}$ be such a sequence. Then since $A\bar{x} = 0$, for each $\lambda \geq 0$ one obtains

$$
\begin{aligned}
\langle A^T a, z + \lambda \bar{x} \rangle &= 2\langle A^T a, d \rangle - \langle A^T a, x \rangle + \lambda \langle A^T a, \bar{x} \rangle \\
&= 2\beta - \langle A^T a, x \rangle > \beta,
\end{aligned}
$$

showing that the ray $\Delta = \{z + \lambda \bar{x} : \lambda \geq 0\}$ does not meet C. Therefore, since $\bar{x} \in C_\infty$, it follows from Proposition 2.4.1 that C is not a continuous set, leading to the desired contradiction. To prove the last part of the proposition, assume now that the linear map is surjective. Since $(\ker(A^T))^\perp = A(\mathbb{R}^n) = \mathbb{R}^m$, it follows that $\ker(A^T) = \{0\}$. Furthermore, from Proposition 1.3.3(e) we have $\sigma_{A(C)}(z) = \sigma_C(A^T z)$. Since $A^T z \neq 0$, $\forall z \neq 0$, and since C is continuous, this implies from Definition 2.4.1 that $\sigma_{A(C)}$ is continuous on $\mathbb{R}^m \setminus \{0\}$, and hence $A(C)$ is continuous. $\square$

Proposition 2.4.3 *For any nonempty closed convex and unbounded set $C \subset \mathbb{R}^n$, $C \neq \mathbb{R}^n$, the following statements are equivalent:*
(a) C is continuous.
(b) $\mathrm{dom}\,\sigma_C \setminus \{0\} = \mathrm{int}(\mathrm{dom}\,\sigma_C) \neq \emptyset$.
(c) $\arg\sup\sigma_C(d)$ is a nonempty compact set for all $d \in \mathrm{dom}\,\sigma_C \setminus \{0\}$.

Proof. We first show that $(a) \implies (b)$. If C is continuous, since C is assumed unbounded then $\mathrm{dom}\,\sigma_C \neq \mathbb{R}^n$ and $\mathrm{int}\,\mathrm{dom}\,\sigma_C \subset \mathrm{dom}\,\sigma_C \setminus \{0\}$. Furthermore, since $C \neq \mathbb{R}^n$, one has $\mathrm{dom}\,\sigma_C \setminus \{0\} \neq \emptyset$. Now let $0 \neq d \in \mathrm{dom}\,\sigma_C$. Then $d \in \mathrm{int}\,\mathrm{dom}\,\sigma_C$. Indeed, in the contrary case there would exist a sequence $\{d_k\}$ converging to d and such that $\sigma_C(d_k) = +\infty$, which is impossible, since $\sigma_C(d) = \lim_{k\to\infty} \sigma_C(d_k)$. To prove the reverse implication $(b) \implies (a)$, note that by Theorem 1.2.3, σ_C is continuous on $\mathrm{dom}\,\sigma_C \setminus \{0\}$, hence everywhere on $\mathbb{R}^n \setminus \{0\}$, implying that the set C is continuous. Finally, by Proposition 1.2.18 one has $\arg\sup\sigma_C(d) = \partial\sigma_C(d)$. But, by Proposition 1.2.16 the subdifferential of σ_C at d is nonempty and compact if and only if $d \in \mathrm{int}\,\mathrm{dom}\,\sigma_C$, and this proves the equivalence between (b) and (c). $\square$

2.5   Asymptotic Functions

Building on the concept of asymptotic cone, we are now interested in understanding the behavior in the large of functions $f : \mathbb{R}^n \to \mathbb{R} \cup \{+\infty\}$.

The epigraph plays a key role. Let F be a nonempty closed set in $\mathbb{R}^{n+1}$ satisfying the property

$$(x, \mu) \in F \implies (x, \mu') \in F, \ \forall \mu' > \mu. \tag{2.16}$$

Then clearly, there exists one and only one function g such that $\mathrm{epi}\, g = F$, defined by

$$g(x) := \inf\{\mu \mid (x, \mu) \in F\}.$$

Now let $f : \mathbb{R}^n \to \mathbb{R} \cup \{+\infty\}$ be proper. Then $\mathrm{epi}\, f$ is nonempty, and therefore $(\mathrm{epi}\, f)_\infty$ is a nonempty closed cone in $\mathbb{R}^{n+1}$ satisfying (2.16); i.e.,

$$(x, \mu) \in (\mathrm{epi}\, f)_\infty \implies (x, \mu') \in (\mathrm{epi}\, f)_\infty, \ \forall \mu' > \mu.$$

Indeed, from the definition of the asymptotic cone given in Definition 2.1.2 and applied to the set $\mathrm{epi}\, f$ we have

$$\exists (x_k, \mu_k) \in \mathrm{epi}\, f, \ \exists t_k \to \infty : \ t_k^{-1}(x_k, \mu_k) \to (x, \mu).$$

Set $\mu_k' := \mu_k + (\mu' - \mu)t_k$. Since $\mu' > \mu$, we have $\mu_k' > \mu_k \geq f(x_k)$ and $t_k^{-1}(x_k, \mu_k') \to (x, \mu')$, showing that $(x, \mu') \in (\mathrm{epi}\, f)_\infty, \ \forall \mu' > \mu$.
The above discussion leads us to introduce the following concept.

Definition 2.5.1 *For any proper function $f : \mathbb{R}^n \to \mathbb{R} \cup \{+\infty\}$, there exists a unique function $f_\infty : \mathbb{R}^n \to \mathbb{R} \cup \{+\infty\}$ associated with f, called the asymptotic function, such that $\mathrm{epi}\, f_\infty = (\mathrm{epi}\, f)_\infty$.*

This definition indicates that the epigraph of an asymptotic function is in fact a closed cone. This can be further elaborated through the use of lower semicontinuity and positively homogeneous functions; cf. Definition 1.3.2. The asymptotic function f_∞ enjoys the following basic properties.

Proposition 2.5.1 *For any proper function $f : \mathbb{R}^n \to \mathbb{R} \cup \{+\infty\}$, we have:*
(a) f_∞ is lsc and positively homogeneous.
(b) $f_\infty(0) = 0$ or $f_\infty(0) = -\infty$.
(c) If $f_\infty(0) = 0$, then f_∞ is proper.

Proof. (a) Since by Definition (2.5.1) $\mathrm{epi}\, f_\infty = (\mathrm{epi}\, f)_\infty$ and $(\mathrm{epi}\, f)_\infty$ is a closed set by definition, it follows that f_∞ is lsc. First, note that $0 \in \mathrm{dom}\, f_\infty$. So, let $x \in \mathrm{dom}\, f_\infty$. Since $\mathrm{epi}\, f_\infty$ is a cone, we have

$$(x, f_\infty(x)) \in \mathrm{epi}\, f_\infty \implies (\lambda x, \lambda f_\infty(x)) \in \mathrm{epi}\, f_\infty, \ \forall \lambda > 0,$$

i.e., $f_\infty(\lambda x) \leq \lambda f_\infty(x)$. Likewise, one has $(\lambda x, f_\infty(\lambda x)) \in \mathrm{epi}\, f_\infty, \ \forall x \in \mathrm{dom}\, f_\infty, \ \forall \lambda > 0$, and hence $(x, \lambda^{-1} f_\infty(\lambda x)) \in \mathrm{epi}\, f_\infty$ by the cone property. Therefore, $\lambda f_\infty(x) \leq f_\infty(\lambda x)$, and (a) is proved whenever $x \in \mathrm{dom}\, f$. Finally, if $x \notin \mathrm{dom}\, f_\infty$, then $\lambda x \notin \mathrm{dom}\, f_\infty, \ \forall \lambda > 0$, and hence $f_\infty(\lambda x) = \lambda f_\infty(x) = +\infty$.

(b) Since f is proper, then epi f is nonempty, and hence either $f_\infty(0)$ is finite or $f_\infty(0) = -\infty$. If $f_\infty(0)$ is finite, then by (a), $f_\infty(0) = \lambda f_\infty(0)$, $\forall \lambda > 0$, and hence $f_\infty(0) = 0$.

(c) Suppose f_∞ is not a proper function. Then there exists x such that $f_\infty(x) = -\infty$. Now let $\{\lambda_k\} \subset \mathbb{R}_{++}$ be a positive sequence converging to 0. Then $\lambda_k x \to 0$, and under our assumption $f_\infty(0) = 0$, the lower semicontinuity of f_∞, and property (a), we obtain

$$0 = f_\infty(0) \le \liminf_{k \to \infty} f_\infty(\lambda_k x) = \liminf_{k \to \infty} \lambda_k f_\infty(x) = -\infty,$$

leading to a contradiction. □

We can now give a fundamental analytic representation of the asymptotic function f_∞.

Theorem 2.5.1 *For any proper function $f : \mathbb{R}^n \to \mathbb{R} \cup \{+\infty\}$ the asymptotic function f_∞ is given by*

$$f_\infty(d) = \liminf_{\substack{d' \to d \\ t \to +\infty}} \frac{f(td')}{t}, \tag{2.17}$$

or equivalently

$$f_\infty(d) = \inf \left\{ \liminf_{k \to \infty} \frac{f(t_k d_k)}{t_k} \mid t_k \to +\infty, d_k \to d \right\}, \tag{2.18}$$

where $\{t_k\}$ and $\{d_k\}$ are sequences in $\mathbb{R}$ and $\mathbb{R}^n$, respectively.

Proof. The equivalence of the two formulas above is clear. Let $g(d)$ denote the right-hand side of (2.18). We will first show that $(\text{epi } f)_\infty \subset \text{epi } g$. Let $(d, \mu) \in (\text{epi } f)_\infty$. Then by Definition 2.1.2, $\exists t_k \to \infty$, $(d_k, \mu_k) \in \text{epi } f$ such that $t_k^{-1}(d_k, \mu_k) \to (d, \mu)$. As a consequence, since $f(d_k) \le \mu_k$ can be written as $t_k^{-1} f(t_k^{-1} d_k \cdot t_k) \le t_k^{-1} \mu_k$, passing to the limit, it follows that $g(d) \le \mu$ and hence $(d, \mu) \in \text{epi } g$. Conversely, let $(d, \mu) \in \text{epi } g$. By definition of g, there exist sequences $\{d_k\} \subset \mathbb{R}^n$ and $\{t_k\} \in \mathbb{R}$ such that

$$g(d) = \lim_{k \to \infty} \frac{f(t_k d_k)}{t_k}, \ t_k \to \infty, \ d_k \to d,$$

and since $(d, \mu) \in \text{epi } g$, it follows from the definition of the limit that $\forall \varepsilon > 0$ and all $k \in \mathbb{N}$ sufficiently large, we have $f(t_k d_k) \le (\mu + \varepsilon) t_k$ and hence $z_k := t_k(d_k, \mu + \varepsilon) \in \text{epi } f$. Since $t_k^{-1} z_k \to (d, \mu + \varepsilon)$, it follows that $(d, \mu + \varepsilon) \in (\text{epi } f)_\infty$, and therefore since $(\text{epi } f)_\infty$ is a closed set and $\varepsilon > 0$ was arbitrary, we also have $(d, \mu) \in (\text{epi } f)_\infty$. □

Corollary 2.5.1 *For a nonempty set $C \subset \mathbb{R}^n$ one has $(\delta_C)_\infty = \delta_{C_\infty}$.*

Proof. This is an immediate consequence of the definition of the asymptotic cone and the previous theorem.

Corollary 2.5.2 *Let* $f : \mathbb{R}^n \to \mathbb{R} \cup \{+\infty\}$ *be a proper function such that* dom f^* *is nonempty. Then* $f_\infty(0) = 0$.

Proof. Indeed, in the contrary case $f_\infty(0) = -\infty$, but by hypothesis there exist $v \in \text{dom } f^*$, $\alpha \in \mathbb{R}$ such that

$$f(x) \geq \langle v, x \rangle - \alpha \quad \forall x.$$

Then from the above inequality we get $t_k^{-1} f(t_k d_k) \geq \langle v, d_k \rangle - \alpha t_k^{-1}$. Passing to the limit with $d_k \to 0$, $t_k \to \infty$, it follows from Theorem 2.5.1 that $f_\infty(0) \geq 0$, which is impossible.

When f is a convex function, there are simpler ways to express the asymptotic function, which is also called in this case the recession function. We will keep the terminology asymptotic function in both the convex and nonconvex cases.

Proposition 2.5.2 *Let* $f : \mathbb{R}^n \to \mathbb{R} \cup \{+\infty\}$ *be a proper, lsc, convex function. The asymptotic function is a positively homogeneous, lsc, proper convex function, and for any* $d \in \mathbb{R}^n$, *one has*

$$f_\infty(d) = \sup\{f(x + d) - f(x) \mid x \in \text{dom } f\} \tag{2.19}$$

and

$$f_\infty(d) = \lim_{t \to +\infty} \frac{f(x + td) - f(x)}{t} = \sup_{t > 0} \frac{f(x + td) - f(x)}{t}, \quad \forall x \in \text{dom } f. \tag{2.20}$$

Proof. By Proposition 2.5.1(a), the asymptotic function f_∞ is lsc and positively homogeneous, while the convexity of f_∞ follows from the convexity of f. Since f_∞ is lsc, that f_∞ is proper follows from (2.19), which we now prove. By definition, the asymptotic function f_∞ associated with f is uniquely determined via epi $f_\infty = (\text{epi } f)_\infty$. Using Proposition 2.1.5 via the representation F of the asymptotic cone, one has $(d, \mu) \in (\text{epi } f)_\infty$ if and only if for all $(x, \alpha) \in \text{epi } f$ it holds that $(x, \alpha) + (d, \mu) \in \text{epi } f$, namely if and only if $f(x + d) \leq \alpha + \mu$. The latter inequality is clearly equivalent to $f(x + d) - f(x) \leq \mu$, $\forall x \in \text{dom } f$, proving formula (2.19). To verify (2.20), let $x \in \text{dom } f$. Then again using Proposition 2.1.5 but via the representation D of the asymptotic cone, one has for any $x \in \text{dom } f$,

$$(\text{epi } f)_\infty = \{(d, \mu) \in \mathbb{R}^n \times \mathbb{R} \mid (x, f(x)) + t(d, \mu) \in \text{epi } f, \ \forall t > 0\},$$

and hence $(d, \mu) \in (\text{epi } f)_\infty$ if and only if for any $x \in \text{dom } f$ we have

$$f(x + td) \leq f(x) + t\mu, \quad \forall t > 0,$$

which means exactly that

$$g(d) := \sup_{t>0} \frac{f(x+td) - f(x)}{t} \le \mu,$$

and hence $(\text{epi } f)_\infty = \text{epi } g$, $\forall x \in \text{dom } f$, proving the formula (2.20). That the limit in t in the first part of formula (2.20) coincides with the supremum in $t > 0$ simply follows by recalling that the convexity of f implies that for fixed $x, d \in \mathbb{R}^n$, for any $\tau > 0$, the function $\tau \to (\tau)^{-1}(f(x+\tau d) - f(x))$ is nondecreasing. $\square$

The next result gives a useful simplification of the formula for f_∞.

Corollary 2.5.3 *For any lsc, proper, and convex function $f : \mathbb{R}^n \to \mathbb{R} \cup \{+\infty\}$, one has*

$$f_\infty(d) = \lim_{t\to 0^+} tf(t^{-1}d), \ \forall d \in \text{dom } f.$$

If $0 \in \text{dom } f$, the formula holds for every $d \in \mathbb{R}^n$.

Proof. When $0 \in \text{dom } f$ one has $f(0) < \infty$, and the formula above is immediate from (2.20). Consider now the case where $0 \notin \text{dom } f$. Using (2.20) with $x := d \in \text{dom } f$, one has for any $d \in \text{dom } f$,

$$f_\infty(d) = \lim_{t\to+\infty} t^{-1}(f((1+t)d) - f(d)),$$

from which the desired formula for f_∞ follows. $\square$

Example 2.5.1 *(Some useful asymptotic functions.)* (a) Let Q be a symmetric $n \times n$ positive semidefinite matrix and $f(x) := (1 + \langle x, Qx \rangle)^{\frac{1}{2}}$. Then $f_\infty(d) = \langle d, Qd \rangle^{\frac{1}{2}}$.
(b) For the quadratic function associated with Q positive semidefinite and given by $f(x) := \frac{1}{2}\langle x, Qx \rangle + \langle c, x \rangle + s$, $c \in \mathbb{R}^n, s \in \mathbb{R}$, one has

$$f_\infty(d) = \begin{cases} \langle c, d \rangle & \text{if } Qd = 0, \\ +\infty & \text{if } Qd \ne 0. \end{cases}$$

For a nonconvex quadratic function, that is, with Q not positive semidefinite, one needs to apply (2.17), and in that case one obtains

$$f_\infty(d) = \begin{cases} -\infty & \text{for } d \text{ with } d^T Qd \le 0, \\ +\infty & \text{for } d \text{ with } d^T Qd > 0. \end{cases}$$

Thus, the asymptotic function of a proper function, and in that case even of a finite function, can be improper.
(c) Let $f(x) = \sum_{j=1}^n e^{x_j}$. Then $f_\infty(d) = \delta_{\mathbb{R}^n_-}$.
(d) Let $f(x) = \log \sum_{j=1}^n e^{x_j}$, $n > 1$. Then $f_\infty(d) = \max_{1\le j\le n} d_j$.

Further examples of asymptotic functions will be given in Sections 2.6–2.7.

Proposition 2.5.3 *For any proper function* $f : \mathbb{R}^n \to \mathbb{R} \cup \{+\infty\}$ *and any* $\alpha \in \mathbb{R}$ *such that* $\mathrm{lev}(f, \alpha) \neq \emptyset$ *one has* $(\mathrm{lev}(f, \alpha))_\infty \subset \mathrm{lev}(f_\infty, \alpha)$, *i.e.,*

$$\{x \mid f(x) \leq \alpha\}_\infty \subset \{d \mid f_\infty(d) \leq 0\}.$$

Equality holds in the inclusion when f *is lsc, proper, and convex.*

Proof. Since $\mathrm{lev}(f, \alpha) \neq \emptyset$, take $d \in (\mathrm{lev}(f, \alpha))_\infty$. We have to show that $f_\infty(d) \leq 0$, $\forall d$. By definition, $d \in (\mathrm{lev}(f, \alpha))_\infty$ means that

$$\exists x_k \in \mathrm{lev}(f, \alpha), \; \exists t_k \to +\infty : \lim_{k \to \infty} t_k^{-1} x_k = d.$$

Set $d_k = t_k^{-1} x_k$. Then $d_k \to d$, and since $x_k \in \mathrm{lev}(f, \alpha)$, we have $t_k^{-1} f(t_k d_k) = t_k^{-1} f(x_k) \leq t_k^{-1} \alpha \to 0$, which implies by the fundamental formula (2.18) of an asymptotic function that $f_\infty(d) \leq 0$, $\forall d$. In the case that f is assumed proper lsc convex, to prove the reverse inclusion, let d be such that $f_\infty(d) \leq 0$. Then for each $x \in \mathrm{lev}(f, \alpha)$ and $\lambda > 0$ one has $f(x + \lambda d) - f(x) \leq \lambda f_\infty(d) \leq 0$, so that $x + \lambda d \in \mathrm{lev}(f, \alpha)$, which by Proposition 2.1.5 means that $d \in (\mathrm{lev}(f, \alpha))_\infty$. $\qquad\square$

Corollary 2.5.4 *Let* $f_i : \mathbb{R}^n \to \mathbb{R} \cup \{+\infty\}$, $i \in I$ *a collection of proper functions and, let* $S \subset \mathbb{R}^n$ *with* $S \neq \emptyset$. *Define* $C := \{x \in S \mid f_i(x) \leq 0, \; \forall i \in I\}$. *Then*

$$C_\infty \subset \{d \in S_\infty \mid (f_i)_\infty(d) \leq 0, \; \forall i \in I\}.$$

The inclusion holds as an equation when $C \neq \emptyset$, S *is closed convex in* $\mathbb{R}^n$, *and each* f_i *is lsc convex.*

Proof. Let $C_i := \{x \in \mathbb{R}^n : f_i(x) \leq 0\}$, so that $C = S \cap \bigcap_{i \in I} C_i$. Invoking Proposition 2.1.9, one has $C_\infty \subset S_\infty \cap_{i \in I} (C_i)_\infty$, and thus using Proposition 2.5.3, the inclusion statement of the theorem is proved. Furthermore, under the additional assumptions given here, by the same Proposition 2.1.9, the previous inclusion holds as an equation. $\qquad\square$

Example 2.5.2 (a) Let Q be an $n \times n$ symmetric positive semidefinite matrix, $a \in \mathbb{R}^n$, and $\beta \in \mathbb{R}$. Let C be the set defined by

$$C = \{x \in \mathbb{R}^n \mid (x - a)^T Q(x - a) \leq \beta\}.$$

Then using the examples given in (2.5.1) one has

$$C_\infty = \{d \in \mathbb{R}^n \mid d^T Qa \leq 0, \; Qd = 0\}.$$

(b) Let $f : \mathbb{R}_{++} \to \mathbb{R}$ defined by $f(t) = t \sin t^{-1}$ and consider the set

$$\text{Gr}(f) = \{(x_1, x_2) \in \mathbb{R}^2 \mid x_2 = f(x_1),\ x_1 > 0\}.$$

Then $(\text{Gr}(f))_\infty = \{(d_1, d_2) \in \mathbb{R}^2 \mid 0 \le d_2 \le d_1\}$.

The next result gives a useful relation between the asymptotic cone of the subdifferential of a convex function and the normal cone of its domain.

Proposition 2.5.4 *Let $f : \mathbb{R}^n \to \mathbb{R} \cup \{+\infty\}$ be proper and convex. Then for any $z \in \text{dom } f$ such that $\partial f(z) \ne \emptyset$ one has*

$$(\partial f(z))_\infty = N_{\text{dom } f}(z).$$

Proof. Let $z \in \text{dom } f$ be such that $\partial f(z) \ne \emptyset$. Then the subdifferential of f is a nonempty closed convex set that can be described as the infinite intersection of closed half-spaces. More precisely, one has

$$\partial f(z) = \bigcap_{y \in \text{dom } f} \{u \in \mathbb{R}^n \mid \langle u, y - z \rangle \le f(y) - f(z)\} \equiv \bigcap_{y \in \text{dom } f} C_y.$$

Applying Proposition 2.1.9(a) to the closed convex sets C_y one obtains

$$(\partial f(z))_\infty = \bigcap_{y \in \text{dom } f} (C_y)_\infty.$$

But by Corollary 2.5.4 one obtains immediately that $(C_y)_\infty = \{u \mid \langle u, y - z \rangle \le 0\}$, and thus it follows that

$$(\partial f(z))_\infty = \bigcap_{y \in \text{dom } f} \{u \mid \langle u, y - z \rangle \le 0\} = N_{\text{dom } f}(z),$$

where the last equality follows from the definition of the normal cone of the nonempty convex set $\text{dom } f$ at z. □

Proposition 2.5.5 *Let $f : \mathbb{R}^n \to \mathbb{R} \cup \{+\infty\}$ be lsc, proper, and convex. Then f is Lipschitz on $\mathbb{R}^n$, i.e.,*

$$|f(x) - f(y)| \le L\|x - y\|,\ \forall x, y \in \mathbb{R}^n,$$

if and only if f_∞ is finite everywhere on $\mathbb{R}^n$. The Lipschitz constant L is given by

$$L = \sup\{f_\infty(d) \mid \|d\| = 1\}.$$

Proof. The asymptotic function f_∞, being finite everywhere on $\mathbb{R}^n$, is continuous, and therefore $\sup\{f_\infty(d) \mid \|d\| = 1\} = L < \infty$. Since f_∞ is positively homogeneous, $f_\infty(d) \leq L\|d\|$, $\forall d$. Invoking the formula (2.19) one thus has

$$f(x+d) - f(x) \leq L\|d\|, \; \forall x \in \text{dom } f, \; d \in \mathbb{R}^n,$$

and since $f(x+d) < \infty$, then dom $f = \mathbb{R}^n$ and L is the Lipschitz constant for f. Conversely, let f be Lipschitz on $\mathbb{R}^n$ and such that $x \in \mathbb{R}^n$. Then

$$f(x+td) - f(x) \leq tL\|d\|, \; \forall t > 0, \; d \in \mathbb{R}^n,$$

which implies by (2.20) that $f_\infty(d) \leq L\|d\|$, $\forall d$, so that dom $f_\infty = \mathbb{R}^n$ and $\sup\{f_\infty(d) \mid \|d\| = 1\} = L$. □

To prove the next result, we need the following obvious technical fact on one-dimensional convex functions, the proof of which is left to the reader.

Lemma 2.5.1 *Let $\psi : \mathbb{R} \to \mathbb{R} \cup \{+\infty\}$ be a proper, lsc, convex function and let S denote the set of optimal solutions of the problem $\inf\{\psi(t) : t \in \mathbb{R}\}$. Then only one of the following two statements holds:*
(a) S is nonempty and compact.
(b) The function ψ is a monotone function on $\mathbb{R}$.

Theorem 2.5.2 *Let $f : \mathbb{R}^n \to \mathbb{R} \cup \{+\infty\}$ be a proper, lsc, convex function, and for any $x \in \text{dom } f$ and $0 \neq d \in \mathbb{R}^n$ let $\psi(t) := f(x+td)$.*
(a) If $f_\infty(d) \leq 0$, then $\limsup_{t \to +\infty} \psi(t) < +\infty$.
(b) If there exists some $z \in \text{dom } f$ such that $\liminf_{t \to +\infty} f(z+td) < +\infty$, then ψ is decreasing on $\mathbb{R}$, which is equivalent to saying that $f_\infty(d) \leq 0$.

Proof. From (2.20) we have $t^{-1}(f(x+td) - f(x)) \leq f_\infty(d) \leq 0$, which implies that $f(x+td) \leq f(x)$, $\forall t > 0$, proving (a). To prove (b), let $z \in \text{dom } f$ satisfy the hypothesis of the theorem. Then one has $f_\infty(d) \leq 0$. Indeed, suppose the contrary. Then again using (2.20) there would exist $t_0 > 0$ and $\alpha > 0$ such that $f(z+td) \geq f(z) + t\alpha$, $\forall t \geq t_0$. Taking the limit as $t \to +\infty$ in the latter inequality implies that $\lim_{t \to +\infty} f(z+td) = +\infty$, which contradicts the assumption made in (b). Since for each $x \in \text{dom } f$, one has $\psi_\infty(1) = f_\infty(d) \leq 0$, invoking Lemma 2.5.1, it follows that the optimal solution set of the problem minimizing ψ is noncompact, and thus ψ is monotone. We can suppose that ψ is nonconstant. Then ψ is decreasing. Indeed, in the opposite case there would exist $t_0 \in \text{dom } \psi$ such that $\psi(t) \leq \psi(t_0)$, $\forall t \leq t_0$, and then it would follow that $\psi_\infty(-1) \leq 0$. Since $\psi_\infty(0) = 0$, it follows from the convexity of ψ_∞ that $\psi_\infty(-1) = \psi_\infty(1) = 0$ and then that ψ is constant. Finally, if ψ is decreasing on $\mathbb{R}$, then $f_\infty(d) \leq 0$ follows from (2.20). □

Definition 2.5.2 *Given $f : \mathbb{R}^n \to \mathbb{R} \cup \{+\infty\}$ a proper convex function, we define:*
(a) The asymptotic cone of f by

$$\mathcal{K}_f = \{d \in \mathbb{R}^n \mid f_\infty(d) \leq 0\} = (\text{epi } f)_\infty \cap \{(d, 0) \mid d \in \mathbb{R}^n\}.$$

This is a closed convex cone containing zero. A vector $d \in \mathcal{K}_f$ is called an asymptotic direction of f.
(b) The constancy space of f by

$$\mathcal{C}_f = \{d \in \mathbb{R}^n \mid f_\infty(d) = f_\infty(-d) = 0\} = (-\mathcal{K}_f) \cap \mathcal{K}_f.$$

(c) The lineality space of f by $L_f = \{d \in \mathbb{R}^n \mid f_\infty(-d) = -f_\infty(d)\}$.

Theorem 2.5.3 *Let $f : \mathbb{R}^n \to \mathbb{R} \cup \{+\infty\}$ be a proper, lsc, convex function. The following conditions on $d \in \mathbb{R}^n$ and $\mu \in \mathbb{R}$ are equivalent:*
(a) $f(x + td) = f(x) + t\mu, \; \forall x \in \text{dom } f, t \in \mathbb{R}$.
(b) $(d, \mu) \in -(\text{epi } f)_\infty \cap (\text{epi } f)_\infty$.
(c) $-f_\infty(-d) = f_\infty(d) = \mu$.
(d) Furthermore, the vector d satisfies (a)–(c) with $\mu = f_\infty(d)$ if there exists one $x \in \text{dom } f$ such that $f(x + td)$ is an affine function of t.
(e) $d \in \mathcal{C}_f$ if and only if f is constant along the direction d.

Proof. Suppose (a) holds, i.e., $f(x+td) = f(x) + t\mu$, $\forall x \in \text{dom } f$, $\forall t \in \mathbb{R}$. Thus in particular, using formula (2.19) one has $\mu = f_\infty(d)$ and $-\mu = f_\infty(-d)$, proving the implication $(a) \implies (c)$. Now if (c) holds, this means that both (d, μ) and $(-d, -\mu)$ are in $\text{epi } f_\infty$, and hence (b) holds. We end the proof by showing that $(b) \implies (a)$. If (b) holds, then for any $t \in \mathbb{R}$ one has $\text{epi } f = \text{epi } f - t(\mu, d)$. Take $x \in \text{dom } f$. Then $(x, f(x)) \in \text{epi } f$, and if we define $h(x) = f(x+td) - t\mu$, then $\text{epi } h = \text{epi } f - t(\mu, d)$, and hence (a) holds. Furthermore, (d) follows from the formula (2.20), and finally, since $f_\infty(0) = 0 \leq f_\infty(d) + f_\infty(-d)$, it follows that $f_\infty(d) = f_\infty(-d) = 0$ if and only if $d \in \mathcal{C}_f$, i.e., if and only if f is constant on the direction d by using the equivalence between (a) and (c). □

There exists an interesting interplay between a proper, lsc, convex function and the convex sets associated with it, together with its conjugate and the corresponding asymptotic function.

Theorem 2.5.4 *Let $f : \mathbb{R}^n \to \mathbb{R} \cup \{+\infty\}$ be a proper convex function, and f^* its conjugate. The following relations hold:*
(a) $(f^*)_\infty = \sigma_{\text{dom } f}$.
(b) If f is also assumed lsc, then

$$f_\infty = \sigma_{\text{dom } f^*}, \quad (f_\infty)^* = \delta_{\text{cl dom } f^*}.$$

Proof. (a) Using the definition of f_∞ given by the formula (2.19) and that of the conjugate function, one has

$$
\begin{aligned}
(f^*)_\infty(d) &= \sup\{f^*(x+d) - f^*(x) : x \in \operatorname{dom} f^*\} \\
&= \sup_{x \in \operatorname{dom} f^*} \{ \sup_{u \in \operatorname{dom} f} \{\langle u, x+d\rangle - f(u)\} - f^*(x)\} \} \\
&= \sup_{u \in \operatorname{dom} f} \{ \sup_{x \in \operatorname{dom} f^*} \{\langle u, x\rangle - f^*(x)\} + \langle u, d\rangle - f(u)\} \} \\
&= \sup_{u \in \operatorname{dom} f} \{\langle u, d\rangle + f^{**}(u) - f(u)\}, \\
&\leq \sigma_{\operatorname{dom} f}(d), \quad \text{since } f^{**} \leq f.
\end{aligned}
$$

We now prove the reverse inequality $(f^*)_\infty(d) \geq \sigma_{\operatorname{dom} f}(d)$. Clearly, for any $d \notin (\operatorname{dom} f^*)_\infty$ there is nothing to prove. Thus, let any $d \in (\operatorname{dom} f^*)_\infty$ and take $\mu = (f^*)_\infty(d)$. Then one has

$$
f^*(z + td) - f^*(z) \leq \mu t, \ \forall z \in \operatorname{dom} f^*, \ \forall t > 0.
$$

On the other hand, by the definition of the conjugate, one has

$$
f(x) + f^*(z + td) \geq \langle x, z\rangle + t\langle x, d\rangle, \ \forall t > 0,
$$

which combined with the inequality above implies

$$
f(x) \geq \langle x, z\rangle - f^*(z) + t(\langle x, d\rangle - \mu), \ \forall t > 0.
$$

Therefore, as $t \to +\infty$ it follows that $\langle x, d\rangle \leq \mu$, $\forall x \in \operatorname{dom} f$, and hence

$$
\sup_{x \in \operatorname{dom} f} \langle x, d\rangle \leq \mu \text{ and thus } \sigma_{\operatorname{dom} f}(d) \leq (f^*)_\infty(d).
$$

(b) We now assume that f is lsc, which implies that $f^{**} = f$. Then replacing f by f^* in the formula proven in (a) yields $(f^{**})_\infty = \sigma_{\operatorname{dom} f^*} = f_\infty$, proving the first formula of (b). Taking conjugates in that formula, and recalling that for any set $C \subset \mathbb{R}^n$, $\sigma_C = \sigma_{\operatorname{cl} C}$ and $\sigma_C = \delta_C^*$ we obtain

$$
(f_\infty)^* = (\sigma_{\operatorname{cl} \operatorname{dom} f^*})^* = \delta_{\operatorname{cl} \operatorname{dom} f^*}^{**} = \delta_{\operatorname{cl} \operatorname{dom} f^*},
$$

proving the second formula in (b). $\qquad\square$

Corollary 2.5.5 *Let $f : \mathbb{R}^n \to \mathbb{R} \cup \{+\infty\}$ be a proper convex function and suppose that $0 \in E := \operatorname{aff} \operatorname{dom} f$, i.e., E is a subspace. Then $\mathcal{C}_{f^*} = L_{f^*} = E^\perp$.*

Proof. From the definition of the lineality space and from Theorem 2.5.3 one has $d \in L_{f^*}$ if and only if the function $\langle \cdot, d\rangle$ is constant on $\operatorname{dom} f$, or equivalently on $E = \operatorname{aff} \operatorname{dom} f$, and since $0 \in E$, this constant is 0, and one has $L_{f^*} = \mathcal{C}_{f^*}$ and $d \in E^\perp$ if and only if $d \in \mathcal{C}_{f^*}$. $\qquad\square$

The characterization of the asymptotic function of a convex function given in terms of the support function of its conjugate in the previous theorem is of fundamental importance, and as we shall see, will be used very often in the following chapters. An interesting immediate application of this formula is the following dual characterization of the constancy subspace $C_f = \{d \in \mathbb{R}^n \mid f_\infty(d) = f_\infty(-d) = 0\}$, which by replacing f_∞ by $\sigma_{\text{dom } f^*}$ (when f is lsc) can thus be simply written as

$$C_f = \{d \in \mathbb{R}^n \mid \langle d, v \rangle = 0, \ \forall v \in \text{dom } f^*\}. \tag{2.21}$$

Lemma 2.5.2 *Let $f : \mathbb{R}^n \to \mathbb{R} \cup \{+\infty\}$ be a proper convex function. Then for all $y \in \mathbb{R}^n$,*

$$f^*(y) = \sigma_{\text{epi } f}(y, -1) = \sup\{\langle y, x \rangle - r \mid (x, r) \in \text{epi } f\}.$$

As a consequence the following formula holds for the epigraph of f:

$$\sigma_{\text{epi } f}(y, -t) = \begin{cases} tf^*(\frac{y}{t}) & \text{if } t > 0, \\ (f^*)_\infty(y) & \text{if } t = 0, \\ +\infty & \text{if } t < 0. \end{cases}$$

Proof. The first formula in the theorem follows from the definition of the epigraph and conjugate of f. In fact, one has

$$\sigma_{\text{epi } f}(y, -1) = \sup_{f(x) \le r} \{\langle x, y \rangle - r\} = \sup_x \sup_{r \ge f(x)} \{\langle x, y \rangle - r\}$$
$$= \sup_x \{\langle x, y \rangle - f(x)\} = f^*(y).$$

We now prove the second formula. First, if $t < 0$, one has $\sigma_{\text{epi } f}(y, -t) = +\infty$. Suppose that $t > 0$. Then using the first formula just proved, we have

$$\sigma_{\text{epi } f}(y, -t) = t \sup_{f(x) \le r} \{\langle x, y/t \rangle - r\} = t\sigma_{\text{epi } f}(y/t, -1) = tf^*(y/t).$$

Finally, if $t = 0$, then

$$\sigma_{\text{epi } f}(y, 0) = \sup_{f(x) \le r} \{\langle x, y \rangle - r \times 0\},$$
$$= \sup\{\langle x, y \rangle \mid (x, r) \in \text{epi } f \text{ for some } r\},$$
$$= \sigma_{\text{dom } f}(y) = (f^*)_\infty(y),$$

where the last equation follows from Theorem 2.5.4(a). □

For each function $f : \mathbb{R}^n \to \mathbb{R} \cup \{+\infty\}$ that is lsc and proper with $0 < f(0) < +\infty$, the positive hull, denoted by $\mathrm{pos}\, f$, is defined through its epigraph via

$$\mathrm{epi}(\mathrm{pos}\, f) = \mathrm{pos}(\mathrm{epi}\, f),$$

which is equivalent to (since $f(0) > 0$, i.e., $(0,0) \notin \mathrm{epi}\, f$)

$$(\mathrm{pos}\, f)(x) = \inf_{\lambda > 0} \lambda f(\lambda^{-1} x). \tag{2.22}$$

Proposition 2.5.6 *Let* $f : \mathbb{R}^n \to \mathbb{R} \cup \{+\infty\}$ *be lsc, and proper, with* $0 < f(0) < +\infty$. *Then:*
(a) $\mathrm{cl}(\mathrm{pos}\, f)(x) = \min\{(\mathrm{pos}\, f)(x), f_\infty(x)\} = \min\{\inf_{\lambda > 0} \lambda f(\lambda^{-1} x), f_\infty(x)\}$.

(b) *If in addition* f *is assumed convex, then* $\mathrm{pos}\, f$ *is lsc, convex, proper, and positively homogeneous.*

Proof. (a) By the definition of the closure and the positive hull of a function, one has

$$\mathrm{epi}(\mathrm{cl}(\mathrm{pos}\, f)) = \mathrm{cl}(\mathrm{epi}(\mathrm{pos}\, f)) = \mathrm{cl}(\mathrm{pos}(\mathrm{epi}\, f)).$$

Using Corollary 2.3.3 one thus obtains

$$
\begin{aligned}
\mathrm{cl}(\mathrm{pos}(\mathrm{epi}\, f)) &= \mathrm{pos}(\mathrm{epi}\, f) \cup (\mathrm{epi}\, f)_\infty \\
&= \mathrm{epi}(\mathrm{pos}\, f) \cup \mathrm{epi}\, f_\infty \\
&= \mathrm{epi}(\min\{\mathrm{pos}\, f, f_\infty\}) \text{ (using Proposition 1.2.1)},
\end{aligned}
$$

which establishes the first formula in (a), while the second formula follows from (2.22).
(b) Since f is convex and $0 \in \mathrm{dom}\, f$, one has $f_\infty(x) = \lim_{\lambda \to 0^+} \lambda f(\lambda^{-1} x)$. Furthermore, since $\lambda f(\lambda^{-1} x) \leq f_\infty(x) - \lambda f(0)$ and $f(0) > 0$, it follows that $\mathrm{pos}\, f\ (x) = \inf_{\lambda > 0} \lambda f(\lambda^{-1} x) \leq f_\infty(x)$, and by part (a) we obtain $\mathrm{cl}\, \mathrm{pos}\, f = \mathrm{pos}\, f$, which proves that $\mathrm{pos}\, f$ is lsc. Since $\mathrm{epi}\, f$ is convex, it follows that $\mathrm{pos}(\mathrm{epi}\, f)$ is convex, and then $\mathrm{pos}\, f$ is convex. Furthermore, since $(\mathrm{pos}\, f)(0) = \inf_{\lambda > 0} \lambda f(0) = 0$, one thus has that $\mathrm{pos}\, f$, which is lsc, is also proper. Finally, since $\mathrm{epi}(\mathrm{pos}\, f)$ is a cone, one thus has that $\mathrm{pos}\, f$ is positively homogeneous. $\square$

Remark 2.5.1 In the convex case, from part (b) of Proposition 2.5.6 we always have $\inf_{\lambda > 0} \lambda f(\lambda^{-1} d) \leq f_\infty(d)$. This inequality can be strict. For example, with $f(x) := x^2 + 1$ we obtain for $d = 1$, $\inf_{\lambda > 0} \lambda f(\lambda^{-1}) < +\infty$ and $f_\infty(1) = +\infty$.

When f is lsc, proper, and convex, the corresponding asymptotic function can be viewed as the closure of another function associated with f.

Definition 2.5.3 *For any proper function* $f : \mathbb{R}^n \to \mathbb{R} \cup \{+\infty\}$, *we associate the function* p_f *defined by*

$$p(t, x) = \begin{cases} tf(t^{-1}x) & \text{if } t > 0, \\ 0 & \text{if } t = 0, x = 0, \\ +\infty & \text{otherwise}, \end{cases} \tag{2.23}$$

with $\operatorname{dom} p = \{(0,0) \cup \{t(1,x) \mid t > 0, \ f(x) < \infty\} = \{(0,0)\} \cup \mathbb{R}_+(\{1\} \times \operatorname{dom} f)$.

Proposition 2.5.7 *Let* $f : \mathbb{R}^n \to \mathbb{R} \cup \{+\infty\}$ *be lsc, proper, and convex. Then the function* $p(t, x)$ *given in (2.23) is proper and jointly convex in* (t, x), *and for any* $d \in \mathbb{R}^n$ *one has*

$$(clp)(t, d) = \sigma_{\operatorname{epi} f^*}(y, -t) = \begin{cases} tf(t^{-1}x) & \text{if } t > 0, \\ f_\infty(d) & \text{if } t = 0, \\ +\infty & \text{if } t < 0. \end{cases}$$

Proof. The joint convexity of $p(t, x)$ follows directly from applying the definition of the convexity of f by observing that for any $\alpha \in [0, 1]$, $\beta = 1 - \alpha$, and any $t_1, t_2 > 0$, $x_1, x_2 \in \mathbb{R}^n$ one has

$$
\begin{aligned}
p(\alpha t_1 + \beta t_2, \alpha x_1 + \beta x_2) &= (\alpha t_1 + \beta t_2)f\left(\frac{\alpha t_1 x_1}{(\alpha t_1 + \beta t_2)t_1} + \frac{\beta t_2 x_2}{(\alpha t_1 + \beta t_2)t_2}\right) \\
&\leq \alpha t_1 f\left(\frac{x_1}{t_1}\right) + \beta t_2 f\left(\frac{x_2}{t_2}\right).
\end{aligned}
$$

Now define the function

$$g(s, x) = \begin{cases} f(x) & \text{if } s = 1, \\ +\infty & \text{if } s \neq 1. \end{cases}$$

Clearly, g is a proper lsc convex function, and since for any x, $g(0, x) = +\infty$, using (2.22) one has

$$(\operatorname{pos} g)(s, x) = \inf_{\lambda > 0} \lambda g(\lambda^{-1}s, \lambda^{-1}x) = \begin{cases} sf(s^{-1}x) & \text{if } s > 0, \\ +\infty & \text{if } s = 0, x = 0, \\ +\infty & \text{if } s < 0. \end{cases}$$

The point $(0, 0)$ is not in the relative interior of p and $\operatorname{pos} g$, respectively, and therefore the functions p and $\operatorname{pos} g$ coincide on their relative interiors, and hence their closures coincide. Applying Proposition 2.5.6(a) one thus obtains

$$\operatorname{cl} p(t, x) = \operatorname{cl}(\operatorname{pos} g)(t, x) = \min\{(\operatorname{pos} g)(t, x), g_\infty(t, x)\},$$

and since here $g_\infty(t, d) = f_\infty(d)$ for $t = 0$ and is $+\infty$ whenever $t \neq 0$, the desired formula for $\operatorname{cl} p(t, x)$ follows using Lemma 2.5.2. $\qquad\square$

Corollary 2.5.6 *Let $C \subset \mathbb{R}^n$ be a nonempty closed convex set with $0 \in C$. Then the gauge function of C, denoted by γ_C and defined by*

$$\gamma_C = \inf\{\lambda \geq 0 \mid x \in \lambda C\}$$

is nonnegative, lsc, and positively homogeneous. Furthermore, one has:
(a) $C = \{x \mid \gamma_C(x) \leq 1\}$.
(b) $C_\infty = \{x \mid \gamma_C(x) = 0\}$.
(c) $\operatorname{pos} C = \operatorname{dom} \gamma_C$.

Proof. Let $h(x) := \delta(x|C) + 1$. Then $1 = h(0) > 0$ and $\lambda h(\lambda^{-1}x) = \delta(x|\lambda C) + \lambda$, $\forall \lambda > 0$. Therefore, by definition one has $\gamma_C(x) = \operatorname{pos} h(x)$ and $\gamma_C \geq 0$. Hence, by Proposition 2.5.6(b) it follows that γ_C is lsc, convex, proper, and positively homogeneous. Moreover, (a) is satisfied, and then $C_\infty = \{d \mid (\gamma_C)_\infty(d) \leq 0\}$. Since γ_C is positively homogeneous, lsc convex, we have $(\gamma_C)_\infty = \gamma_C$, and since $\gamma_C \geq 0$, we obtain formula (b). Finally, using (a) and once more the fact that γ_C is positively homogeneous, we obtain $\operatorname{pos} C = \operatorname{dom} \gamma_C$. □

Domains of proper, lsc, convex functions can be usefully characterized via the use of asymptotic functions.

Proposition 2.5.8 *Let $h : \mathbb{R}^n \to \mathbb{R} \cup \{+\infty\}$ be lsc, proper, and convex. Let z be a fixed vector and let $f(x) = h(x) - \langle x, z \rangle$. Then:*
(a) $z \in \operatorname{aff}(\operatorname{dom} h^*) \iff f_\infty(d) = 0$, $\forall d$ such that $-f_\infty(-d) = f_\infty(d)$.
(b) $z \in \operatorname{ri}(\operatorname{dom} h^*) \iff f_\infty(d) > 0$, $\forall d$ except those satisfying $-f_\infty(-d) = f_\infty(d) = 0$.
(c) $z \in \operatorname{int}(\operatorname{dom} h^*) \iff f_\infty(d) > 0$, $\forall d \neq 0$.
(d) $z \in \operatorname{cl}(\operatorname{dom} h^*) \iff f_\infty(d) \geq 0$, $\forall d$.

Proof. We have $f^*(y) = \sup_x\{\langle x, y + z \rangle - h(x)\} = h^*(y + z)$, so that $\operatorname{dom} f^* = \operatorname{dom} h^* - z$. Thus, $z \in \operatorname{cl} \operatorname{dom} h^*$ is equivalent to $0 \in \operatorname{cl} \operatorname{dom} f^*$, and similarly for each of the other statements in the theorem. By Theorem 2.5.4(b), one has $\sigma_{\operatorname{dom} f^*} = f_\infty$. Therefore, statements (a)–(d) correspond exactly to those of Theorem 1.3.2 stated for the support function of $\operatorname{dom} f^*$. □

2.6   Differential Calculus at Infinity

We develop here some useful formulas for computing asymptotic functions.

Proposition 2.6.1 *Let $f_i : \mathbb{R}^n \to \mathbb{R} \cup \{+\infty\}$, $i = 1, \ldots, p$, be a collection of proper functions, $f := \sum_{i=1}^{p} f_i$, and suppose that f is proper; i.e.,*

$\mathrm{dom}\, f = \cap_{i=1}^{p} \mathrm{dom}\, f_i \neq \emptyset$. *Then:*
(a) If the functions are all lsc, f is also lsc.
(b) $f_\infty(d) \geq \sum_{i=1}^{p} (f_i)_\infty(d)$ for all d satisfying the following condition:
if $(f_i)_\infty(d) = +\infty$ (respectively $-\infty$) for some i, then $(f_j)_\infty(d) > -\infty$
respectively $(< +\infty)$ for $j \neq i$.
If in addition all the functions are lsc and convex, then equality holds in
the inequality.

Proof. For arbitrary extended real valued functions g_i, one has for $y \in \mathbb{R}^n$,

$$\lim_{x \to y} \inf (g_1(x) + \cdots + g_p(x)) \geq \lim_{x \to y} \inf g_1(x) + \cdots + \lim_{x \to y} \inf g_p(x),$$

as long as the right side of the inequality has a meaning, and hence (a)
follows immediately by using the definition of lower semicontinuity. The
inequality of assertion (b) follows also from the above inequality by using
the definition of the asymptotic function (2.18), while equality in the case
of lsc convex functions follows from formula (2.20). □

Proposition 2.6.2 *For a collection $\{f_i\}_{i \in I}$ of proper functions $f_i : \mathbb{R}^n \to$*
$\mathbb{R} \cup \{+\infty\}$ one has

$$\left(\sup_{i \in I} f_i \right)_\infty \geq \sup_{i \in I} (f_i)_\infty \quad and \quad \left(\inf_{i \in I} f_i \right)_\infty \leq \inf_{i \in I} (f_i)_\infty.$$

The first relation is an equality when f_i are lsc, proper, and convex. The
second relation is an equality for a finite index set I.

Proof. Apply Proposition 2.1.9 to the corresponding epigraphs, recalling
the epigraph algebraic properties of Proposition 1.2.1. □

Proposition 2.6.3 *Let $g : \mathbb{R}^m \to \mathbb{R} \cup \{+\infty\}$ be a proper function, let A*
be a linear map from $\mathbb{R}^n$ to $\mathbb{R}^m$ with $A(\mathbb{R}^n) \cap \mathrm{dom}\, g \neq \emptyset$, and let $f(x) =$
$g(Ax)$ be the proper composite function. The following properties hold for
the function f:
(a) If g is convex and there exists some y such that $Ay \in \mathrm{ri\, dom}\, g$, then for
any x, $\mathrm{cl}\, f(x) = \mathrm{cl}\, g(Ax)$.
(b) If g is lsc, then f is lsc and one has

$$f_\infty(d) \geq g_\infty(Ad) \quad \forall d.$$

If in addition g is convex (so is f), this holds also as an equality.

Proof. (a) Since $\mathrm{dom}\, f = A^{-1} \mathrm{dom}\, g$, using Proposition 1.1.6(c) we get
$\mathrm{ri\, dom}\, f = A^{-1}(\mathrm{ri\, dom}\, g)$, which is equivalent to saying that $y \in \mathrm{ri\, dom}\, f$ if

and only if $Ay \in \mathrm{ri\,dom}\,g$. Let y be such that $Ay \in \mathrm{ri\,dom}\,g$ (which exists by hypothesis). Then, since $y \in \mathrm{ri\,dom}\,f$, by Proposition 1.2.6 we have

$$\mathrm{cl}\,f(x) = \lim_{t \to 0^+} f(x + t(y-x)) = \lim_{t \to 0^+} g(Ax + t(Ay - Ax)).$$

By the same proposition, we also have

$$\mathrm{cl}\,g(Ax) = \lim_{t \to 0^+} g(Ax + t(Ay - Ax)),$$

so that $\mathrm{cl}\,f(x) = \mathrm{cl}\,g(Ax)$.

(b) If g is lsc, obviously f is also lsc. Furthermore, using the representation of the asymptotic function we obtain for each d

$$f_\infty(d) = \inf_{\substack{d_k \to d \\ t_k \to +\infty}} \liminf_{k \to \infty} t_k^{-1} g(t_k A d_k) \ge \liminf_{\substack{d' \to Ad \\ t \to \infty}} t^{-1} g(td') = g_\infty(Ad).$$

If in addition g is convex, then obviously f is also convex. Let $x_0 \in \mathrm{dom}\,f$. Then since f is lsc, proper, and convex, we have

$$f_\infty(d) = \lim_{\lambda \to \infty} \frac{g(Ax_0 + \lambda Ad) - g(Ax_o)}{\lambda} = g_\infty(Ad).$$

$\square$

It is interesting and often useful to have a formula for the asymptotic function defined as a composition of a nondecreasing function with a convex function.

Proposition 2.6.4 (*Composition rule*) *Let $f : \mathbb{R}^n \to \mathbb{R} \cup \{+\infty\}$ be lsc, proper, and convex, and let $\psi : (-\infty, b) \to \mathbb{R}$ with $0 \le b \le +\infty$ be a convex nondecreasing function with $\psi_\infty(1) > 0$, and with $\mathrm{dom}\,\psi \cap f(\mathbb{R}^n) \ne \emptyset$. Consider the composite function*

$$g(x) = \begin{cases} \psi(f(x)) & \text{if } x \in \mathrm{dom}\,f, \\ +\infty & \text{otherwise.} \end{cases}$$

Then g is a proper, lsc, convex function, and one has

$$g_\infty(d) = \begin{cases} \psi_\infty(f_\infty(d)) & \text{if } d \in \mathrm{dom}\,f_\infty, \\ +\infty & \text{otherwise.} \end{cases}$$

Proof. Clearly, the composite function g is an lsc convex function. Let $x \in \mathrm{dom}\,f$ be such that $f(x) \in \mathrm{dom}\,\psi$. For every $s < f_\infty(d)$ there exists τ such that $f(x + td) \ge f(x) + ts$, for $t \ge \tau$, and since ψ is nondecreasing, we obtain

$$\frac{g(x + td) - g(x)}{t} \ge \frac{\psi(f(x) + ts) - \psi(f(x))}{t},$$

and passing to the limit with $t \to +\infty$ we deduce $g_\infty(d) \geq \psi_\infty(s)$. If $f_\infty(d) = +\infty$, using $\psi(1) > 0$ and letting $s \to +\infty$ we get $g_\infty(d) = +\infty$. In the other case, $f_\infty(d) < +\infty$, letting $s \to f_\infty(d)$ and since ψ_∞ is lsc, we deduce $g_\infty(d) \geq \psi_\infty(f_\infty(d))$. On the other hand, since $f(x + td) \leq f(x) + tf_\infty(d)$, using the monotonicity of ψ we also get

$$
\begin{aligned}
g_\infty(d) &= \lim_{t \to +\infty} \frac{g(x + td) - g(x)}{t} \\
&\leq \lim_{t \to +\infty} \frac{\psi(f(x) + tf_\infty(d)) - \psi(f(x))}{t} = \psi_\infty(f_\infty(d)).
\end{aligned}
$$

$\square$

The next set of results concerns arbitrary proper functions.

Proposition 2.6.5 *Let $f_i : \mathbb{R}^n \to \mathbb{R} \cup \{+\infty\}$ be lsc proper functions, $i = 1, \ldots, m$, such that closed sets $\{\mathrm{epi}\, f_i \mid i = 1, \ldots, m\}$ are supposed to be in relative general position (see Definition 2.3.3) and let $h := \mathrm{conv}\{f_i \mid i = 1, \ldots, m\}$. Then for all $x \in \mathrm{dom}\, \mathrm{cl}\, h$,*

$$
\exists x_i \in \mathrm{dom}\, f_i,\ y_j \in \mathrm{dom}(f_j)_\infty,\ \lambda \in \mathrm{int}\, \Delta_p,\ i \in [1, p],\ j \in [1, q],
$$

with $p + q \leq n + 1$, $p \geq 1$, such that

$$
\mathrm{cl}\, h(x) = \sum_{i=1}^{p} \lambda_i f_i(x) + \sum_{j=1}^{q} (f_j)_\infty(y_j),
$$

with $x = \sum_{i=1}^{p} \lambda_i x_i + \sum_{j=1}^{q} y_j$.

Proof. Apply Theorem 2.3.5 to the corresponding epigraphs. $\square$

As in the case for sets where from Theorem 2.3.5 we have obtained an interesting corollary concerning the convex hull of a set and its closure we will now show how the convex hull of a function and its closure are related. For that purpose we need the following definition.

Definition 2.6.1 *Let $f : \mathbb{R}^n \to \mathbb{R} \cup \{+\infty\}$ be a proper function. Then f is said to be semibounded if its epigraph is semibounded, i.e., if $\mathrm{epi}\, f_\infty$ is pointed.*

Proposition 2.6.6 *Let $f : \mathbb{R}^n \to \mathbb{R} \cup \{+\infty\}$ be a proper, lsc, semibounded function. Then for all $x \in \mathrm{dom}(\mathrm{cl}\, \mathrm{conv}\, f)$,*

$$
\exists x_i \in \mathrm{dom}\, f,\ y_j \in \mathrm{dom}(f)_\infty,\ \lambda \in \mathrm{int}\, \Delta_p,\ i \in [1, p],\ j \in [1, q],
$$

with $p + q \leq n + 1$, $p \geq 1$ such that

$$\text{cl conv } f(x) = \sum_{i=1}^{p} \lambda_i f(x_i) + \sum_{j=1}^{q} f_\infty(y_j),$$

with $x = \sum_{i=1}^{p} \lambda_i x_i + \sum_{j=1}^{q} y_j$.

Proof. Use Proposition 2.6.5 with $m = n + 1$, $f = f_i \ \forall i = 1, \ldots, p$. □

An example of a function where Proposition 2.6.6 becomes simpler concerns cofinite functions.

Definition 2.6.2 *Let $f : \mathbb{R}^n \to \mathbb{R} \cup \{+\infty\}$ be a proper lsc function. Then f is said to be cofinite if*

$$f_\infty(0) = 0, \quad \text{and} \quad f_\infty(y) = +\infty \quad \forall y \neq 0.$$

Corollary 2.6.1 *Let f be a cofinite function. Then $\text{cl conv } f = \text{conv } f$ and*

$$\forall x \in \text{dom}(\text{conv } f), \ \exists x_i \in \text{dom } f, \ \lambda \in \Delta_{n+1}, \ i \in [1, n+1],$$

such that

$$\text{conv } f(x) = \sum_{i=1}^{n+1} \lambda_i f(x_i),$$

with $x = \sum_{i=1}^{n+1} \lambda_i x_i$.

Proof. As an immediate consequence of Corollary 2.3.6, since $\text{epi } f_\infty = \{0\}$, we obtain $\text{cl conv } f = \text{conv } f$. From the definition of a cofinite function it follows that f is semibounded, and the claimed formula is obtained by using Proposition 2.6.6. □

Proposition 2.6.7 *Let $f : \mathbb{R}^n \to \mathbb{R} \cup \{+\infty\}$ be a proper lsc function. Then f is cofinite if and only if $\text{dom } f^* = \mathbb{R}^n$.*

Proof. Suppose that f is cofinite and there exists some $v \notin \text{dom } f^*$, i.e., $f^*(v) = +\infty$. Then we have

$$\inf\{f(x) - \langle v, x \rangle \mid x \in \mathbb{R}^n\} = -\infty,$$

and there exists a minimizing sequence $\{x_k\}$ such that $f(x_k) - \langle v, x_k \rangle \to -\infty$. Let j be an arbitrary integer. Then for k sufficiently large we have

$$f(x_k) - \langle v, x_k \rangle \leq -j. \tag{2.24}$$

Thus there are two cases:
(i) The sequence $\{x_k\}$ is bounded. Therefore, it has at least one cluster

point, and for each cluster point x, since f is lsc, passing to the limit in the last inequality it follows that

$$f(x) - \langle v, x \rangle \leq -j,$$

and taking $j \to \infty$ we get a contradiction.

(ii) Suppose now that $\{x_k\}$ is unbounded and without loss of generality suppose that

$$\|x_k\| \to \infty, \qquad \frac{x_k}{\|x_k\|} \to y \neq 0.$$

Then dividing both members of formula (2.24) by $\|x_k\|$ we obtain

$$\|x_k\|^{-1} f\left(\frac{x_k}{\|x_k\|}\|x_k\|\right) - \langle v, x_k\|x_k\|^{-1}\rangle \leq -j\|x_k\|^{-1},$$

and passing to the limit we obtain

$$f_\infty(y) - \langle v, y \rangle \leq 0,$$

a contradiction to the fact that f is cofinite. We prove now the reverse statement. Suppose that $\mathrm{dom}\, f^* = \mathbb{R}^n$. Then from Corollary 2.5.2 $f_\infty(0) = 0$, and f_∞ is proper. Suppose that f is not cofinite. Then there exists some $y \neq 0$ with $f_\infty(y) := \alpha \in \mathbb{R}$ such that we can find two sequences $\{y_k\} \to y$ and $\{t_k\} \to \infty$ for which

$$\lim_{k \to \infty} \frac{f(t_k y_k)}{t_k} \to \alpha.$$

Let $s \in \mathbb{R}^n$. Then we have $t_k^{-1}(\langle s, t_k y_k \rangle - f(t_k y_k)) \leq t_k^{-1} f^*(s)$, and passing to the limit, since $f^*(s) \in \mathbb{R}$, we get $\langle s, y \rangle - \alpha \leq 0 \quad \forall s$, which is obviously impossible.

$\square$

We end this section with some interesting facts and properties of *one-dimensional* convex functions and their relations with the corresponding asymptotic function.

Lemma 2.6.1 *Let* $\psi : (-\infty, b) \to \mathbb{R}$ *with* $0 \leq b \leq +\infty$ *be a convex nondecreasing function with* $\psi_\infty(1) > 0$. *Then* ψ^{-1}, *the inverse of* ψ *exists and is a concave function, and the following relations hold:*
(a)

$$(-\psi^{-1})^*(-s) = \begin{cases} s\psi^*(s^{-1}) & \text{if } s > 0, \\ +\infty & \text{otherwise} . \end{cases}$$

(b) $s \in \mathrm{dom}\, \psi^* \iff -s^{-1} \in \mathrm{dom}(-\psi^{-1})^*$ *and* $\mathrm{dom}\, \psi^* \subset \mathbb{R}_+$.
(c) $\psi^{-1}(t) = \inf_{s>0} \{st + s\psi(s^{-1})\}$.
(d) $(-\psi^{-1})_\infty(t) = -\inf\{ts^{-1} : s \in \mathbb{R}_{++} \cap \mathrm{dom}\, \psi^*\}$.

Proof. The strict monotonicity and continuity of ψ implies that ψ^{-1} exists, and thus from the definition of the convexity of ψ it follows that ψ^{-1} is concave. By definition of the conjugate of the convex function $-\psi^{-1}$ one has

$$
\begin{aligned}
(-\psi^{-1})^*(-s) &= \sup\{-ts + \psi^{-1}(t) : t \in \operatorname{dom}\psi^{-1}\} = \sup_u\{-s\psi(u) + u\} \\
&= \sup_u\{s(us^{-1} - \psi(u)\} \\
&= \begin{cases} s\sup_u\{us^{-1} - \psi(u)\} & \text{if } s > 0, \\ +\infty & \text{otherwise,} \end{cases}
\end{aligned}
$$

proving the desired formula in (a). The relation between the domains in (b) follows as a direct consequence. Furthermore, since $\operatorname{ri}\operatorname{dom}\psi^* \subset \operatorname{rge}\partial\psi \subset \operatorname{dom}\psi^*$, and we assumed that ψ is nondecreasing, it follows that $\operatorname{rge} \subset \partial\psi\mathbb{R}_{++}$, so that $\operatorname{ri}\operatorname{dom}\psi^* \subset \mathbb{R}_{++}$, and hence $\operatorname{dom}\psi^* \subset \mathbb{R}_+$, as stated. To prove the infimal representation of the inverse function ψ in (c) we use the fact that $-\psi^{-1}$ is lsc proper convex, so that $-\psi^{-1}(t) = (-\psi^{-1})^{**}(t)$, and with the help of the conjugate formula derived in (a) we then have

$$
\begin{aligned}
(-\psi^{-1})^{**}(t) = -\psi^{-1}(t) &= \sup_{s<0}\{st - (-\psi^{-1})^*(s)\} \\
&= \sup_{s<0}\{st + s\psi^*(-s^{-1})\} = -\inf_{s>0}\{st + s\psi^*(s^{-1})\}.
\end{aligned}
$$

Finally, using the first formula in (b) of the same theorem, and (b) proven above, we have

$$
\begin{aligned}
(-\psi^{-1})_\infty(t) &= \sup\{st : s \in \operatorname{dom}(-\psi^{-1})^*\} \\
&= -\inf\{st : -s \in \operatorname{dom}(-\psi^{-1})^*\} \\
&= -\inf\{ts^{-1} : s \in \mathbb{R}_{++} \cap \operatorname{dom}\psi^*\},
\end{aligned}
$$

proving (d). $\qquad\qquad\qquad\qquad\qquad\qquad\qquad\qquad\qquad\qquad\qquad\qquad\square$

Note that if in the above lemma we suppose in addition that $\psi_\infty(1) = +\infty$, $\psi_\infty(-1) = 0$, then the function $t \to -\psi^{-1}(-t)$ is convex and nondecreasing with domain the interval $(-\infty, b)$ and such that $-\psi^{-1}(1) = 0$, $-\psi^{-1}(-1) = +\infty$.

2.7   Application I: Semidefinite Optimization

For an arbitrary closed convex cone $K \subset \mathbb{R}^n$ we define a partial ordering $\succeq$ for vectors $x, y \in \mathbb{R}^n$ via

$$
x \succeq y \iff x - y \in K.
$$

The partial ordering $\succeq$ satisfies the following conditions:
(i) $x \succeq x, \ \forall x$.
(ii) $x \succeq y \Longrightarrow -y \succeq -x$.
(iii) $x \succeq y \Longrightarrow \lambda x \succeq \lambda y, \ \forall \lambda \geq 0$.
(iv) $x \succeq y$ and $x' \succeq y' \Longrightarrow x + x' \geq y + y'$.
(v) $x_k \succeq y_k, \ x_k \to x, \ y_k \to y \Longrightarrow x \succeq y$.
The additional antisymmetric property (vi) $x \succeq y$ and $y \succeq x \Longrightarrow x = y$
holds if and only if K is in addition a *pointed* cone.
The standard case, already mentioned in Chapter 1, is for vector inequalities when $K = \mathbb{R}^n_+$, the nonnegative orthant, which is a closed convex pointed cone. In that case $\succeq$ reduces simply to the usual coordinatewise inequalities

$$x \succeq y \iff x_j - y_j \geq 0, \ \forall j = 1, \ldots, n.$$

Another interesting and important partial ordering that fit the conditions $(i) - (v)$ is for the space of square real symmetric matrices. Let M_n be the space of real square matrices of order n equipped with the inner product

$$\langle A, B \rangle := \sum_{i,j=1}^n a_{ij} b_{ij} = \operatorname{tr} AB,$$

where tr stands for the trace operator of a matrix $C \in M_n$, which is the sum of the diagonal elements of C.
Of particular importance is the space of symmetric real matrices of order n, denotes by S_n. This fact can be viewed as a linear subspace of M_n with dimension $n(n+1)/2$ and can be treated as a Euclidean vector space with the trace inner product.
Let $K \subset S_n$ be a given cone. Then as in the vector case we may define the dual cone

$$K^\circ = \{ Y \in S_n \mid \langle Y, X \rangle \geq 0, \ \forall X \in K \}.$$

Among all the cones in S_n, one of particular interest is the cone of positive semidefinite matrices, denoted by S_n^+ and defined by

$$S_n^+ := \{ A \in S_n \mid x^T A x \geq 0, \ \forall x \in \mathbb{R}^n \}.$$

It is easy to verify that S_n^+ is a closed convex pointed cone. The partial ordering associated with S_n^+ is quantified by

$$A \succeq B \iff A - B \text{ positive semidefinite}$$

and obeys the rules $(i) - (vi)$.
The cone S_n^+ is self dual, i.e., $(S_n^+)^\circ = S_n^+$. The interior of S_n^+ consists of positive definite matrices, namely

$$S_n^{++} := \{ A \in S_n \mid x^T A x > 0, \ \forall x \neq 0 \}.$$

Another particularly interesting partial ordering is the Lorentz cone (also called the second-order or ice-cream cone), defined by

$$L := \{x \in \mathbb{R}^{n+1} \mid x_{n+1} \geq \|x\|\}.$$

Spectrally Defined Matrix Functions

Spectrally defined functions arise in various areas of mathematical sciences and have recently gained strong interest within the area of matrix optimization problems, called semidefinite programming.

Recall that a function f on $\mathbb{R}^n$ is called symmetric if

$$f(Px) = f(x)$$

for all $n \times n$ permutation matrices P.

Definition 2.7.1 *The function* $\Phi : S_n \to \mathbb{R} \cup \{+\infty\}$ *is said to be spectrally defined if there exists a symmetric function* $f : \mathbb{R}^n \to \mathbb{R} \cup \{+\infty\}$ *such that*

$$\Phi(X) = \Phi_f(X) := f(\lambda(X)), \quad \forall X \in S_n,$$

where $\lambda(X) := (\lambda_1(X), \ldots, \lambda_n(X))^T$ *is the vector of eigenvalues of* X *in nondecreasing order.*

The function Φ is spectrally defined if and only if Φ is orthonormally invariant, that is,

$$\Phi(U^T A U) = \Phi(A), \quad \forall U \in \mathcal{U}_n,$$

with $\mathcal{U}_n :=$ the set of $n \times n$ orthogonal matrices. The symmetric function f in the definition is then given by

$$f(x) = \Phi(\operatorname{diag} x), \quad \forall x \in \mathbb{R}^n,$$

where $\operatorname{diag} x$ is the diagonal matrix with diagonal elements $x_1, \ldots x_n$.

Of particular interest is the class of spectrally defined functions associated with symmetric functions f that are proper convex and lsc, leading to convex matrix functions Φ_f. Most of these are in fact represented via $\operatorname{tr} f_s(A)$, for a particular choice of f. We recall that for a smooth scalar real-valued function f and a symmetric matrix A with real eigenvalues $\lambda_i(A)$, $i = 1, \ldots, n$, one defines $f_s(A) := V^T \operatorname{diag}(f(\lambda_1(A)), \ldots, f(\lambda_n(A)))V$, with $\lambda_i(A) \in \operatorname{dom} f$, $i = 1, \ldots, n$, arranged in nondecreasing order, and the trace of $f_s(A)$ is thus given by

$$\operatorname{tr} f_s(A) = \sum_{i=1}^{n} f(\lambda_i(A)). \tag{2.25}$$

Example 2.7.1 *(i) Let*

$$f(\lambda) = \begin{cases} -\sum_{i=1}^{n} \log \lambda_i & \text{if } \lambda > 0, \\ +\infty & \text{otherwise.} \end{cases}$$

Then for $X \in S_n$ one has

$$\Phi_f(X) = \begin{cases} -\log \det X & \text{if } X \succ 0, \\ +\infty, & \text{otherwise,} \end{cases}$$

where $\det X$ is the determinant of the matrix X.
(ii) Let $f(\lambda) = \max_{1 \le i \le n} \lambda_i$. Then $\Phi_f(X) = \lambda_{\max}(X)$, the largest eigenvalue of X.
(iii) Let

$$f(\lambda) = \begin{cases} -\sum_{i=1}^{n} \frac{1}{\lambda_i} & \text{if } \lambda > 0, \\ +\infty & \text{otherwise.} \end{cases}$$

Then for $X \in S_n$ one has

$$\Phi_f(X) = \begin{cases} -\operatorname{tr}(X^{-1}) & \text{if } X \succ 0, \\ +\infty, & \text{otherwise.} \end{cases}$$

(iv) Let

$$f(\lambda) = \begin{cases} -(\Pi_{i=1}^{n} \lambda_i)^{1/n} & \text{if } \lambda > 0, \\ +\infty & \text{otherwise.} \end{cases}$$

Then for $X \in S_n$ one has

$$\Phi_f(X) = \begin{cases} -(\det X)^{1/n} & \text{if } X \succeq 0, \\ +\infty, & \text{otherwise.} \end{cases}$$

Convex analysis tools in $\mathbb{R}^n$ can be translated to S_n. Most of the interesting properties of Φ_f can be deduced directly from those of f, and therefore convex analysis of spectrally defined convex functions can be easily developed.
The effective domain of Φ is defined by

$$\operatorname{dom} \Phi := \{X \in S_n \mid \Phi(X) < \infty\}.$$

The conjugate Φ^* of Φ is defined by

$$\Phi^*(Y) := \sup\{\langle X, Y \rangle - \Phi(X) \mid X \in S_n\}, \ \forall Y \in S_n,$$

and the subdifferential of Φ at $X \in \operatorname{dom} \Phi$ is

$$\begin{aligned} \partial \Phi(X) &= \{Y \in S_n \mid \Phi(Z) \ge \Phi(X) + \langle Z - X, Y \rangle, \ \forall Z \in S_n\} \\ &= \{Y \in S_n \mid \Phi^*(Y) + \Phi(X) = \langle X, Y \rangle\}. \end{aligned}$$

An important and useful inequality is the *trace inequality*

$$\langle X, Y \rangle \leq \langle \lambda(X), \lambda(Y) \rangle, \quad \forall X, Y \in S_n, \tag{2.26}$$

and equality holds if the matrices X, Y are simultaneously diagonalizable, i.e., if and only if there exists a matrix $U \in \mathcal{U}_n$ such that

$$U^T X U = \operatorname{diag} \lambda(X) \text{ and } U^T Y U = \operatorname{diag} \lambda(Y).$$

Theorem 2.7.1 *Let $f : \mathbb{R}^n \to \mathbb{R} \cup \{+\infty\}$ be a symmetric proper convex and lsc function and $\Phi_f : S_n \to \mathbb{R} \cup \{+\infty\}$ the induced convex spectral function. Then f^* is symmetric, and the following relations hold:*
(a) $\Phi_f^* = \Phi_{f^*}$.
(b) Φ_f is a proper, lsc, convex function.
(c) $\Phi_f^*(Y) + \Phi_f(X) \geq \langle \lambda(X), \lambda(Y) \rangle, \quad \forall X, Y \in S_n$.
(d) $Y \in \partial \Phi_f(X) \iff \langle \lambda(X), \lambda(Y) \rangle = \langle X, Y \rangle$ and $\lambda(Y) \in \partial f(\lambda(X))$.
Furthermore, X and Y are simultaneously diagonalizable.

Proof. The symmetry of f^* follows immediately from the fact that for any invertible matrix A, with $h(x) = f(Ax)$ one obtains $h^*(y) = f^*((A^T)^{-1}y)$, and thus if A is an orthogonal matrix (in particular a permutation matrix) one has $A^{-1} = A^T$ and hence $f^*(y) = f^*(Ay)$. To prove (a), clearly, we have $\Phi_f^*(Y) = \Phi^*(\operatorname{diag} \lambda(Y))$. Using the definition of the conjugate function one thus has

$$\begin{aligned} \Phi_f^*(Y) = \Phi^*(\operatorname{diag} \lambda(Y)) &= \sup_{X \in S_n} \{\langle X, \operatorname{diag} \lambda(Y) \rangle - \Phi_f(X)\}, \\ &\geq \sup_{x \in \mathbb{R}^n} \{\langle \operatorname{diag} x, \operatorname{diag} \lambda(Y) \rangle - \Phi_f(\operatorname{diag} x))\}, \\ &= \sup\{\langle x, \lambda(Y) \rangle - f(x)\} = f^*(\lambda(Y)) = \Phi_{f^*}(Y). \end{aligned}$$

The proof of the reverse inequality $\Phi_f^*(Y) \leq \Phi_{f^*}(Y)$ is simply obtained by combining the Fenchel–Young inequality $f^*(\lambda(Y)) + f(\lambda(X)) \geq \langle \lambda(X), \lambda(Y) \rangle$ with the trace inequality (2.26). The statements (b), (c), (d) follow as easy consequences of the formula proven in (a), noting that in (d) the last statement follows from the trace equality condition given after (2.26). $\square$

As a consequence of Theorem 2.7.1, the optimization problems $\min\{\Phi_f(X) \mid X \in S_n\}$ and $\min\{f(x) \mid x \in \mathbb{R}^n\}$ are equivalent. In fact, one has

$$\inf_{X \in S_n} \Phi_f(X) = -\Phi_f^*(0) = -\Phi_{f^*}(0) = -f^*(0) = \inf f.$$

Moreover,

$$X \in \operatorname{argmin} \Phi_f \iff \lambda(X) \in \operatorname{argmin} f,$$

with $\operatorname{argmin} \Phi := \{X \in S_n \mid \Phi(X) = \inf \Phi\}$.

Following Proposition 2.5.2, the asymptotic function of the proper convex lsc function $\Phi : S_n \to \mathbb{R} \cup \{+\infty\}$ is defined by

$$\Phi_\infty(D) = \sup_{t>0} \frac{\Phi(A + tD) - \Phi(A)}{t} \quad \forall D \in S_n,$$

where $A \in \operatorname{dom} \Phi$. Using Theorem 2.5.4 one has, in fact,

$$\Phi_\infty(D) = \sup\{\langle B, D, \rangle : B \in \operatorname{dom} \Phi^*\} \quad \forall D \in S_n.$$

We remark that if f is symmetric, lsc, convex, and proper, by using Proposition 2.5.2 so is f_∞.

Theorem 2.7.2 *Let $f : \mathbb{R}^n \to \mathbb{R} \cup \{+\infty\}$ be symmetric, lsc, proper, and convex with induced spectral function Φ_f. Then,*

$$(\Phi_f)_\infty(D) = \Phi_{f_\infty}(D) = f_\infty(\lambda(D)).$$

Proof. For any $D \in S_n$, one has

$$
\begin{aligned}
(\Phi_f)_\infty(D) &= \sup\{\langle Z, D \rangle \mid Z \in \operatorname{dom} \Phi_f^*\} \\
&= \sup\{\langle Z, D \rangle \mid Z \in \operatorname{dom} \Phi_{f^*}\} \\
&= \sup\{\langle Z, D \rangle \mid Z \in \operatorname{cl} \operatorname{dom} \Phi_{f^*}\},
\end{aligned}
$$

where in the second equality we invoke Theorem 2.7.1(a). On the other hand, we also have

$$
\begin{aligned}
\Phi_{f_\infty}(D) &= \sup\{\langle Z, D \rangle \mid Z \in \operatorname{dom} \Phi_{f_\infty}^*\} \\
&= \sup\{\langle Z, D \rangle \mid Z \in \operatorname{dom} \Phi_{(f_\infty)^*}\} \\
&= \sup\{\langle Z, D \rangle \mid Z \in \operatorname{cl} \operatorname{dom} \Phi_{(f_\infty)^*}\}.
\end{aligned}
$$

To prove the desired statement, it thus suffices to verify that $\operatorname{cl} \operatorname{dom} \Phi_f^* = \operatorname{cl} \operatorname{dom} \Phi_{(f_\infty)^*}$. Using again Theorem 2.7.1(a) one has

$$\operatorname{dom} \Phi_{f^*} = \{U \operatorname{diag} x\, U^T \mid U \in \mathcal{U}_n,\ x \in \operatorname{dom} f^*\},$$

and therefore

$$
\begin{aligned}
\operatorname{cl} \operatorname{dom} \Phi_{f^*} &= \{U \operatorname{diag} x\, U^T \mid U \in \mathcal{U}_n,\ x \in \operatorname{cl} \operatorname{dom} f^*\} \\
&= \{U \operatorname{diag} x\, U^T \mid U \in \mathcal{U}_n,\ x \in \operatorname{dom}(f_\infty)^*\} = \operatorname{dom} \Phi_{(f_\infty)^*},
\end{aligned}
$$

where in the second equality we use the fact $\operatorname{dom}(f_\infty)^* = \operatorname{cl} \operatorname{dom} f^*$, which follows from Theorem 2.5.4(b). $\qquad\square$

2.8   Application II: Modeling and Smoothing Optimization Problems

Asymptotic functions are essentially *built-in* within most optimization problems. This can be realized by considering the following general composite optimization model,

$$\inf\{f_0(x) + p(F(x)) \mid F(x) \in \operatorname{dom} p\},$$

where $F(x) = (f_1(x), \ldots, f_m(x))^T$ with all f_i real-valued, and $p : \mathbb{R}^m \to \mathbb{R} \cup \{+\infty\}$ is some given function. The composite model is rich enough to encompass most of the interesting formulations of nonlinear optimization problems and allows for recovering a generic class of both smooth and nonsmooth problems. In fact, as illustrated below, a viable choice for the function p is simply to pick the support function of some given subset $Y \subset \mathbb{R}^m$, i.e., to set $p(u) := \sup_{y \in Y} \langle y, u \rangle$.

Example 2.8.1 (i) l_1-norm approximation/optimization problems

$$p(u) = \sum_{i=1}^{p} |u_i|, \quad Y = \{y : \|y\|_\infty \le 1\}.$$

(ii) Discrete minimax problems

$$p(u) = \max_{i=1,\ldots,m} u_i, \quad Y = \Delta_m = \left\{ y \in \mathbb{R}^m \mid \sum_{i=1}^{m} y_i = 1, y \ge 0 \right\}.$$

(iii) l_2-norm nonsmooth problems

$$p(u) = \|u\|, \quad Y = \{y \mid \|y\| \le 1\}.$$

Now suppose that there exists an lsc proper convex function H with conjugate H^* and such that $Y = \operatorname{dom} H^*$. Then by Theorem 2.5.4 one has $\sigma_Y = \sigma_{\operatorname{dom} H^*} = H_\infty$, and the composite model can be represented in terms of asymptotic functions and takes the form

$$(\text{CM}) \quad v = \inf\{\phi(x) \mid x \in \mathbb{R}^n\}$$

with

$$\phi(x) = \begin{cases} f_0(x) + H_\infty(f_1(x), \ldots, f_m(x)) & \text{if } x \in \bigcap_{i=1}^{m} \operatorname{dom} f_i, \\ +\infty & \text{otherwise.} \end{cases}$$

Now there are two key observations that can be made:
(i) The model above is in fact fairly generic, since it contains all usual constrained optimization problems. Indeed, for $p(u) = \delta_{\mathbb{R}^m_-}(u)$, problem (CM) is equivalent to

$$\inf\{f_0(x) \mid f_i(x) \le 0, \ i = 1, \ldots, m\}.$$

(ii) By Corollary 2.5.3 the asymptotic function H_∞ of a given lsc proper convex function H can be *approximated* via

$$H_\infty(y) = \lim_{r \to 0^+} \{H_r(y) := rH(r^{-1}y)\}, \quad \forall y \in \operatorname{dom} H,$$

and thus leads one naturally to consider as an approximate problem for (CM) the problem

$$(\mathrm{CM})_r \quad v_r = \inf\{\phi_r(x) \mid x \in \mathbb{R}^n\}$$

with

$$\phi_r(x) = \begin{cases} f_0(x) + H_r(f_1(x), \dots, f_m(x)) & \text{if } x \in \bigcap_{i=1}^m \operatorname{dom} f_i, \\ +\infty & \text{otherwise.} \end{cases}$$

Thus, with H smooth enough, the above mechanism might lead to generic smooth approximation–type algorithms, such as penalty–barrier methods, which are often used as basic algorithms in the numerical treatment of optimization problems. Thus these methods can be analyzed within a unified framework based on the properties of asymptotic functions and where here $r > 0$ plays the role of *smoothing* parameter. The natural questions which thus arise are: Does there exist a solution x_r of problem $(\mathrm{CM})_r$? If it exists, do we have $x_r \to x$ as a solution of (CM) and $v_r \to v$? This type of questions can be efficiently handled via calculus rules of asymptotic functions, as illustrated below and in Chapter 5 later on.

Before formalizing the above approach, let us go back first to some of the examples above. For the corresponding function $p = H_\infty$ in the three examples, we can easily verify that the corresponding functions H are respectively given by the smooth functions

$$H(y) = \sum_{i=1}^m \sqrt{1 + y_i^2}, \quad H(y) = \log \sum_{i=1}^m e_i^y, \quad H(y) = \sqrt{\|y\|^2 + 1}.$$

In the general case of constrained optimization problems it turns out that there exists a wide class of functions H such that $H_\infty = \delta_{\mathbb{R}_-^m}$.

Proposition 2.8.1 *Let $\theta : \mathbb{R} \to \mathbb{R} \cup \{+\infty\}$ be an lsc, proper convex and nondecreasing function with $\operatorname{dom} \theta = (-\infty, b)$ and $b \in [0, \infty)$, $\theta_\infty(-1) = 0$, $\theta_\infty(1) = +\infty$ and set $H(y) = \sum_{i=1}^m \theta(y_i)$. Then*

$$H_\infty(d) = \delta_{\mathbb{R}_-^m}(d), \quad \forall d \in \mathbb{R}^m.$$

Proof. By Proposition 2.6.1 one has $H_\infty(d) = \sum_{i=1}^m \theta_\infty(d_i), \quad \forall d \in \mathbb{R}^m$. Under the hypothesis on the function θ we have

$$\theta_\infty(s) = \begin{cases} s\theta_\infty(1) & \text{if } s > 0, \\ -s\theta_\infty(-1) & \text{if } s \leq 0, \end{cases}$$

which proves the desired formula for H_∞.   □

Particularly interesting choices for the functions θ are

$$
\begin{aligned}
\theta_1(u) &= \exp(u), \ \mathrm{dom}\,\theta = \mathbb{R}, \\
\theta_2(u) &= -\log(1-u), \ \mathrm{dom}\,\theta = (-\infty,1), \\
\theta_3(u) &= \frac{u}{1-u}, \ \mathrm{dom}\,\theta = (-\infty,1), \\
\theta_4(u) &= -\log(-u), \ \mathrm{dom}\,\theta = (-\infty,0), \\
\theta_5(u) &= -u^{-1}, \ \mathrm{dom}\,\theta = (-\infty,0).
\end{aligned}
$$

Another interesting example is provided by the class of semidefinite optimization problems introduced in the previous section.

Example 2.8.2 Semidefinite programming
Consider the following semidefinite optimization problem

$$\text{(SDP)} \quad \inf \ c^T x : \text{ subject to } B(x) \preceq 0,$$

with $B(x) = B_0 + \sum_{i=1}^m x_i B_i$. The problem data are the vector $c \in \mathbb{R}^m$ and the $(m+1)$ symmetric matrices $B_0, B_1, \ldots, B_m$ of order $n \times n$. This special class of convex problems can also be handled through the general composite model by a simple adjustment of the involved finite-dimensional setting. Let us replace the set $Y \subset \mathbb{R}^m$ with $Y \subset S_n$, the space of $n \times n$ symmetric matrices. More precisely, recalling that the indicator of the negative cone in $\mathbb{R}^n$ is a symmetric function, (cf. Section 2.7), we can associate to it the spectral function $p : S_n \to \mathbb{R} \cup \{+\infty\}$ defined by $p(U) = \delta_{\mathbb{R}^n_-}(\lambda(U))$, where $\lambda(U)$ is the vector of eigenvalues of the symmetric matrix U in nondecreasing order. For any symmetric matrix U one has $\delta_{S_n^-}(U) = \delta_{\mathbb{R}^n_-}(\lambda(U))$, where S_n^- denotes the space of symmetric negative definite matrices, and thus one can write

$$p(U) = \delta_{S_n^-}(U) = \sup\{\langle Z, U \rangle \mid Z \in S_n^+\}.$$

Therefore, problem (SDP) can be written as the composite optimization model

$$\inf\{c^T x + p(B(x)) \mid B(x) \in \mathrm{dom}\,p\}.$$

From here, we can thus proceed as in Example 2.8.1. Let $h(y) = \sum_{i=1}^n \theta(y_i)$ with $\theta : \mathbb{R} \to \mathbb{R} \cup \{+\infty\}$ a function satisfying Proposition 2.8.1. Then one has $h_\infty = \delta_{\mathbb{R}^n_-}$. Invoking Theorem 2.7.2 (once again recall that the indicator of the negative cone in $\mathbb{R}^n$ is a symmetric function), it follows that for any $D \in S_n$ one has $H_\infty(D) = \delta_{S_n^-}(D)$ with $H(D) = h(\lambda(D))$. Thus problem (SDP) can be written in the form of the composite model (CM)

$$\inf_{x \in \mathbb{R}^m} \{c^T x + H_\infty(B(x))\},$$

with corresponding approximate model given by

$$(SDP_r) \quad \inf_{x \in \mathbb{R}^m} \{c^T x + H_r(B(x))\},$$

where $H_r(D) = r^{-1} H(rD) = r^{-1} h(r\lambda(D))$.

Using for example the functions θ given above we thus obtain for any symmetric matrix D the following corresponding functions H

$\quad H_1(D) = \text{tr}(\exp D)$,

$\quad H_2(D) = -\log(\det(I - D))$ for $D \prec I$, $+\infty$ otherwise,

$\quad H_3(D) = \text{tr}((I - D)^{-1} D)$ for $D \prec I$, $+\infty$ otherwise,

$\quad H_4(D) = -\log(\det(-D))$ for $D \prec 0$, $+\infty$ otherwise,

$\quad H_5(D) = \text{tr}(-D^{-1})$ for $D \prec 0$, $+\infty$ otherwise.

We note that for these examples each H is C^∞ on the interior of its domain.

To formalize the above approach we need to make some minimal hypothesis on functionals involved in the structural representation of the problem (CM) $\inf\{\phi(x) \mid x \in \mathbb{R}^n\}$ and its corresponding approximation defined in $(CM)_r$ for every $r > 0$.

Definition 2.8.1 *A function $H : \mathbb{R}^m \to \mathbb{R} \cup \{+\infty\}$ that is a proper, lsc, convex function with asymptotic function H_∞ is called a generating asymptotic approximation kernel if the following conditions hold:*
(a) H_∞ is isotone, i.e.,

$$y_i \leq z_i \quad \forall i = 1, 2 \ldots, m \implies H_\infty(y) \leq H_\infty(z).$$

(b) $\lim_{r \to 0+} H_r(y) = H_\infty(y)$, $\forall y \in \text{ri dom } H_\infty$.
(c) $\text{ri dom } H_\infty \subset \text{ri dom } H_r$, $\forall r > 0$.
(d) The constancy space of H_∞ satisfies $\mathcal{C}_{H_\infty} = \{0\}$.

It can be verified that the functions H defined through Proposition 2.8.1, and based on the choice of the functions θ outlined above, are generating asymptotic approximation kernels. In the case of semidefinite programming given in Example 2.8.2, a similar analysis can be developed to verify that claim. This can be done by either identifying the space S_n with $\mathbb{R}^{(n+1)n/2}$ or in a direct way. We refer the reader to the notes and references for details.

Lemma 2.8.1 *For any generating asymptotic approximation kernel H one has*

$$\lim_{\lambda \to +\infty} H_\infty(a_1, \ldots, a_{i-1}, a_i + \lambda, a_{i+1}, \ldots, a_m) = +\infty, \ \forall a \in \mathbb{R}^m.$$

Proof. By definition, H_∞ is convex and isotone. Suppose the result does not hold. Then one has

$$\lim_{\lambda \to +\infty} \inf \psi(\lambda) := H_\infty(a_1, \ldots, a_{i-1}, a_i + \lambda, a_{i+1}, \ldots, a_m) < +\infty,$$

and by Theorem 2.5.2, the function $\psi(\lambda)$ is decreasing, and since H_∞ is assumed isotone, it follows that H_∞ is constant along each direction $d = e_i$, where e_i, $i = 1, \ldots, m$, denotes the canonical basis of $\mathbb{R}^m$. This, however, contradicts the condition $C_{H_\infty} = \{0\}$ (cf. Definition 2.8.1) imposed on a generating asymptotic kernel.   □

In the remainder of this section, for the problem's data (CM) we now make the following minimal assumptions, which are needed to guarantee that (CM) is well-defined.
(i) The functions $f_i : \mathbb{R}^n \to \mathbb{R} \cup \{+\infty\}$, $i = 0, 1, \ldots, m$, are closed, proper and satisfy

$$(f_i)_\infty(d) > -\infty \quad \forall d.$$

(ii) There exists $x_0 \in \mathrm{dom}\, f_0$ such that $F(x_0) \in \mathrm{dom}\, H_\infty$, where $F(x) := (f_1(x), f_2(x), \ldots, f_m(x))$. When $0 \notin \mathrm{dom}\, H$ we suppose in addition that $F(x_0) \in \mathrm{ri}\,\mathrm{dom}\, H_\infty$.
Note that the assumption (i) on f_i is always satisfied when f_i is convex. Note also that whenever H is isotone, then the isotonicity of H_∞ follows at once, and thus if f_0 is in addition convex, then ϕ and ϕ_r are convex, and (CM) and (CM)$_r$ are convex problems.

Lemma 2.8.2 *Consider problems* (CM) *and* $(CM)_r$ *with a generating asymptotic approximation kernel* H. *Then:*
(a) The function ϕ *is lsc, proper, and the same holds for any* ϕ_r *with* $r > 0$ *if* H *is isotone.*
(b) If $0 \in \mathrm{dom}\, H$, *one has* $x \in \mathrm{dom}\, \phi \implies x \in \mathrm{dom}\, \phi_r$.

Proof. (a) We prove only that ϕ is lsc, since with the same arguments we can prove that ϕ_r is also lsc. Let x be arbitrary and let $\{x_n\}$ be a sequence converging to x. We have to prove that $\liminf_{n\to\infty} \phi(x_n) \geq \phi(x)$. As a consequence we have only to consider the case where $x_n \in \mathrm{dom}\, \phi$. Let $\epsilon > 0$ be arbitrary. Since the functions f_i are lsc, then for n sufficiently large we have

$$f_i(x_n) \geq f_i(x) - \epsilon, \quad \forall i = 0, 1 \ldots, m.$$

Now, since H_∞ is isotone, it follows for n sufficiently large that

$$\phi(x_n) \geq f_0(x) + H_\infty(f_1(x) - \epsilon, \ldots, f_m(x) - \epsilon) - \epsilon.$$

Passing to the limit as $n \to \infty$ we obtain

$$\liminf_{n\to\infty} \phi(x_n) \geq f_0(x) + H_\infty(f_1(x) - \epsilon, \ldots, f_m(x) - \epsilon) - \epsilon.$$

Now let $\epsilon \to 0^+$. Since H_∞ is lsc, we get

$$\phi(x) \leq \liminf_{n\to\infty} \phi(x_n),$$

and it follows that ϕ is lsc.

Since f_0, H_∞, H are proper, it follows that ϕ and ϕ_r never take the value $-\infty$. The hypothesis on the problem's data that there exists $x_0 \in \mathrm{dom}\, f_0$ with $F(x_0) \in \mathrm{dom}\, H_\infty$ implies that $\phi(x_0)$ is finite. If $0 \in \mathrm{dom}\, H$, then from Proposition 2.5.2, it follows that $\phi_r(x_0)$ is also finite. In the other case, since $\mathrm{ri}\,\mathrm{dom}\, H_\infty$ is a cone, it follows that $r^{-1}F(x_0) \in \mathrm{ri}\,\mathrm{dom}\, H_\infty$, and then from Definition 2.8.1(c) it follows that $\phi_r(x_0)$ is finite, proving that ϕ and ϕ_r are proper. To prove (b) we suppose now that $0 \in \mathrm{dom}\, H$ and let $x \in \mathrm{dom}\, \phi$. Then $F(x) \in \mathrm{dom}\, H_\infty$, $f_0(x)$ is finite, and from Corollary 2.5.2 it follows that $\phi_r(x)$ is finite. $\qquad\square$

We arrive now with a key asymptotic formula that can be used in the analysis of the approximation problem $(\mathrm{CM})_r$.

Proposition 2.8.2 *Consider the problem (CM) with a given generating asymptotic approximation kernel H. Let*

$$\widetilde{\phi}_\infty(d) = (f_0)_\infty(d) + H_\infty(F_\infty(d)) \ \textit{if}\ d \in \bigcap_{i=1}^{m} \mathrm{dom}(f_i)_\infty, +\infty \ \textit{otherwise},$$

where $F_\infty(d) = ((f_1)_\infty(d), \dots, (f_m)_\infty(d))$. Then

$$\phi_\infty(d) \geq \widetilde{\phi}_\infty(d) \quad \forall d \in \mathbb{R}^n,$$

with equality when the functions f_i are convex.

Proof. Let $a_i < (f_i)_\infty(d)$ for $i = 0, 1, \dots, m$ and $d_n \to d, t_n \to +\infty$ with $\phi_\infty(d) = \liminf_{n \to +\infty} t_n^{-1}\phi(t_n d_n)$. Then using the fundamental analytical formula of an asymptotic function (cf. Theorem 2.5.1), for n sufficiently large we have for each i

$$f_i(t_n d_n) \geq a_i t_n \iff F(t_n d_n) \geq a t_n \ \text{with}\ a = (a_1, \dots, a_m).$$

Furthermore, since asymptotic functions are positively homogeneous, it follows that

$$\frac{\phi(t_n d_n)}{t_n} = \frac{f_0(t_n d_n)}{t_n} + H_\infty\left(\frac{F(t_n d_n)}{t_n}\right).$$

Since H_∞ is assumed isotone, using the inequalities above we get

$$\frac{\phi(t_n d_n)}{t_n} \geq a_0 + H_\infty(a).$$

Now let $a_i \to (f_i)_\infty(d)$. Then since H_∞ is lsc, using the definition of the asymptotic function and passing to the limit in the above formula, we obtain

$$\phi_\infty(d) \geq (f_0)_\infty(d) + H_\infty(F_\infty(d)) \ \text{if}\ d \in \bigcap_{i=1}^{m} \mathrm{dom}(f_i)_\infty,$$

proving the result for such a direction. Otherwise, the result remains valid by invoking Lemma 2.8.1. Now, if we suppose that for each $i = 0, 1, \ldots, m$ the functions f_i are convex, since H_∞ is sublinear and assumed isotone, it follows that

$$\frac{\phi(x + \lambda d) - \phi(x)}{\lambda} \leq (f_0)_\infty(d) + H_\infty(F_\infty(d)) \text{ if } d \in \bigcap_{i=1}^m \text{dom}(f_i)_\infty.$$

Passing to the limit as $\lambda \to +\infty$ we get $\phi_\infty(d) \leq \tilde{\phi}_\infty(d)$, and thus the desired equality in the convex case follows.   □

We end this section by indicating that other approximating schemes can be naturally generated via the analytical asymptotic function formula. Indeed, consider a scalar function $\alpha : \mathbb{R}_+ \to \mathbb{R}$ satisfying

$$\alpha(r) > 0, \ \forall r > 0, \ \lim_{r \to 0_+} \alpha(r) = 0, \ \lim_{r \to 0_+} r^{-1}\alpha(r) = +\infty.$$

Consider now the class of functions $\theta : \mathbb{R} \to \mathbb{R}_+$, that are nondecreasing, convex, and such that

$$\lim_{u \to -\infty} \theta(u) = 0, \ 0 < \theta_\infty(1) < +\infty.$$

Since here $\theta_\infty(1) \neq +\infty$, we no longer have the representation $\delta_{\mathbb{R}^m_-}(u) = \lim_{r \to 0_+} r\theta(r^{-1}u)$. However, it is possible to write now that $\delta_{\mathbb{R}^m_-}(u) = \lim_{r \to 0_+} \alpha(r)\theta(r^{-1}u)$. Thus, with the functions α and θ as above, we can now construct the approximation kernel $H_r(y) = \alpha(r)H(r^{-1}y)$, where $H(y) = \sum_{i=1}^m \theta(y_i)$, and we have $\lim_{r \to 0_+} H_r(y) = \delta_{\mathbb{R}^m_-}(y)$. Within this class we then construct the approximate models $(CM)_r$ and $(SDP)_r$ as before. Two particular examples of functions θ that can be used to generate the approximate models are

$$\theta_6(u) = \log(1 + e^u), \ \theta_7(u) = 2^{-1}(u + \sqrt{u^2 + 4}).$$

For any symmetric matrix D, the corresponding functions H used for approximating the semidefinite program (SDP) then take the forms

$$H_6(D) = \log \det(I + \exp(D)), \ H_7(D) = 2^{-1} \text{tr} \left(D + \sqrt{D^2 + 4I} \right).$$

2.9   Notes and References

The concept of asymptotic cone and many of its properties seem to have appeared first in the literature in the work of Steinitz in 1913 [126]. Properties of unbounded convex point sets have been also studied in 1940 by

Stoker [127]. These concepts can also be found in Bourbaki [39] and Choquet [47]. For convex sets, the importance of asymptotic cones were already realized in the work of Dieudonné in 1966 [60] and for closed convex sets, in Rockafellar [119], who uses the terminology of *recession* cones. The generalization of the concept of asymptotic cone of convex sets to closed sets in an arbitrary topological vector space was considered in the works of Dedieu [58], [59]. Section 2.1 includes the fundamental results on asymptotic cones in both the convex and nonconvex cases, most of which can be found in the book of Rockafellar [119], while in the nonconvex case several results outlined here appear in the references cited above, but in greater details in the recent book of Rockafellar–Wets [123]. The dual characterization given in Section 2.2 can be found in the book of Aubin–Ekeland [4]. Many results on closedness criteria have been scattered in the literature, and our intent here was to present some of the most useful criteria. Most results were originally concerned with sufficient conditions that ensure the closedness of the image of a closed set. The conditions are expressed in terms of some properties shared by the map and the asymptotic cone of the set in question. In the convex case, such results can be found in Fenchel [72], Choquet [47], and Rockafellar [117]. The weaky coercive case described by Corollary 2.3.2(b) can be found in [119]. It seems that Dedieu [57] was the first to give the corresponding result to Corollary 2.3.2(a) in the nonconvex coercive case. The necessary and sufficient condition that ensures that the image of a closed set under a linear map remains closed and given in Theorem 2.3.1 is due to Auslender [13]. The concept of asymptotic linear sets and their properties were introduced by Auslender [18]. Theorem 2.3.2, stating that the image of the asymptotic cone of a set coincides with the asymptotic cone of its image, is new. However, such a result was proven earlier in the coercive case by Zalinescu [135] and can be found for the convex case in Rockafellar's book [119]. Specific results ensuring the closedness of the sum of closed sets and the set of convex combinations of a finite collection of closed sets have been given for the convex case in [119] and extended to the nonconvex case in [13]. Results concerning semibounded sets can be found in an article of Fadeev and Fadeva published in the collected works [100]. Continuous convex sets originated in the work of Gale and Klee [75], and the results given here follow this work. Propositions 2.4.2 and 2.4.3 are due to Auslender–Coutat [11]. Building on the concept of asymptotic cone, one can analyze the behavior in the large of real-valued functions through their epigraphs. This has been done by Rockafellar [119] for convex functions, who gave the representation formula of Proposition 2.5.2. For the nonconvex case, the notion of asymptotic function and the representation formula given in Theorem 2.5.1 were first given by Dedieu [58] and later on by Biaocchi, Butazzo, Gastaldi, and Tomeralli [22]. Most of the results presented in Section 2.5 and Section 2.6 are classical in the convex case and can be found in [119] as well as in the more recent book of Rockafellar and Wets [123], which also includes results for nonconvex functions. Some other less classical or not so well known results include Proposition 2.6.6, due to Benoist and Hirriart-Urruty [26]; Corollary 2.6.1, due to Valadier [128]; Proposition 2.6.5 from Auslender [13]. Proposition 2.6.7, (well known is convex analysis) is new, and the formula on the composite of a convex function in Proposition 2.6.4 given in Auslender, Cominetti, and Haddou [14], while the results given in Lemma 2.6.1 on one-dimensional convex functions is from Ben-Tal, Ben-Israel, and Teboulle [27]. Systems of matrix inequalities and related optimization problems, today called *Semidefinite Optimization* originated in 1963 with the work of Bellman and Fan [24]. This class of problems today forms an active research area; see for example, the recent handbook of semidefinite programming [134], which includes about one thousand references. We gave in Section 2.7 some basic notions on convex functions of matrices first characterized via their eigenvalues by Davis [55] and further studied and extended by Lewis in [87] and Seeger [124], on which our presentation is based. The formula for the associated asymptotic function given in Theorem 2.7.2 can be found in Seeger [124], with a slightly different proof given here. Section 2.8 considers smoothing and approximation of optimization problems. Initial studies on that topic can be found in Bertsekas in [33] and in [34], who proposed and studied in particular the logarithmic sum of exponentials. Related numerical algorithms based on this approach such as penalty–barrier and Lagrangians methods can be found in [34] and the more recent book of Berstekas [36], which also includes recent developments in the field together with the relevant references. The idea of smoothing optimization via asymptotic functions was originally proposed by Ben-Tal and Teboulle [28] for finite convex functions, with some applications that can be found in Ben-Tal, Teboulle, and Hang [29]. More recently, Auslender [17] proposed a unified framework within asymptotic functional representation of optimization problems to handle and analyze the more important case of extended real-valued functions, thus allowing the modeling of general problems such as nonlinear constrained optimization and semidefinite programs via the use of Propositions 2.8.1 and 2.8.2 proven in [17]. The presentation here follows the work developed in [17].

3
Existence and Stability in Optimization Problems

A central question in optimization is to know whether an optimal solution exists. Associated with this question is the stability problem. A classical result states that a lower semicontinuous function attains its minimum over a compact set. The compactness hypothesis is often violated in the context of extremum problems, and thus the need for weaker and more realistic assumptions. This chapter develops several fundamental concepts and tools revolving around asymptotic functions to derive existence and stability results for general and convex minimization problems.

3.1 Coercive Problems

A central question in minimization problems is to find appropriate conditions that will ensure the existence of a minimizer, namely that $f : \mathbb{R}^n \to \mathbb{R} \cup \{+\infty\}$ will attain its minimum over $\mathbb{R}^n$. A classical calculus result providing an answer to this question is the well-known Weierstrass theorem, which guarantees that a continuous function defined on a compact subset of $\mathbb{R}^n$ attains its minimum (as well as its maximum). Both conditions required in that theorem, compactness and continuity, are rather stringent and often not present in the context of a given minimization problem. Nevertheless, the Weierstrass theorem provides the underlying idea for deriving existence results under weaker hypotheses on the problem's data. The standard starting point is to realize that solving the general optimization problem

$$(P) \quad \inf\{f(x) \mid x \in \mathbb{R}^n\}$$

in equivalent to minimizing f over one of its level sets $\text{lev}(f, \lambda) = \{x \mid f(x) \leq \lambda\}$, for $\lambda > \inf f$, and thus one should look at "Weierstrass-type" theorems on that set.

Let us begin by first recalling the following basic result, which reveals that continuity can be replaced by lower semicontinuity.

Proposition 3.1.1 *Let $f : \mathbb{R}^n \to \mathbb{R} \cup \{+\infty\}$ be an lsc, proper function and let C be a nonempty compact set in $\mathbb{R}^n$. Then the optimal solution set of the optimization problem*

$$m := \inf\{f(x) \mid x \in C\}$$

is nonempty and compact.

Proof. Clearly, if $f(x) = \infty$ for every $x \in C$, then any $x \in C$ attains the minimum of f over C. We can thus assume that $\inf_C f < \infty$. Let $\{x_k\} \in C$ be a minimizing sequence, i.e.,

$$\lim_{k \to \infty} f(x_k) = \inf\{f(x) \mid x \in C\}.$$

Since C is compact, there exists a subsequence $\{x_{k_l}\}$ converging to some $x \in C$, and since f is lsc, it follows that $f(x) \leq \liminf_{l \to \infty} f(x_{k_l}) = m$, and the theorem is proved, since $f(x) \geq m$. □

A similar result for the existence of maxima of f over C can be stated by assuming that f is upper semicontinuous on the compact set C.

Going back to the general optimization problem (P), since the optimal solution set of (P) is the same as that of minimizing f on a level set of f, we obtain as an immediate consequence of Proposition 3.1.1 the following result.

Corollary 3.1.1 *Let $f : \mathbb{R}^n \to \mathbb{R} \cup \{+\infty\}$ be lsc, and proper, and suppose that $\text{lev}(f, \lambda)$ is bounded for some $\lambda > \inf f$. Then the optimal solution set of problem (P) is nonempty and compact.*

Proof. Since f is proper, $\inf f < \infty$. Thus for any $\lambda > \inf f$, the level set $\text{lev}(f, \lambda)$ is nonempty and closed, since f is lsc, hence compact, by assumption. We can thus apply Proposition 3.1.1 with $C := \text{lev}(f, \lambda)$. □

The fundamental role played by the asymptotic function in existence questions for minimization problems then comes naturally into action, since the boundedness requirement of a given set can be translated geometrically as asking for its associated asymptotic cone to reduce to the singleton $\{0\}$ (cf. Proposition 2.1.2). The set in question is here the level set of f, and thus for any $\lambda > \inf f$, we have that $\text{lev}(f, \lambda)$ is nonempty and bounded if and only if $\text{lev}(f, \lambda)_\infty$ is reduced to $\{0\}$. Furthermore, as a consequence of the definition of the asymptotic function f_∞ we know that (cf. Proposition 2.5.3)

$$\mathrm{lev}(f, \lambda)_\infty \subset \{d \mid f_\infty(d) \le 0\}, \tag{3.1}$$

with equality replacing the inclusion when f is also assumed convex. This naturally leads one to define the following concepts.

Definition 3.1.1 *The function $f : \mathbb{R}^n \to \mathbb{R} \cup \{+\infty\}$ is called*
(a) level bounded if for each $\lambda > \inf f$, the level set $\mathrm{lev}(f, \lambda)$ is bounded,
(b) coercive if $f_\infty(d) > 0 \;\; \forall d \ne 0$.

As an immediate consequence of the definition we remark that f is level bounded if and only if

$$\lim_{\|x\| \to \infty} f(x) = +\infty, \tag{3.2}$$

which means that the values of $f(x)$ cannot remain bounded on any subset of $\mathbb{R}^n$ that is not bounded.

This level boundedness property as given above is widely used in the literature (often called coercivity in classical analysis textbooks). Note that a coercive function is always level bounded, i.e., its level set $\mathrm{lev}(f, \lambda)$ is bounded and nonempty for any $\lambda > \inf f$. The following proposition gives a useful criterion for recognizing a coercive function.

Proposition 3.1.2 *Let $f : \mathbb{R}^n \to \mathbb{R} \cup \{+\infty\}$ be lsc and proper, and define*

$$\alpha = \liminf_{\|x\| \to \infty} \frac{f(x)}{\|x\|}, \quad \beta = \inf \{f_\infty(d) \mid \|d\| = 1\}. \tag{3.3}$$

Then the infimum β is attained and $\alpha = \beta$. Furthermore, f is coercive if and only if there exist $\gamma > 0$, $\delta \in \mathbb{R}$ such that

$$f(x) \ge \gamma \|x\| + \delta, \quad \forall x. \tag{3.4}$$

Proof. Since for any proper function f, its asymptotic function f_∞ is lsc, using (3.3) there exists x such that

$$\|x\| = 1, \quad f_\infty(x) = \beta. \tag{3.5}$$

Then from the analytic representation of f_∞ (see Theorem 2.5.1) there exists a sequence d_k converging to d, and a sequence t_k converging to $+\infty$ such that $\lim_{t_k \to +\infty} t_k^{-1} f(t_k d_k) = \beta$. As a consequence we get $\alpha \le \beta$. Now let us prove that $\alpha \ge \beta$. From the definition of α there exists a sequence $\{x_k\}$ such that

$$\|x_k\| \to \infty, \quad x_k \|x_k\|^{-1} \to \bar{x}, \; \|\bar{x}\| = 1, \quad \lim_{k \to \infty} \frac{f(x_k \|x_k\|^{-1} \|x_k\|)}{\|x_k\|} = \alpha.$$

Once again, using the analytic representation of f_∞ (cf. (2.18)), and passing to the limit, it follows that $f_\infty(\bar{x}) \le \alpha$, which implies that $\alpha \ge \beta$, thus proving the first part of the proposition. Suppose now that (3.4) holds. Then it obviously follows that $\alpha \geq \gamma > 0$. Since $\alpha = \beta$ and since $f_\infty(\cdot)$ is positively homogeneous, this implies coercivity. Conversely, suppose that f is coercive. Then $\alpha = \beta > 0$, and suppose that (3.4) does not hold. Then, for any $\delta \in \mathbb{R}$, $\gamma > 0$, there exists x such that $f(x) < \gamma\|x\| + \delta$. Pick $\delta = -k$, $\gamma = k^{-1}$, $k \in \mathbb{N}$. Then there exists x_k such that $f(x_k) < k^{-1}\|x_k\| - k$. If x_k is bounded, it follows that $f(x_k) \to -\infty$, which is impossible. If x_k is unbounded, then from the last inequality it follows that

$$\liminf_{\|x\|\to+\infty} \frac{f(x)}{\|x\|} \leq \liminf_{k\to+\infty} \frac{f(x_k)}{\|x_k\|} \leq 0,$$

which contradicts the coercivity of f. $\square$

Definition 3.1.2 *For any function $f : \mathbb{R}^n \to \mathbb{R}\cup\{+\infty\}$, directions $d \in \mathbb{R}^n$ such that $f_\infty(d) \leq 0$ are called asymptotic directions, or recession directions in the case of f convex.*

In the convex case, the above concepts and relations are in fact all equivalent.

Proposition 3.1.3 *Let $f : \mathbb{R}^n \to \mathbb{R} \cup \{+\infty\}$ be lsc and proper. If f is coercive, then it is level bounded. Furthermore, if f is also convex, then the following statements are equivalent:*
(a) f is coercive.
(b) f is level bounded.
(c) The optimal set $\{x \in \mathbb{R}^n \mid f(x) = \inf f\}$ is nonempty and compact.
(d) $0 \in \operatorname{int} \operatorname{dom} f^*$.

Proof. Since $\operatorname{lev}(f, \lambda)_\infty \subset \{d \mid f_\infty(d) \leq 0\}$, then whenever f is coercive, namely $f_\infty(d) > 0$, $\forall 0 \neq d$, it follows that f is level bounded. When f is convex, the inclusion above becomes an equality, and since $\operatorname{lev}(f, \lambda)$ is compact, it follows that $\{d \mid f_\infty(d) \leq 0\}$ is reduced to 0, so that f is coercive, and this proves the equivalence between (a) and (b). Moreover, by Theorem 1.3.2(c) one has $0 \in \operatorname{int} \operatorname{dom} f^*$ if and only if $\sigma_{\operatorname{dom} f^*}(d) > 0$, $\forall d \neq 0$, and since by Theorem 2.5.4(a) $\sigma_{\operatorname{dom} f^*} = f_\infty$, the statements (a) and (d) are equivalent. Finally, since the set of optimal solutions can be written as $\{x \in \mathbb{R}^n \mid f(x) = \inf f\} = \partial f^*(0)$, then from Proposition 1.2.16, (c) and (d) are equivalent. $\square$

Using asymptotic rules of calculus, the coercivity condition can translate into useful tests for optimization problems described via explicit constraints.

Corollary 3.1.2 *Let $f_i : \mathbb{R}^n \longrightarrow \mathbb{R} \cup \{+\infty\}$ $i = 0, 1 \ldots, m$ be lsc, proper functions. Let $C := \{x \mid f_i(x) \leq 0 \quad i = 1, \ldots, m\}$, $\quad f = f_0 + \delta_C(\cdot)$, and consider the constrained optimization problem*

$$(P) \qquad \inf\{f_0(x) \mid x \in C\},$$

with $\operatorname{dom} f_0 \cap C \neq \emptyset$. *Suppose that* $(f_0)_\infty(d) > -\infty \ \forall d \neq 0$. *If the functions* $f_i, \ i = 0, 1, \ldots, m$ *have no common nonzero asymptotic direction, i.e.,*

$$(f_i)_\infty(d) \leq 0 \ \forall i = 0, \ldots, m \implies d = 0,$$

then the objective function f is coercive (and thus the optimal set of (P) is nonempty and compact). When the the functions $f_i, \ i = 0, 1, \ldots, m$ are convex, the converse statement holds.

Proof. By asymptotic calculus rules one has $f_\infty(d) \geq (f_0)_\infty(d) + \delta_{C_\infty}(d)$, which implies $f_\infty(d) \geq (f_0)_\infty(d)$, $\forall d \in C_\infty$. Since $C_\infty \subset \cap_{i=1}^m \mathcal{K}_{f_i}$, the asymptotic assumption implies that f is coercive. Conversely, if the functions $\{f_i, \ i = 0, \ldots, m\}$ are assumed convex, then the above inequalities and the inclusion become equalities thanks to Proposition 2.1.9(a) and Proposition 2.6.1, respectively. As a result, the functions $f_i, \ i = 0, \ldots, m$, have no common nonzero asymptotic directions. $\qquad \square$

Level-boundedness is also linked with the so-called Palais–Smale condition.

Definition 3.1.3 *Suppose that $f : \mathbb{R}^n \to \mathbb{R}$ is a C^1 function (continuously differentiable). Then:*
(a) A sequence $\{x_k\}$ such that $\nabla f(x_k) \to 0$ and $\{f(x_k)\}$ is bounded is said to be a Palais Smale [PS] sequence for f.
(b) f is said to satisfy the Palais–Smale condition if each [PS] sequence for f is bounded.

Clearly, if f is level bounded, then f satisfies the Palais–Smale condition. The converse holds if in addition f is assumed bounded below, and this will be proved as a corollary of Ekeland's variational principle given in Section 4.1.

3.2   Weak Coercivity

For many theoretical purposes in convex analysis, coercivity is a notion that is often very stringent and thus replaced by *weak coercivity*. In fact, weak coercivity is a property that emerges from a wide class of functions called *asymptotically directionally constant*.

Definition 3.2.1 *Let* $f : \mathbb{R}^n \to \mathbb{R} \cup \{+\infty\}$ *be a proper and lsc function. Then the function f is said to be:*
(a) asymptotically directionally constant (adc) if

$$f_\infty(d) = 0 \Longrightarrow f(x + \rho d) = f(x), \quad \forall x \in \text{dom } f, \ \forall \rho \in \mathbb{R}, \qquad (3.6)$$

(b) weakly coercive if in addition to (3.6) one has

$$f_\infty(d) \geq 0 \quad \forall d \neq 0. \qquad (3.7)$$

Note that a coercive function is obviously adc. The next example shows that there is a wide class of (not necessarily convex) functions that are adc and weakly coercive.

Example 3.2.1 Let $g : \mathbb{R}^m \to \mathbb{R} \cup \{+\infty\}$ be a proper, lsc, and coercive function, and let A be an $m \times n$ real matrix with $\ker A \neq \{0\}$, and $b \in -\text{dom } g$. Consider the function $x \to f(x) := g(Ax - b)$. Clearly, this function has unbounded level sets. Indeed,

$$\lim_{k \to \infty} f(t_k d) = g(-b), \quad \text{for } 0 \neq d \in \ker A, \quad \text{with } \lim_{k \to \infty} t_k = +\infty.$$

Furthermore, for such d since $\lim_{k \to \infty} t_k^{-1} f(t_k d_k) = 0$, it follows by the analytic representation of $f_\infty(d)$ that $f_\infty(d) \leq 0$. Now using again the analytical representation of the asymptotic function we obtain

$$f_\infty(d) \geq \inf_{\substack{d_k \to Ad \\ t_k \to \infty}} \left\{ \liminf_{k \to \infty} t_k^{-1} g(t_k d_k) \right\} = g_\infty(Ad).$$

Then if $Ad \neq 0$, it follows from the coercivity of g that $f_\infty(d) > 0$, and we conclude that $f_\infty(d) \leq 0$ if and only if $d \in \ker A$, so that f is adc. Furthermore, if $g_\infty(0) = 0$, which is the case when g is convex, then (3.7) is satisfied and f is weakly coercive.

In the convex case there are several classes of functions that are adc.

Proposition 3.2.1 *Let* $f : \mathbb{R}^n \to \mathbb{R} \cup \{+\infty\}$ *be a proper, lsc, and convex function such that f is affine on each line for which the direction is an asymptotic direction. Then f is adc. Quadratic convex functions, polynomial convex functions, and functions for which $\text{dom } f^*$ is an affine set belong to this class.*

Proof. By hypothesis, f is affine on each line for which the direction d is an asymptotic direction, i.e., satisfying $f_\infty(d) \leq 0$. Invoking Theorem 2.5.3 we get $f_\infty(d) = -f_\infty(-d)$. Then assuming that $f_\infty(d) = 0$, it follows by parts (a)–(c) of the same theorem that $f(x + \rho d) = f(x)$, $\forall x \in \text{dom } f$, $\rho \in \mathbb{R}$. Thus f is constant on such a direction, and f is adc. Consider now the class of convex quadratic functions. Using the formula of the corresponding asymptotic function f_∞ given in Example 2.5.1(b), one easily see that f is adc for such functions. More generally, if f is a convex polynomial, then for each x and y the function $h : \lambda \to f(x + \lambda y)$ is a convex polynomial of the single variable $\lambda \in \mathbb{R}$. Since the coefficient associated with the term of highest degree is positive for a nonaffine convex polynomial, then it follows that h is affine or level bounded, so that f is adc. Finally, we consider the class of functions for which dom f^* is an affine set. By Theorem 2.5.4(b), f_∞ is the support functional of dom f^*. Then if d is an asymptotic direction of f, it follows that the linear functional $u \to \langle d, u \rangle$ is bounded above by 0 on dom f^*. Since every functional that is bounded above on an affine set is constant, it follows that $-\sigma_{\mathrm{dom}\, f^*}(-d) = \sigma_{\mathrm{dom}\, f^*}(d)$, which is equivalent to $-f_\infty(-d) = f_\infty(d)$. By Theorem 2.5.3 this means that f is affine along the direction d, and thus f is adc. $\qquad \square$

We give now further important equivalent ways for characterizing weak coercivity of convex functions.

Theorem 3.2.1 Let $f : \mathbb{R}^n \to \mathbb{R} \cup \{+\infty\}$ be an lsc, proper, convex function. Then

$$f_\infty(d) \geq 0 \ \ \forall 0 \neq d \iff 0 \in \mathrm{cl}\,\mathrm{dom}\, f^*. \tag{3.8}$$

Furthermore, the following statements are equivalent:
(a) f is weakly coercive.
(b) $f_\infty(d) \geq 0 \ \ \forall\, 0 \neq d$ and $\{d \mid f_\infty(d) = 0\}$ is a subspace, i.e., $f_\infty(d) = 0 \implies f_\infty(-d) = 0$.
(c) $f_\infty(d) > 0, \ \forall d \neq 0, \ d \in \mathcal{C}_f^\perp = \{d \mid f_\infty(d) = f_\infty(-d) = 0\}^\perp$.
(d) $0 \in \mathrm{ri}\,\mathrm{dom}\, f^*$.

Proof. By Theorem 2.5.4(b) we have $f_\infty = \sigma_{\mathrm{dom}\, f^*} = \sigma_{\mathrm{cl}\,\mathrm{dom}\, f^*}$, and then by Theorem 1.3.2, one has $0 \in \mathrm{cl}\,\mathrm{dom}\, f^*$ if and only if $\sigma_{\mathrm{cl}\,\mathrm{dom}\, f^*}(d) \geq 0$, for every d, thus proving the relation (3.8). The equivalence between (a) and (b) is a direct consequence of Theorem 2.5.3. The equivalence between (c) and (d) is a consequence of Theorem 2.5.4 and Proposition 2.5.8(b). Indeed, by the latter proposition one has $0 \in \mathrm{ri}\,\mathrm{dom}\, f^*$ if and only if $f_\infty(d) > 0$ for all $d \notin \mathcal{C}_f$, so that (d) clearly implies (c). Conversely, let P_C be the projection operator on C. Then any $d \in \mathbb{R}^n$ can be decomposed as $d = d_1 + d_2$ with $d_1 = P_{\mathcal{C}_f}(d), \ \ d_2 = P_{\mathcal{C}_f^\perp}(d)$. But by Theorem 2.5.4 one obtains

$$\begin{aligned}
f_\infty(d) = f_\infty(d_1 + d_2) &= \sup_{u \in \mathrm{dom}\, f^*} \{\langle u, d_1 \rangle + \langle d_2, u \rangle\} \\
&= \sup_{u \in \mathrm{dom}\, f^*} \langle d_2, u \rangle = f_\infty(d_2),
\end{aligned}$$

where in the third equality we use the fact (cf. (2.21)) $d_1 \in \mathcal{C}_f = \{y : \langle y, v \rangle = 0, \ \ \forall v \in \mathrm{dom}\, f^*\}$. Therefore, we have proved that for any $d \notin \mathcal{C}_f, \ d_2 \neq 0, \ f_\infty(d) = f_\infty(d_2) > 0$, proving that assertion (c) implies (d).

Finally, that (b) implies (c) is obvious, while (d) implies (b) follows at once from Proposition 2.5.8(b) and the fact that $f_\infty(d) = f_\infty(d_2)$.   □

The following definition used for characterizing general weakly coercive functions will be of particular interest in the next section, devoted to the existence of optimal solutions.

Definition 3.2.2 *Let $f : \mathbb{R}^n \to \mathbb{R} \cup \{+\infty\}$ be any function and define by S_f the set of sequences $\{x_k\}$ satisfying*

$$\|x_k\| \to +\infty, \quad \{f(x_k)\} \ \text{is bounded above and} \quad x_k\|x_k\|^{-1} \to \bar{x} \in \ker f_\infty. \tag{3.9}$$

Then f is said to satisfy Assumption (A) if either one of the following two conditions is satisfied:
(a) S_f is empty,
(b) for each sequence $\{x_k\} \in S_f$ there exist $\varepsilon \in (0, \frac{1}{2})$, $\rho_k \in ((1+\varepsilon)\|x_k\|, (2-\varepsilon)\|x_k\|)$ such that for k sufficiently large we have

$$f(x_k - \rho_k \bar{x}) \leq f(x_k). \tag{3.10}$$

Proposition 3.2.2 *Let $f : \mathbb{R}^n \to \mathbb{R} \cup \{+\infty\}$ be an lsc proper convex function satisfying $f_\infty(d) \geq 0$, $\forall d \neq 0$. Then f is weakly coercive if and only if f satisfies Assumption (A).*

Proof. We have only to prove the equivalence between Assumption (A) of Definition 3.2.2 and the relation (3.6). Suppose that relation (3.6) holds for some $\bar{x}$. Then from the equivalence of (a) and (b) in Theorem 3.2.1 it follows that $f(x_k - \rho_k \bar{x}) = f(x_k)$ for each ρ_k and (3.10) is satisfied. Conversely, suppose that f satisfies assumption (A) and let $\bar{x}$ be such that $\|\bar{x}\| = 1$ and $f_\infty(\bar{x}) = 0$. We have only to prove that $f_\infty(-\bar{x}) = 0$. Let $x \in \operatorname{dom} f$, and set $x_k = x + k\bar{x}$. Since $k^{-1}(f(x_k) - f(x)) \leq f_\infty(\bar{x}) = 0$, it follows that $\{f(x_k)\}$ is bounded above by $f(x)$, and then from Assumption (A) there exists $\rho_k \in ((1 + \varepsilon)\|x_k\|, (2 - \varepsilon)\|x_k\|)$ satisfying (3.10) for k sufficiently large. Furthermore, since $f_\infty(\bar{x}) \geq 0$, we obtain $\lim_{k\to\infty} \|x_k\|^{-1} f(x_k) = 0$. Without loss of generality we may suppose that $\rho_k\|x_k\|^{-1} \to \alpha > 1$. Let $y_k = \|x_k\|^{-1}(x_k - \rho_k \bar{x})$, then $y_k \to (1 - \alpha)\bar{x}$. Using (3.10) we obtain

$$0 = \lim_{k\to\infty} \frac{f(x_k)}{\|x_k\|} \geq \liminf_{k\to\infty} \frac{f(\|x_k\|y_k)}{\|x_k\|} \geq f_\infty((\alpha - 1)(-\bar{x}))$$
$$= (\alpha - 1)f_\infty(-\bar{x}) \geq 0,$$

and it follows that $f_\infty(-\bar{x}) = 0$.   □

Very often, the functions that are to be used appear as a result of applying different operations to other simpler functions, for example, the infimal convolution operation. It is therefore important to give criteria for verifying when the resulting function is weakly coercive, and this will be done by characterizing weakly coercive functions via images of coercive functions under linear transformations. To achieve this task, we first introduce a useful device as follows. To establish existence results we encounter conditions requiring that a point belong to the interior of the domain of a convex function. When the domain is of full dimensionality, the interior is guaranteed to be nonempty. However, it often occurs that the domain of the function under study is not fully dimensional, in which case the interior might be empty. It turns out that we can easily manage this kind of situation by employing a simple and useful dimension-reducing argument.

Let $h : \mathbb{R}^n \to \mathbb{R} \cup \{+\infty\}$ be a convex function with $0 \in \operatorname{dom} h$ and let E be the affine hull of its domain, i.e., $E = \operatorname{aff}(\operatorname{dom} h)$. Note that since $0 \in \operatorname{dom} h$, then E is in fact a vector subspace of $\mathbb{R}^n$. Given h, we associate the function h_E defined by

$$h_E(x) = (h \circ P_E)(x) = h(P_E(x)),$$

where P_E denotes the orthogonal projection operator from $\mathbb{R}^n$ onto E.

Proposition 3.2.3 *Let* $h : \mathbb{R}^n \to \mathbb{R} \cup \{+\infty\}$ *be a proper convex function with* $0 \in \operatorname{dom} h$ *and* $E = \operatorname{aff}(\operatorname{dom} h)$. *The following properties hold:*
(a) $\operatorname{dom} h_E = \operatorname{dom} h + E^\perp$.
(b) $\operatorname{int} \operatorname{dom} h_E = \operatorname{ri} \operatorname{dom} h + E^\perp$.
(c) $\partial h_E(x) \subset E$ *and* $\partial h_E(x) = \partial h_E(y)$ *if* $y - x \in E^\perp$.
(d) $h^*(y) = h_E^*(P_E(y))$.
(e) $\partial h(x) = \partial h_E(x) + E^\perp$ *if* $x \in E$; *otherwise,* $\partial h(x) = \emptyset$.
(f) $\partial h_E(x) = \partial h(P_E(x)) \cap E$.

Proof. (a) Follows from the definition of h_E. To verify (b), use Proposition 1.1.8 and recall that for any subspace M one has $\operatorname{ri} M = M$. Then since here $\operatorname{ri} \operatorname{dom} h_E = \operatorname{int} \operatorname{dom} h_E$, (b) follows from (a). To prove (c), consider $w \in \partial h_E(x)$ and take $v \in E^\perp$. By (a), $x + \lambda v \in \operatorname{dom} h_E$ for each $\lambda \in \mathbb{R}$, and we have

$$h_E(x) = h_E(x + \lambda v) \geq h_E(x) + \lambda \langle w, v \rangle,$$

so that $w \in E$. Furthermore, $h_E(x) = h_E(y)$ when $x - y \in E^\perp$, by definition of h_E. Then $w \in \partial h_E(x)$ if and only if for each $z \in \mathbb{R}^n$ we have

$$h_E(z) \geq h_E(x) + \langle w, z - x \rangle = h_E(y) + \langle w, z - y \rangle + \langle w, y - x \rangle,$$

but $w \in E$, so that $\langle w, y - x \rangle = 0$, and the second part of (c) is proved. To prove (d), using the definition of the conjugate we obtain

$$h^*(y) \;=\; \sup_{x \in E} \{ \langle x, y \rangle - h(x) \} = \sup_{x \in \mathbb{R}^n} \{ \langle y, P_E(x) \rangle - h(P_E(x)) \}$$

$$= \sup_{x \in \mathbb{R}^n} \{\langle x, P_E(y)\rangle - h_E(x)\} = h_E^*(P_E(y)).$$

We now prove (e) and (f). Let $y = u + v$ with $u \in \partial h_E(x)$, $v \in E^\perp$, and let us prove that $y \in \partial h(x)$, i.e.,

$$h(z) - h(x) \geq \langle y, z - x\rangle, \quad \forall z \in \mathrm{dom}\, h.$$

But since $h(z) = h_E(z)$, $h(x) = h_E(x)$, the above inequality follows from

$$h_E(z) - h_E(x) \geq \langle u, z - x\rangle = \langle u + v, z - x\rangle.$$

Conversely, suppose that $y \in \partial h(x)$ and let us prove that $P_E y \in \partial h_E(x)$ so that (e) will hold. For any $z \in \mathbb{R}^n$ we have $h(z) \geq h_E(x) + \langle y, z - x\rangle$, and then for any $z \in \mathbb{R}^n$ we get

$$\begin{aligned}
h(P_E(z)) &= h_E(z) \geq h_E(x) + \langle y, P_E(z) - x\rangle \\
&= h_E(x) + \langle P_E(y), P_E(z) - x\rangle \\
&= h_E(x) + \langle P_E(y), z - x\rangle.
\end{aligned}$$

From (c) and (e) we obtain (f) via the relations

$$\begin{aligned}
\partial h_E(x) = [\partial h_E(x) + E^\perp] \cap E &= [\partial h_E(P_E(x)) + E^\perp] \cap E \\
&= \partial h(P_E(x)) \cap E.
\end{aligned}$$

$\square$

Armed with Proposition 3.2.3, we are then able to analyze existence results by working with the function h_E, instead of h, since h_E is continuous on the interior of its domain, which is always nonempty. The following result shows that weakly coercive functions can be characterized as images of coercive functions under linear transformations.

Corollary 3.2.1 *Let $f : \mathbb{R}^n \to \mathbb{R} \cup \{+\infty\}$ be lsc, proper, and convex with $\inf f > -\infty$. Then*

$$f(x) = (f^* \circ P_E)^*(P_E(x)), \quad with \ E = \mathrm{aff}(\mathrm{dom}\, f^*), \tag{3.11}$$

and f is weakly coercive if and only if $(f^* \circ P_E)^*$ is coercive.

Proof. Since $-f^*(0) = \inf f$, then $0 \in \mathrm{dom}\, f^*$, and we can apply Proposition 3.2.3 with $h := f^*$. Then formula (3.11) is nothing else than (d) in Proposition 3.2.3. By Theorem 3.2.1(d), f is weakly coercive if and only if $0 \in \mathrm{ri}\,\mathrm{dom}\, f^*$, which by Proposition 3.2.3(b) is equivalent to saying that $0 \in \mathrm{int}\,\mathrm{dom}(f^* \circ P_E)$, which in turn is equivalent to saying that $(f^* \circ P_E)^*$ is coercive by Proposition 3.1.3(d).

$\square$

Proposition 3.2.4 *Let $f, g : \mathbb{R}^n \to \mathbb{R} \cup \{+\infty\}$ be lsc, proper, and convex with g weakly coercive, and let $h(x) = (f \square g)(x) = \inf_u \{f(u) + g(x - u)\}$ be their infimal convolution. The following hold:*
(a) If f is weakly coercive, then so is h.
(b) Conversely, if g is cofinite, i.e., $\operatorname{dom} g^* = \mathbb{R}^n$ and h is weakly coercive, then f is weakly coercive.
(c) The constancy space of h satisfies $\mathcal{C}_h \supset \mathcal{C}_f + \mathcal{C}_g$.

Proof. From the equality $h^* = f^* + g^*$ (cf. Proposition 1.2.10(a)), we get $\operatorname{dom} h^* = \operatorname{dom} f^* \cap \operatorname{dom} g^*$, and therefore using Proposition 1.1.7 one has $\operatorname{ri} \operatorname{dom} h^* = \operatorname{ri} \operatorname{dom} f^* \cap \operatorname{ri} \operatorname{dom} g^*$, from which the first two claims (a)–(b) of the proposition follow from Theorem 3.2.1. Property (c) follows directly from this last formula and from the characterization given in (2.21) for $\mathcal{C}_h = \{d \mid \langle d, v \rangle = 0, \ \forall v \in \operatorname{dom} h^*\}$, as well as the characterization for $\mathcal{C}_f$ and $\mathcal{C}_g$. □

Corollary 3.2.2 *Let $f : \mathbb{R}^n \to \mathbb{R} \cup \{+\infty\}$ be lsc, proper, convex, and let $E = \operatorname{aff}(\operatorname{dom} f^*)$. Then the optimization problem $\inf\{f(x) \mid x \in \mathbb{R}^n\}$ has a nonempty optimal set of the form $S = K + E^\perp$ with $K = \partial(f^*)_E(0)$ a nonempty convex compact set if and only if f is weakly coercive.*

Proof. We are in the case where $0 \in \operatorname{dom} f^*$. Since $S = \partial f^*(0)$, it follows from Proposition 3.2.3 (e) that $\partial f^*(0) = \partial(f^*)_E(0) + E^\perp$, and by Proposition 3.1.3 we have that $\partial(f^*)_E(0)$ is nonempty and compact if and only if $0 \in \operatorname{int} \operatorname{dom}(f^*)_E$, i.e., from Proposition 3.2.3(b) if and only if $0 \in \operatorname{ri} \operatorname{dom} f^*$. □

Proposition 3.2.5 *Let $f_i : \mathbb{R}^n \longrightarrow \mathbb{R} \cup \{+\infty\}$, $i = 0, 1 \ldots, m$ be lsc, proper, convex functions. Let $C := \{x \mid f_i(x) \leq 0 \quad i = 1, 2 \ldots, m\}$, $f = f_0 + \delta_C(\cdot)$, and consider the constrained convex optimization problem*

$$\text{(P)} \qquad \inf\{f_0(x) \mid x \in C\},$$

with $\operatorname{dom} f_0 \cap C \neq \emptyset$. Then f is weakly coercive if and only if $(f_0)_\infty(d) > 0$, $\forall 0 \neq d$ satisfying $(f_i)_\infty(d) \leq 0 \ \forall i = 1, \ldots, m$, $d \in L^\perp$, with $L = \{d : (f_i)_\infty(d) = (f_i)_\infty(-d) = 0, \ i = 1, \ldots, m\}$. Furthermore, $L = \mathcal{C}_f$, the constancy space of f.

Proof. Since $f_\infty(d) = (f_0)_\infty(d)$ if $d \in C_\infty$ and $f_\infty(d) = +\infty$ otherwise, then $f_\infty(-d) = (f_0)_\infty(-d)$ whenever $-d \in C_\infty$. Thus, it follows that $d \in \mathcal{C}_f$, the constancy space of f, if and only if $d \in \mathcal{C}_{f_0}$ and

$$d \in C_\infty \cap -C_\infty = \cap_{i=1}^m (\mathcal{K}_{f_i} \cap -\mathcal{K}_{f_i}) = \cap_{i=1}^m \mathcal{C}_{f_i}.$$

To end the proof it then suffices to remark that $C_\infty = \{d \mid (f_i)_\infty(d) \leq 0,\ i = 1, \ldots, m\}$ and use Theorem 3.2.1 together with the first statement proved at the beginning.

□

The following proposition proves that in some cases weak coercivity of a function is a kind of stability property with respect to sequences of functions.

Proposition 3.2.6 *For a nondecreasing sequence $\{f_k\}$ of lsc, proper and convex functions on $\mathbb{R}^n$ with $f := \sup_k f_k$ the following properties hold:*
(a) The sequence $\{f_k^*\}$ is nonincreasing and $f^* = \operatorname{cl}(\inf_k f_k^*)$. Moreover, the linear manifolds $E_k = \operatorname{aff}(\operatorname{dom} f_k^*)$ coincide for k sufficiently large with $E = \operatorname{aff}(\operatorname{dom} f^*)$.
(b) If f is weakly coercive, then f_k is weakly coercive for k sufficiently large.

Proof. (a) Since Fenchel conjugacy reverses inequalities ($g \leq h \Rightarrow g^* \geq h^*$), it follows at once that f_k^* is nonincreasing. Now by direct computation one has

$$\left(\inf_k f_k^*\right)^* = \sup_k f_k^{**} = \sup_k f_k = f,$$

so that by taking the Fenchel conjugate we deduce $f^* = \operatorname{cl}(\inf_k f_k^*)$. From this characterization and noting that a convex function and its closure have the same affine hull of their corresponding domains, we deduce

$$E = \operatorname{aff}(\operatorname{dom}(\inf_k f_k^*)) = \operatorname{aff}(\cup_k \operatorname{dom} f_k^*) = \cup_k \operatorname{aff}(\operatorname{dom} f_k^*) = \cup_k E_k.$$

Since f_k^* is nonincreasing, the sets E_k form a nondecreasing sequence of linear manifolds contained in E and whose union gives the entire set E. The conclusion then follows easily: All the sets E_k must coincide with E for k large enough.

(b) Since a convex function and its closure have the same relative interior of their respective domains, we deduce from (a),

$$\operatorname{ri} \operatorname{dom} f^* = \operatorname{ri}(\cup_k \operatorname{dom} f_k^*).$$

Thus, f weakly coercive implies $0 \in \operatorname{ri}(\cup_k \operatorname{dom} f_k^*)$, which can also be written as

$$E = \mathbb{R}_+(\cup_k \operatorname{dom} f_k^*) = \cup_k \mathbb{R}_+(\operatorname{dom} f_k^*).$$

Again, the linear manifold E has been expressed as a monotone union of the family of convex cones $\mathbb{R}_+(\operatorname{dom} f_k^*)$, so that for k large enough we must have the equality $\mathbb{R}_+(\operatorname{dom} f_k^*) = E = E_k$, which amounts precisely to $0 \in \operatorname{ri} \operatorname{dom} f_k^*$, that is, f_k is weakly coercive.

□

3.3   Asymptotically Level Stable Functions

As will be seen in the next section, coercivity and weak coercivity will ensure the existence of optimal solutions. We want now to exhibit the widest class of functions for which existence of optimal solutions is ensured. Such functions will be called *asymptotically level stable* functions. Coercive, weakly coercive, and asymptotically directionally constant functions belong to this class, and our purpose here is to show that this class is very large. In fact, the importance of this class goes beyond the questions of existence of optimal solutions. Indeed, we will see in Section 3.5 that if we suppose in addition convexity, the alternative theorems and closedness criteria can be obtained with such functions under minimal assumptions.

In order to prove that this class covers a large spectrum we now introduce different kinds of functions that rely on the concept of asymptotic linear sets introduced in Chapter 2, in Definition 2.3.1. Recall that a nonempty closed set C of $\mathbb{R}^n$ is asymptotically linear (als) if for each $\rho > 0$ and each sequence $\{x_k\}$ satisfying

$$\{x_k\} \in C, \quad ||x_k|| \to \infty \quad x_k ||x_k||^{-1} \to \bar{x}, \tag{3.12}$$

there exists $k_0 \in \mathbb{N}$ such that

$$x_k - \rho \bar{x} \in C \quad \forall k \geq k_0.$$

Proposition 3.3.1 *Let D be a nonempty closed convex set of $\mathbb{R}^n$ supposed asymptotically linear. For $i = 1, \ldots, m$, let $f_i : \mathbb{R}^n \to \mathbb{R} \cup \{+\infty\}$ be convex asymptotically directionally constant, and let $\lambda_i \in \mathbb{R}$. Consider the set*

$$C = D \bigcap_{i=1}^{m} \mathrm{lev}(f_i, \lambda_i),$$

supposed nonempty. For each $i = 1, \ldots, m$ define

$$C_i = D \bigcap_{j \neq i}^{m} \mathrm{lev}(f_j, \lambda_j),$$

and suppose that f_i is bounded below on C_i. Then C is asymptotically linear.

Proof. Since all the sets are closed convex sets, we have

$$C_\infty = D_\infty \bigcap_{i=1}^{m} \mathrm{lev}(f_i, \lambda_i)_\infty \ , \ (C_i)_\infty = D_\infty \bigcap_{j \neq i}(\mathrm{lev}(f_j, \lambda_j))_\infty, \ \forall i = 1, \ldots, m.$$

Since f_i is bounded below on C_i, it follows that

$$y \in D_\infty, \ (f_j)_\infty(y) \leq 0 \ \forall j \neq i \Longrightarrow (f_i)_\infty(y) \geq 0. \tag{3.13}$$

Now let $\{x_k\}$ be a sequence satisfying (3.12). Then $\bar{x} \in D_\infty$, and since

$$\|x_k\|^{-1} f_j \left(\frac{x_k \|x_k\|}{\|x_k\|} \right) \leq \|x_k\|^{-1} \lambda_j,$$

passing to the limit, it follows from the formula of the asymptotic function of a convex function given in (2.18) that $(f_j)_\infty(\bar{x}) \leq 0 \quad \forall j = 1, \ldots, m$. Then from (3.13) it follows that $(f_j)_\infty(\bar{x}) = 0$, and since f_j is adc, we have

$$f_j(x_k - \rho\bar{x}) = f_j(x_k) \leq \lambda_j \quad \forall j = 1, \ldots, m.$$

Furthermore, since D is asymptotically linear, it follows that there exists k_0 such that $x_k - \rho\bar{x} \in D$ for each $k \geq k_0$, so that $x_k - \rho\bar{x} \in C$ for such k, and then C is asymptotically linear. □

Note that Proposition 3.3.1 is particularly useful for $m = 1$.

Definition 3.3.1 *Let $f : \mathbb{R}^n \to \mathbb{R} \cup \{+\infty\}$ be lsc, and proper. Then f is said to be asymptotically linear if* epi f *is an asymptotically linear set.*

Proposition 3.3.2 *Let f be asymptotically linear, then for each λ such that* lev(f, λ) *is nonempty,* lev(f, λ) *is asymptotically linear.*

Proof. Set $C = \text{lev}(f, \lambda) \neq \emptyset$. Let $\rho > 0$, and let $\{x_k\}$ be a sequence satisfying

$$f(x_k) \leq \lambda, \quad \|x_k\| \to \infty, \quad x_k \|x_k\|^{-1} \to \bar{x}.$$

Then $(x_k, \lambda) \in \text{epi } f$, and $\frac{(x_k, \lambda)}{\|(x_k, \lambda)\|} \to (\bar{x}, 0)$, with $(\bar{x}, 0) \in (\text{epi } f)_\infty$. Since epi f is asymptotically linear, it follows that for k sufficiently large $(x_k - \rho\bar{x}, \lambda - \rho \cdot 0) \in \text{epi } f$ is equivalent to $f(x_k - \rho\bar{x}) \leq \lambda$, proving that lev$(f, \lambda)$ is als. □

Now we introduce the fundamental class of asymptotically level stable (als) functions.

Definition 3.3.2 *Let $f : \mathbb{R}^n \to \mathbb{R} \cup \{+\infty\}$ be an lsc, proper function. Then f is said asymptotically level stable (als) if for each $\rho > 0$, each bounded sequence of reals $\{\lambda_k\}$, and each sequence $\{x_k\} \in \mathbb{R}^n$ satisfying*

$$x_k \in \text{lev}(f, \lambda_k), \quad \|x_k\| \to +\infty, x_k \|x_k\|^{-1} \to \bar{x} \in \ker f_\infty, \qquad (3.14)$$

there exists k_0 such that

$$x_k - \rho\bar{x} \in \text{lev}(f, \lambda_k) \qquad \forall k \geq k_0. \qquad (3.15)$$

We remark that if for each bounded sequence $\{\lambda_k\}$ there exists no sequence $\{x_k\}$ satisfying (3.14), then automatically f is als. As a consequence coercive functions are als. In the next proposition we shall prove that asymptotically linear functions, as well as piecewise linear quadratic convex functions are, als. The latter class of functions is particularly wide and enjoys many interesting properties. It is often met in various fields such as stochastic programming and optimal control.

Definition 3.3.3 *A function $f : \mathbb{R}^n \to \mathbb{R} \cup \{+\infty\}$ is called piecewise linear quadratic (plq) if $\operatorname{dom} f$ can be represented as the union of finitely many polyhedral sets, relative to each of which $f(x)$ is given by an expression of the form $\langle x, Ax \rangle + 2\langle a, x \rangle + \alpha$, where $\alpha \in \mathbb{R}$, $a \in \mathbb{R}^n$, and $A \in \mathbb{R}^{n \times n}$ is a symmetric matrix.*

It can be proved (see Notes and References) that the following operations preserve the plq property:
(i) The sum of a finite number of plq functions is plq.
(ii) If $g : \mathbb{R}^m \to \overline{\mathbb{R}}$ is plq and $f(x) = g(Ax + a)$ for some $A \in \mathbb{R}^{m \times n}, a \in \mathbb{R}^m$, then f is plq.
(iii) If $\varphi(x) = f(x, \bar{u})$ with $f : \mathbb{R}^n \times \mathbb{R}^m \to \overline{\mathbb{R}}$ with f plq, then φ is plq.
(iv) If $f : \mathbb{R}^n \to \mathbb{R} \cup \{+\infty\}$ is lsc, proper, and convex, then f is plq if and only if its conjugate f^* is plq.
(v) For a nonempty polyhedral set $Y \subset \mathbb{R}^m$ and a symmetric positive semidefinite matrix $B \in \mathbb{R}^{m \times m}$ the function $\theta_{Y,B} : \mathbb{R} \to \overline{\mathbb{R}}$ defined by

$$\theta_{Y,B}(u) := \sup_{y \in Y} \{ \langle y, u \rangle - \langle y, By \rangle \}$$

is lsc, proper, convex, and plq on $\mathbb{R}^m$.

Proposition 3.3.3 *Let $f : \mathbb{R}^n \to \mathbb{R} \cup \{+\infty\}$ be lsc and proper.*
(a) If f is asymptotically linear, then f is als.
(b) If f is asymptotically directionally constant, then f is als.
(c) Let $g : \mathbb{R}^n \to \mathbb{R} \cup \{+\infty\}$ be als and let C be an asymptotically linear set in $\mathbb{R}^n$ such that $\operatorname{dom} g \cap C \neq \emptyset$ and $g_\infty(d) \geq 0$, $\forall d \neq 0$ $d \in C_\infty$. Then $f := g + \delta_C$ is als.
(d) Let $f : \mathbb{R}^n \to \mathbb{R} \cup \{+\infty\}$ be lsc, proper, and convex such that $\operatorname{dom} f$ can be represented as the union of finitely many asymptotically linear convex sets $D_l, l = 1, \ldots, m$, relative to each of which $f = f_l$, with f_l convex and als. Then f is als. In particular, if f is plq and convex, then f is als.

Proof. (a) Suppose that f is asymptotically linear and let $\{x_k\}$ be a sequence satisfying (3.14). Then since $\{\lambda_k\}$ is bounded, we have

$$(x_k, \lambda_k) \in \operatorname{epi} f \ , \ (\bar{x}, 0) = \lim_{k \to \infty} \frac{(x_k, \lambda_k)}{\|(x_k, \lambda_k)\|} \ , \ (\bar{x}, 0) \in (\operatorname{epi} f)_\infty.$$

Since epi f is asymptotically linear, it follows that for k sufficiently large $(x_k - \rho\bar{x}, \lambda_k - \rho \cdot 0) \in$ epi f, which is equivalent to $f(x_k - \rho\bar{x}) \leq \lambda_k$, i.e., $x_k - \rho\bar{x} \in \text{lev}(f, \lambda_k)$, and by Definition 3.3.2, that f is als.

(b) Let f be asymptotically directionally constant and $\{x_k\}$ be a sequence satisfying (3.14). Then $f_\infty(\bar{x}) = 0$, and since f is adc, it follows that $f(x_k - \rho\bar{x}) = f(x_k)$ for each k, and therefore (3.15) is satisfied.

(c) Let $\{x_k\}$ be a sequence satisfying (3.14). Then for k sufficiently large we have

$$\{x_k\} \in C, \qquad g(x_k) \leq \lambda_k.$$

As a consequence it follows that $\bar{x} \in C_\infty$ and $g_\infty(\bar{x}) \leq 0$, which in turn implies that $g_\infty(\bar{x}) = 0$. Then since C is asymptotically linear and g is als, for k sufficiently large we have

$$x_k - \rho\bar{x} \in C, \quad g(x_k - \rho\bar{x}) \leq \lambda_k,$$

which is equivalent to saying that f is als.

(d) Let $\{x_k\}$ be a sequence satisfying (3.14). For each $l = 1, \ldots, m$ define

$$S(l) = \{k \mid x_k \in D_l\} \quad \text{and} \quad I = \{l \mid S(l) \text{ is nonfinite}\}.$$

Now let $l \in I$ and $y_l \in \text{dom } f \cap D_l$. Since $\lim_{k\to\infty, k\in S(l)} \|x_k\|^{-1}x_k = \bar{x}$, it follows that $\bar{x} \in (D_l)_\infty$, and then $y_l + \lambda\bar{x} \in D_l$ for each $\lambda > 0$ and every $l \in I$. As a consequence we have

$$\frac{f(y_l + t\bar{x}) - f(y_l)}{t} = \frac{f_l(y_l + t\bar{x}) - f_l(y_l)}{t}.$$

Passing to the limit as $t \to +\infty$ it follows that $0 = f_\infty(\bar{x}) = (f_l)_\infty(\bar{x}) \; \forall l \in I$. Since f_l are als and D_l are asymptotically linear, it follows that for each $l \in I$, there exists $k_0(l)$ such that $x_k - \rho\bar{x} \in D_l$, $f_l(x_k - \rho\bar{x}) \leq \lambda_k \quad \forall k \geq k_0(l), k \in S(l)$. As a consequence we have

$$f(x_k - \rho\bar{x}) \leq \lambda_k \quad \forall k \geq k_0(l), k \in S(l).$$

Let $k^* \in \mathbb{N}$ be such that $\mathbb{N} \setminus [1, k^*] \subset \bigcup_{l\in I} S(l)$ and $k_0 = \max_{l\in I}(k^*, k_0(l))$. Then for $k > k_0$ it follows that $x_k - \rho\bar{x} \in \text{lev}(f, \lambda_k)$, which proves that f is als. Finally, the last assertion is immediate, by definition of convex plq functions. □

3.4 Existence of Optimal Solutions

The results developed in the previous sections will now be applied to derived general necessary and sufficient conditions for the existence of minimizers.

Let $f : \mathbb{R}^n \to \mathbb{R} \cup \{+\infty\}$ be an lsc, proper function and consider the optimization problem

$$(P) \quad \inf\{f(x) \mid x \in \mathbb{R}^n\}.$$

We introduce two basic hypotheses under which the optimal set S of (P) is nonempty.

The first hypothesis is simply expressed in terms of the asymptotic function

$$\mathcal{H}_0 \qquad\qquad f_\infty(d) \geq 0 \ \forall d \neq 0 \in \mathbb{R}^n.$$

If $\inf f > -\infty$, by the analytic representation of f_∞ it follows that $\mathcal{H}_0$ holds and appears as a natural necessary condition for the existence of optimal solutions. The second hypothesis is more delicate and is as follows. For any $f : \mathbb{R}^n \to \mathbb{R} \cup \{+\infty\}$ lsc, and proper, recall (cf. (3.9)) that S_f denotes the set of sequences $\{x_k\}$ satisfying

$$||x_k|| \to \infty, \ \{f(x_k)\}\text{bounded above and } x_k||x_k||^{-1} \to \bar{x} \in \ker f_\infty. \quad (3.16)$$

Then we say that $\mathcal{H}_1$ holds if
(a) either S_f is empty, or
(b) for each sequence $\{x_k\} \in S_f$, there exists $z_k, \rho_k \in (0, ||x_k||)$ such that for k sufficiently large one has

$$f(x_k - \rho_k z_k) \leq f(x_k) \qquad \text{and} \ z_k \to z \ \text{with} \ ||\bar{x} - z|| < 1. \quad (3.17)$$

Remark 3.4.1 A coercive function obviously satisfies $\mathcal{H}_0$ and $\mathcal{H}_1$, because in this case S_f is always empty. Indeed, suppose the contrary. Then there would exist a sequence $\{x_k\}$ and a point $\bar{x}$ such that

$$\lim_{k \to \infty} ||x_k|| = +\infty, \quad \lim_{k \to \infty} x_k ||x_k||^{-1} = \bar{x}, \quad f_\infty(\bar{x}) = 0.$$

But since $||\bar{x}|| = 1 \neq 0$, this is impossible.

Theorem 3.4.1 *Let $f : \mathbb{R}^n \to \mathbb{R} \cup \{+\infty\}$ be an lsc, proper function. Then a necessary and sufficient condition for the optimal set S to be nonempty is that $\mathcal{H}_0$ and $\mathcal{H}_1$ are satisfied.*

Proof. Suppose first that $\mathcal{H}_0$ and $\mathcal{H}_1$ hold and let $\{\epsilon_k\}$ be a nonincreasing sequence of positive reals converging to 0. For each $k \in \mathbb{N}$, set

$$f_k(x) = f(x) + \epsilon_k g(x), \quad g(x) = ||x||^2,$$

and consider the optimization problem

$$(P_k) \quad \inf\{f_k(x) \mid x \in \mathbb{R}^n\}.$$

Since $(f_k)_\infty(d) \geq f_\infty(d) + \epsilon_k g_\infty(d)$, then from $\mathcal{H}_0$ it follows that $(f_k)_\infty(d) > 0 \ \forall d \neq 0$. As a consequence f is coercive, and from Corollary 3.1.1 there exists at least an optimal solution x_k of (P_k). Let us prove now that the sequence $\{x_k\}$ is bounded. Suppose the contrary. Then without loss of generality we can suppose that $\|x_k\| \to \infty$ and $x_k\|x_k\|^{-1} \to \bar{x}$, and since $\{f(x_k)\}$ is bounded above by $f(x) + \epsilon_0\|x\|^2$ with $x \in \mathrm{dom}\, f$, dividing $f(x_k)$ by $\|x_k\|$ and passing to the limit, it follows that $f_\infty(\bar{x}) \leq 0$. By Hypothesis $\mathcal{H}_0$, this implies that $\bar{x} \in \ker f_\infty$, and then by Hypothesis $\mathcal{H}_1$ there exist $z_k, \rho_k \in (0, \|x_k\|]$ satisfying (3.17) for k sufficiently large. Therefore, since $x_k \in \mathrm{argmin}\,\{f_k(x) \mid x \in \mathbb{R}^n\}$, we obtain

$$f(x_k) + \epsilon_k\|x_k\|^2 \leq f(x_k - \rho_k z_k) + \epsilon_k\|x_k - \rho_k z_k\|^2 \leq f(x_k) + \epsilon_k\|x_k - \rho_k z_k\|^2,$$
(3.18)

and since $\varepsilon_k > 0$, it follows that $\|x_k - \rho_k z_k\| \geq \|x_k\|$. Thus, if we set $x'_k = x_k\|x_k\|^{-1}$, we have

$$
\begin{aligned}
\|x_k - \rho_k z_k\| &= \|(1 - \rho_k\|x_k\|^{-1})x_k + \rho_k(x'_k - z_k)\| \\
&\leq \|x_k\|(1 - \rho_k\|x_k\|^{-1}) + \rho_k\|x'_k - z_k\| \\
&\leq \|x_k\| + \rho_k(\|x'_k - z_k\| - 1).
\end{aligned}
$$
(3.19)

It follows that $\|x'_k - z_k\| \geq 1$, and passing to the limit we get $\|\bar{x} - z\| \geq 1$, which is impossible, and therefore the sequence $\{x_k\}$ is bounded. Now let $\tilde{x}$ be a limit point of the sequence $\{x_k\}$. Since

$$f(x_k) + \epsilon_k\|x_k\|^2 \leq f(x) + \epsilon_k\|x\|^2 \quad \forall x \in \mathbb{R}^n,$$

passing to the limit it follows that $\tilde{x} \in S$. To prove the converse, let $\tilde{x} \in S$; then writing $\tilde{x} = x_k - \|x_k\|(x_k - \tilde{x})\|x_k\|^{-1}$, the hypothesis $\mathcal{H}_1$ holds with $\rho_k = \|x_k\|, z_k = (x_k - \tilde{x})\|x_k\|^{-1}$.   □

A useful criterion for choosing z_k in (3.17) consists in taking it constant and equal to $\alpha\bar{x}$ with $\alpha \in (0, 2)$. Then for this specific choice, the hypothesis $\mathcal{H}_1$ can be written as follows:
(a) Either S_f defined by (3.16) is empty, or
(b) for each sequence $\{x_k\} \in S_f$ there exists $\varepsilon \in (0, 2)$ such that for k sufficiently large, there exists $\rho_k \in \mathbb{R}$ satisfying

$$\rho_k \in (0, (2 - \varepsilon)\|x_k\|\,], \quad f(x_k - \rho_k\bar{x}) \leq f(x_k).$$
(3.20)

This reformulation of $\mathcal{H}_1$ will be denoted $\mathcal{H}'_1$. As a consequence, we immediately obtain the following corollary.

Corollary 3.4.1 *Let $f : \mathbb{R}^n \to \mathbb{R} \cup \{+\infty\}$ be an lsc, proper function such that $\mathcal{H}_0$ and $\mathcal{H}'_1$ hold. Then the optimal set S of (P) is nonempty.*

Example 3.4.1 *Let $g : [0.5, 1] \to \mathbb{R}$ be continuous and satisfy*

$$g(0.5) = e^{-0.5}, \quad g(1) = e^{-1}, \quad \min\{g(x) \mid x \in [0.5, 1]\} < 0.$$

Consider the continuous function

$$f(x) = \begin{cases} e^{-x^2} & \text{if } x \notin [0.5, 1] \\ g(x) & \text{otherwise.} \end{cases}$$

Then we have $f_\infty(-1) = f_\infty(1) = 0$, so that $\mathcal{H}_0$ holds and $\mathcal{H}'_1$ is obviously satisfied.

Remark 3.4.2 If in addition to hypothesis $\mathcal{H}_0$ and $\mathcal{H}'_1$ the function f is convex, and in (3.20) we suppose that $\varepsilon \in (0, \frac{1}{2})$ and $\rho_k \geq (1 + \varepsilon)||x_k||$, then not only is the optimal set S nonempty, but by Proposition 3.2.2, f is weakly coercive and S is equal to the sum of a compact set and a linear space.

Corollary 3.4.2 *Let* $f : \mathbb{R}^n \to \mathbb{R} \cup \{+\infty\}$ *be asymptotically level stable with* $\inf f > -\infty$. *Then the optimal set of (P) is nonempty.*

Proof. Since $\inf f > -\infty$, it follows that $\mathcal{H}_0$ holds. Furthermore, let $\{x_k\} \in S_f$. Then by definition of S_f the sequence $\{\lambda_k = f(x_k)\}$ is bounded, and then if $\rho > 0$, by definition of an asymptotically level stable function it follows that for k sufficiently large

$$f(x_k - \rho\bar{x}) \leq f(x_k),$$

and $\mathcal{H}'_1$ is satisfied. $\qquad\square$

Corollary 3.4.3 *Let* $g : \mathbb{R}^n \to \mathbb{R} \cup \{+\infty\}$ *be lsc and proper and let* C *be a subset of* $\mathbb{R}^n$ *that is asymptotically linear. Consider the optimization problem*

$$(P_C) \qquad\qquad m = \inf\{g(x) \mid x \in C\}.$$

Suppose that $C \cap \operatorname{dom} g \neq \emptyset$ *and that*

$$g_\infty(d) \geq 0 \quad \forall d \in C_\infty, d \neq 0. \tag{3.21}$$

Suppose also that g *is adc or als with* $m > -\infty$. *Then the optimal set* S *of* (P_C) *is nonempty.*

Proof. Problem (P_C) is equivalent to

$$m = \inf\{f(x) \mid x \in \mathbb{R}^n\},$$

where $f = g + \delta_C$. Since $f_\infty(d) = +\infty$, if $d \notin C_\infty$ and

$$(g + \delta_C)_\infty(d) \geq g_\infty(d) + \delta_{C_\infty}(d) \quad \text{for } d \in C_\infty, \tag{3.22}$$

it follows from (3.21) that $(g)_\infty(d) \geq 0$ for each $d \neq 0$, and $\mathcal{H}_0$ holds. Now let $\{x_k\}$ be a sequence in S_f. Then $\{x_k\} \in C$, and since $f_\infty(\bar{x}) = 0$, it follows from (3.21) and (3.22) that $g_\infty(\bar{x}) = 0$ and $\bar{x} \in C_\infty$. Therefore, if g is adc, we have $g(x_k - \rho\bar{x}) = g(x_k)$ for each $\rho \in \mathbb{R}$. Since C is asymptotically linear, for k sufficiently large we also have $x_k - \rho\bar{x} \in C$, so that $f(x_k - \rho\bar{x}) = f(x_k)$ and $\mathcal{H}'_1$ holds. If g is als and $m > -\infty$, then from Proposition 3.3.3(c), f is also als, and thus from Corollary 3.4.2 the optimal set is nonempty. $\square$

Remark 3.4.3 If we suppose that g and C are convex, then the condition $m > -\infty$ implies that (3.21) holds, since in this case $(g + \delta_C)_\infty = g_\infty + \delta_{C_\infty}$.

3.5   Stability for Constrained Problems

Let $F : \mathbb{R}^n \times \mathbb{R}^m \longrightarrow \mathbb{R} \cup \{+\infty\}$ be an lsc proper function. A very important problem in non linear analysis consists in studying the marginal function associated with F and defined by

$$h(y) = \inf\{F(x, y) \mid x \in \mathbb{R}^n\}, \tag{3.23}$$

and in particular to give simple conditions that imply that the infimum is attained and that h is lsc at some point y. The function $F(x, y)$ stands in fact as a *perturbation* function, parameterized by the vector $y \in \mathbb{R}^m$, and (3.23) can be viewed as a perturbed or parametric optimization problem. A key problem is thus to study the behavior of the optimal value $h(y)$ and the corresponding optimal solution $x(y)$, with respect to changes in the parameter y. This question is called *stability*, and it plays a central role in duality for optimization problems, a topic that will be developed in detail in Chapter 5. In this section we focus on the conditions that guarantee that there exists a minimum in (3.23) and that h is lsc. For this purpose we begin by studying a simpler problem, from which general results will easily follow.

Let A be an $m \times n$ real matrix and $f : \mathbb{R}^n \to \mathbb{R} \cup \{+\infty\}$ an lsc, proper function. Consider the marginal function defined by

$$h(y) := Af(y) := \inf\{f(x) \mid x \in C(y)\}, \tag{3.24}$$

where $C(y) := \{x \in \mathbb{R}^n \mid Ax = y\}$.

We wish to give conditions close to hypotheses $\mathcal{H}_0$ and $\mathcal{H}'_1$ under which it may be proved that h is lsc at a fixed point y. First we remark that

$$h(y) = \inf\{F_y(x) \mid x \in \mathbb{R}^n\}, \tag{3.25}$$

with $F_y(x) = f(x) + \delta_{C(y)}(x)$. Then since $C(y)_\infty = \{d \mid Ad = 0\}$, the hypothesis

$$\mathcal{H}'_0 \qquad\qquad f_\infty(d) \geq 0 \ \ \forall d \in \ker A, \ d \neq 0, \tag{3.26}$$

implies $\mathcal{H}_0$ for F_y and coincides with $\mathcal{H}_0$ when f is convex.
The second hypothesis, which corresponds to $\mathcal{H}_1$, has to take into account perturbations around y. In similarity with the set S_f defined in (3.16), we thus let $\{y_k\}$ be a sequence converging to y and define by $S_f[\{y_k\}, y]$ the set of sequences $\{x_k\}$ satisfying $Ax_k = y_k$ such that

$$||x_k|| \to \infty, \quad \{f(x_k)\} \text{ bounded above}, \quad x_k ||x_k||^{-1} \to \bar{x} \in \ker f_\infty \cap \ker A. \tag{3.27}$$

Then $\mathcal{H}_1'[y]$ holds if for any sequence $\{y_k\}$ converging to y:
(a) either $S[\{y_k\}, y]$ is empty, or
(b) for each $\{x_k\} \in S[\{y_k\}, y]$ there exists $z_k, \rho_k \in (0, ||x_k||]$ such that for k sufficiently large we have

$$f(x_k - \rho_k z_k) \leq f(x_k), \quad Az_k = 0, \quad z_k \to z \text{ with } ||\bar{x} - z|| < 1. \tag{3.28}$$

Theorem 3.5.1 *Suppose that $f : \mathbb{R}^n \to \mathbb{R} \cup \{+\infty\}$ is lsc and proper, and that $\mathcal{H}_0'$ and $\mathcal{H}_1'[y]$ hold. Then h is lsc at y, and when $C(y) \cap \operatorname{dom} f \neq \emptyset$, the optimal set $S(y)$ is nonempty.*

Proof. Suppose that $C(y) \cap \operatorname{dom} f$ is nonempty and set $y_k = y$. To prove that $S(y)$ is nonempty we have only to verify assumption $\mathcal{H}_1$ for F_y. But this is an immediate consequence of $\mathcal{H}_1'[y]$ if we set $y_k = y$ for each k. It remains to prove that h is lsc at y; that is, $h(y) \leq \liminf_{k \to \infty} h(y_k)$. Without loss of generality we can suppose that

$$\liminf_{k \to \infty} h(y_k) = \lim_{k \to \infty} h(y_k) < \infty. \tag{3.29}$$

Then it follows that for each k, $C(y_k) \cap \operatorname{dom} f$ is nonempty. Let $\varepsilon_k \to 0^+$, and set

$$m_k = \begin{cases} h(y_k) + \varepsilon_k & \text{if } h(y_k) > -\infty \\ -k & \text{otherwise}, \end{cases} \tag{3.30}$$

and

$$S_k = \{x \mid f(x) \leq m_k \quad Ax = y_k \}. \tag{3.31}$$

Define x_k as

$$x_k \in \operatorname{argmin}\{f(x) + \varepsilon_k ||x||^2 \mid x \in S_k\}.$$

Thanks to hypothesis $\mathcal{H}_0'$ the objective function $f(\cdot) + \epsilon_k || \cdot ||^2 + \delta_{S_k}(\cdot)$ is coercive, and from Corollary 3.1.1 such an x_k exists. If the sequence $\{x_k\}$ is bounded, there exists a subsequence $\{x_{n_k}\}$ converging to x^*. Passing to the limit, since f is lsc, it follows from (3.31) that $Ax^* = y$ and $f(x^*) \leq \liminf_{k \to \infty} h(y_k)$. Since $x^* \in C(y)$, we then have $h(y) \leq f(x^*)$, which proves that h is lsc at y. Let us prove that the sequence $\{x_k\}$ is indeed bounded. In the contrary case we can suppose that

$$||x_k|| \to \infty, \quad x'_k = x_k ||x_k||^{-1} \to \bar{x} \neq 0. \tag{3.32}$$

Since $f(x_k) \leq m_k$, from (3.29), (3.30), there exists a constant c such that $f(x_k) \leq c$ for each k. Then using (2.18) it follows that $f_\infty(\bar{x}) \leq 0$. Furthermore,

$$A\bar{x} = \lim_{k \to \infty} Ax_k \|x_k\|^{-1} = \lim_{k \to \infty} \frac{y_k}{\|x_k\|} = 0.$$

As a consequence, using condition $\mathcal{H}_0'$ it follows that $A\bar{x} = 0 = f_\infty(\bar{x})$. Now we can use hypothesis $\mathcal{H}_1'[y]$, and thus there exist z_k, $\rho_k \in (0, \|x_k\|]$ such that for k sufficiently large (3.28) holds. As a consequence, $x_k - \rho_k z_k$ belongs to S_k, and by definition of x_k the inequalities (3.18) hold, from which it follows that $\|x_k\| \leq \|x_k - \rho_k z_k\|$. Then using the same arguments as in the proof of Theorem 3.4.1, we obtain the desired contradiction. $\square$

Corollary 3.5.1 *Suppose that f is als and that*

$$f_\infty(d) \geq 0 \quad \forall d \in \ker A, \quad d \neq 0.$$

Suppose also that $\operatorname{cl} h(y) > -\infty$. *Then h is lsc at y, and when $C(y) \cap \operatorname{dom} f \neq \emptyset$, the optimal set $S(y)$ is nonempty.*

Proof. We have only to verify that $\mathcal{H}_1'[y]$ holds. Take a sequence $\{x_k\} \in S_f[\{y_k\}, y]$ with $y_k \to y$ and set $z_k = \bar{x} \; \forall k$. Then since $Ax_k = y_k$, for each $\varepsilon > 0$ we have for k sufficiently large, $f(x_k) > \operatorname{cl} h(y) - \varepsilon$, and it follows that the sequence $\{f(x_k)\}$ is bounded. As a consequence, since f is als, the hypothesis $\mathcal{H}_1'[y]$ is satisfied. $\square$

Let us come back to our original motivation and study the marginal function h of an lsc proper function $F : \mathbb{R}^n \times \mathbb{R}^m : \mathbb{R} \cup \{+\infty\}$, i.e.,

$$h(y) := \inf\{F(x, y) \mid x \in \mathbb{R}^n\}.$$

If we set $A(x, y) = y$, then $h(y) = (AF)(y)$, and one can use Theorem 3.5.1. Condition $\mathcal{H}_0'$ reduces to

$$\mathcal{H}_0'' \qquad F_\infty(x, 0) \geq 0 \;\; \forall x \neq 0,$$

and $\mathcal{H}_1'[y]$ becomes $\mathcal{H}_1''[y]$ as follows: Let $\{y_k\}$ be a sequence converging to y and define $T_F[\{y_k\}, y]$ as the set of sequences $\{x_k\}$ satisfying

$$\|x_k\| \to \infty, \;\; \{F(x_k, y_k)\} \text{ bounded above}, \;\; x_k \|x_k\|^{-1} \to \bar{x} \in \ker F_\infty(\cdot, 0). \tag{3.33}$$

Then $\mathcal{H}_1''[y]$ holds if for any sequence $\{y_k\}$ converging to y:
(a) either $T_F[\{y_k\}, y]$ is empty, or
(b) for each sequence $\{x_k\} \in T_F[\{y_k\}, y]$ there exist $z_k, \rho_k \in (0, \|(x_k, y_k)\|]$ such that for k sufficiently large we have

$$F(x_k - \rho_k z_k, y_k) \leq F(x_k, y_k), \;\; z_k \to z \text{ with } \|\bar{x} - z\| < 1. \tag{3.34}$$

As an immediate consequence of Theorem 3.5.1 we then obtain the following properties for the marginal function h defined in (3.23).

Corollary 3.5.2 *Suppose that $F : \mathbb{R}^n \times \mathbb{R}^m \to \mathbb{R} \cup \{+\infty\}$ is lsc, proper, and that $\mathcal{H}_0''$ and $\mathcal{H}_1''[y]$ hold. Then h is lsc at y and the optimal set $S(y) = \{x \mid F(x,y) = h(y)\}$ is nonempty.*

Corollary 3.5.3 *Suppose that $F : \mathbb{R}^n \times \mathbb{R}^m \to \mathbb{R} \cup \{+\infty\}$ is lsc, proper and that*

$$F_\infty(x,0) \geq 0 \ \forall x \neq 0.$$

Then h is lsc at each y, and the optimal set $S(y) = \{x \mid F(x,y) = h(y)\}$ is nonempty under any one of the following conditions:
(a) $F_\infty(x,0) > 0 \ \forall x \neq 0$.
(b) F is convex and for each x such that $F_\infty(x,0) = 0$ we have $F_\infty(-x,0) = 0$.
(c) $\operatorname{cl} h(y) > -\infty$ and F is asymptotically level stable.

Proof. To prove the validity of the assertion under condition (a) or (b) we use Corollary 3.5.2, and we have only to verify that hypothesis $\mathcal{H}_1''[y]$ holds. This is immediate for case (a), since for each sequence $\{y_k\} \to y$ the set $T_F[\{y_k\}, y]$ is obviously empty. In case (b), let $\{x_k\} \in T_F[\{y_k\}, y]$. Then since F is convex and since for each x such that $F_\infty(x,0) = 0$ we have $F_\infty(-x,0) = 0$, it follows that F is constant on each direction $(x,0)$, so that for k sufficiently large one has $F(x_k - \rho\bar{x}, y_k) = F(x_k, y_k) \ \ \forall \rho > 0$ and (3.34) holds. Finally, as an immediate consequence of Corollary 3.5.1 the assertion under (c) follows. $\qquad\square$

Corollary 3.5.4 *Let $f, g : \mathbb{R}^n \to \mathbb{R} \cup \{+\infty\}$ be proper, convex and polyhedral. Then $f \square g$ is polyhedral, and the infimum is attained in the infimal convolution whenever the latter is proper.*

Proof. Let $F(x,y) := f(x) + g(y - x)$ and let h be the associated marginal function. Then $\operatorname{epi} h = \operatorname{epi} f + \operatorname{epi} g$, so that $\operatorname{epi} h$ is a polyhedral set and thus h is polyhedral and hence also lsc. If we suppose that h is also proper, then $h(y) > -\infty$, which implies that $F_\infty(x,0) \geq 0$ for all $x \neq 0$. Furthermore, by Proposition 1.2.11, F is polyhedral, and then it follows that F is als and one can apply part (c) of Corollary 3.5.3. $\qquad\square$

Now we want to give a formula for the asymptotic function $(Af)_\infty$ in terms of the asymptotic function f_∞. In order to derive such a formula we need to suppose that $\mathcal{H}_1'[y]$ holds for each y, and also that formula (3.27) does not depend on y_k, y. More precisely, we introduce the following hypothesis $\mathcal{H}_2'$ as follows: Let X be the set of sequences $\{x_k\} \in \operatorname{dom} f$ satisfying

$$||x_k|| \to \infty, \quad x_k ||x_k||^{-1} \to \bar{x} \in \ker f_\infty \cap \ker A. \tag{3.35}$$

Then $\mathcal{H}_2'$ holds if:
(a) either X is empty, or
(b) for each sequence $\{x_k\} \in X$ there exists $z_k, \rho_k \in (0, \|x_k\|]$ such that for k sufficiently large (3.28) holds; i.e.,

$$Az_k = 0, \quad f(x_k - \rho_k z_k) \le f(x_k), \quad z_k \to z, \quad \text{with } \|\bar{x} - z\| < 1. \qquad (3.36)$$

Remark 3.5.1 If $\mathcal{H}_2'$ is satisfied, then $\mathcal{H}_1'[y]$ holds for every $y \in \mathbb{R}^n$. Also, note that if f is asymptotically directionally constant, then $\mathcal{H}_2'$ holds.

Theorem 3.5.2 *Let* $f : \mathbb{R}^n \to \mathbb{R} \cup \{+\infty\}$ *be an lsc, proper function and* A *a real* $m \times n$ *matrix. Suppose that there exists* $y \in \mathbb{R}^n$ *and* $x \in \mathrm{dom}\, f$ *such that* $Ax = y$. *Suppose also that* $\mathcal{H}_0'$ *and* $\mathcal{H}_2'$ *hold. Then the marginal function* $h = Af$ *is lsc and proper. Furthermore, the optimal set* $S(y)$ *is nonempty for each* y *such that* $C(y) \cap \mathrm{dom}\, f \ne \emptyset$, *and one has*

$$\forall y \in \mathbb{R}^n : \quad (Af)_\infty(y) = \inf\{f_\infty(d) : Ad = y\}.$$

Proof. Since $\mathcal{H}_2'$ holds, it follows that $\mathcal{H}_1'[y]$ holds for each y, and by Theorem 3.5.1 it follows that for any y, h is lsc and the optimal set $S(y)$ in nonempty whenever $C(y) \cap \mathrm{dom}\, f \ne \emptyset$, so that h is proper, proving the first part of the theorem. Let $\tilde{h}_\infty(y) = \inf\{f_\infty(d) \mid Ad = y\}$. We prove first that $\tilde{h}_\infty \le h_\infty = (Af)_\infty$. Let $s \in \mathbb{R}$ satisfy $h_\infty(y) \le s$, which is equivalent to $(y, s) \in (\mathrm{epi}\, h)_\infty$. Then there exist sequences $\{(y_k, s_k)\}$ and $\{t_k\}$ with $t_k \to \infty$ such that

$$h(y_k) \le s_k, \quad \frac{(y_k, s_k)}{t_k} \to (y, s).$$

Let $S_k = \{x \mid f(x) \le s_k, \ Ax = y_k\}$. Since $h(y_k) < \infty$, it follows from Theorem 3.5.1 that the optimal set $S(y_k)$ is nonempty, and we may define

$$x_k \in \mathrm{argmin}\left\{ f(x) + \varepsilon_k \|x\|^2 \mid x \in S_k \right\}$$

for a sequence $\{\varepsilon_k\} \to 0^+$. Let us show that we cannot have $\lim_{k\to\infty} t_k^{-1}\|x_k\| = \infty$. Suppose the contrary. Then $\lim_{k\to\infty} \|x_k\| = +\infty$, and we may suppose without loss of generality that $\xi_k := x_k \|x_k\|^{-1} \to \bar{x}$ with $\|\bar{x}\| = 1$. Furthermore, one has

$$\frac{(x_k, s_k)}{\|x_k\|} = \left(\xi_k, \frac{s_k}{t_k} \frac{t_k}{\|x_k\|} \right) \to (\bar{x}, 0).$$

Since

$$A\xi_k = \frac{y_k}{\|x_k\|}, \quad \frac{f(\xi_k\|x_k\|)}{\|x_k\|} \le \frac{s_k}{t_k} \frac{t_k}{\|x_k\|},$$

then passing to the limit and using formula (2.18) we get $A\bar{x} = 0$, $f_\infty(\bar{x}) \le 0$. Now, by hypothesis $\mathcal{H}_2'$, there exist z_k and $\rho_k \in (0, \|x_k\|]$ such that

$$f(x_k - \rho_k z_k) \le f(x_k), \quad A(x_k - \rho_k z_k) = y_k, \quad z_k \to z, \quad \text{with } \|\bar{x} - z\| < 1.$$

Therefore, it follows that $x_k - \rho_k z_k \in S_k$ and one has

$$f(x_k) + \varepsilon_k \|x_k\|^2 \le f(x_k - \rho_k z_k) + \varepsilon_k \|x_k - \rho_k z_k\|^2 \le f(x_k) + \varepsilon_k \|x_k - \rho_k z_k\|^2,$$

which implies that $\|x_k - \rho_k z_k\| \ge \|x_k\|$ and then using the same arguments as given at the end of the proof of Theorem 3.4.1 we obtain the desired contradiction. Without loss of generality, we may now suppose that the sequence $\{t_k^{-1}\|x_k\|\}$ is bounded and that $t_k^{-1} x_k \to \bar{x}$. Furthermore, $t_k^{-1} f(x_k) = t_k^{-1} f(t_k^{-1} x_k t_k) \le t_k^{-1} s_k$, $t_k^{-1} A x_k = t_k^{-1} y_k$. Passing to the limit, we get $f_\infty(\bar{x}) \le s$, $A\bar{x} = y$, and consequently $\tilde{h}_\infty(y) \le s$, from which it follows that $\tilde{h}_\infty(y) \le h_\infty(y) = (Af)_\infty(y)$. To prove the reverse inequality, $\tilde{h}_\infty \ge h_\infty$, it is only necessary to prove that for each (y, s) with $\tilde{h}_\infty(y) \le s$ we have $h_\infty(y) \le s$. Let $D = \{(u, v) \mid Au = v\}$ and $F(x, y) = f(x) + \delta_D((x, y))$. Let $P : \mathbb{R}^n \times \mathbb{R}^m \times \mathbb{R} \to \mathbb{R}^m \times \mathbb{R}$ be the projection map $P(x, y, r) = (y, r)$. Then one has $P(\text{epi } F) = \text{epi } h$. Indeed, if $(y, \mu) \in \text{epi } h$, then from Theorem 3.5.1 there exists an x such that $F(x, y) = h(y) \le \mu$, and then $(y, \mu) = P(x, y, \mu) \subset P(\text{epi } F)$. Conversely, let $(y, \mu) = P(x, y, \mu)$ with $(x, y, \mu) \in \text{epi } F$. Then $F(x, y) \le \mu$ and $h(y) \le \mu$. Thus, using epi $h = P(\text{epi } F)$ one has $(y, s) \in \text{epi } h$ if and only if there exists an x such that $Ax = y$ and $f(x) \le s$. As a consequence, one obtains that $(y, s) \in (\text{epi } h)_\infty$ if and only if there exists a sequence $t_k \to +\infty$ and a sequence $\{x_k\}$ with $Ax_k = y_k$, $f(x_k) \le s_k$, $t_k^{-1}(y_k, s_k) \to (y, s)$. Define $E = \{(y, s) \mid \exists x \text{ with } Ax = y, f_\infty(x) \le s\}$ and take $(y, s) \in E$. Then there exist x and sequences $\{t_k\}, \{x_k\}$ satisfying

$$t_k \to \infty, \quad f(x_k) \le s_k, \quad t_k^{-1}(x_k, s_k) \to (x, s), \quad Ax = y.$$

Set $y_k = Ax_k$. Then $t_k^{-1} y_k \to Ax = y$, and it follows that $(y, s) \in (\text{epi } h)_\infty$. Now let $\tilde{h}_\infty(y) \le s$ and $\varepsilon_k \to 0^+$. By definition of $\tilde{h}_\infty$ there exists a sequence $\{x_k\}$ such that $Ax_k = y$, $f_\infty(x_k) \le s + \varepsilon_k$, and thus from the argument above it follows that $(y, s + \varepsilon_k) \in (\text{epi } h)_\infty$, which means that $h_\infty(y) \le s + \varepsilon_k$. Passing to the limit, it thus follows that $h_\infty(y) \le s$, and the proof is complete. $\qquad \square$

We now return to the general case of h defined in (3.23), to derive a formula for its asymptotic function in terms of F_∞. This will be obtained as an immediate consequence of Theorem 3.5.2. For that purpose, we replace the hypothesis $\mathcal{H}_2'$ with $\mathcal{H}_2''$ defined as follows:
Let W be the set of sequences $\{x_k, y_k\} \in \text{dom } F$ with

$$\|(x_k, y_k)\| \to \infty, \quad \frac{(x_k, y_k)}{\|(x_k, y_k)\|} \to (\bar{x}, 0) \in \ker F_\infty.$$

Then $\mathcal{H}_2''$ holds if:
(a) Either W is empty, or
(b) for each sequence $\{x_k, y_k\} \in W$ there exists $z_k, \rho_k \in (0, \|(x_k, y_k)\|]$ such that for k sufficiently large (3.34) holds.

Corollary 3.5.5 *Let $F : \mathbb{R}^n \times \mathbb{R}^m \to \mathbb{R} \cup \{+\infty\}$ be an lsc, proper function. Suppose that $\mathcal{H}_0''$ and $\mathcal{H}_2''$ hold. Then the marginal function $y \to h(y) = \inf\{F(x,y) \mid x \in \mathbb{R}^n\}$ is proper, lsc, and for each $y \in \mathbb{R}^m$,*

$$h_\infty(y) = \inf\{F_\infty(x,y) : \quad x \in \mathbb{R}^n\}. \tag{3.37}$$

Proof. Set $A(x,y) = y$, so that $h(y) = AF(y)$ and apply Theorem 3.5.2. $\square$

Using the same argument as in Corollary 3.5.3 we obtain from Corollary 3.5.5 the following results.

Corollary 3.5.6 *Let $F : \mathbb{R}^n \times \mathbb{R}^m \to \mathbb{R} \cup \{+\infty\}$ be an lsc, proper function and suppose that one of the following conditions holds:*
(a) $F_\infty(x,0) > 0 \ \forall x \neq 0$.
(b) F is convex , $F_\infty(x,0) \geq 0 \ \forall x \neq 0$, and for each x such that $F_\infty(x,0) = 0$ we have $F_\infty(-x,0) = 0$.
Then the marginal function $y \to h(y) = \inf\{F(x,y) \mid x \in \mathbb{R}^n\}$ is lsc, proper, and for each y, the infimum is attained. Furthermore, one has for each y,

$$h_\infty(y) = \inf\{F_\infty(x,y) \mid x \in \mathbb{R}^n\}.$$

Proof. Since $\mathcal{H}_0''$ holds in cases (a) and (b), we have only to verify the hypothesis $\mathcal{H}_2''$ and use Corollary 3.5.5. In case (a), $\mathcal{H}_2''$ holds since W is empty. In case (b), let $(x_k, y_k) \in W$. Then since F is convex and since for each x such that $F_\infty(x,0) = 0$ we have $F_\infty(-x,0) = 0$, it follows that F is constant on $(x,0)$, and (3.34) holds. $\square$

Corollary 3.5.7 *Let $f : \mathbb{R}^n \to \mathbb{R} \cup \{+\infty\}$ be lsc, proper, and let A be a linear map from $\mathbb{R}^n$ to $\mathbb{R}^m$. Consider the function Af, defined by (3.24). Suppose that $Af \not\equiv \infty$ (i.e., $\mathrm{dom}\, f \cap A^{-1}(\mathbb{R}^m) \neq \emptyset$). Then, under either of the following conditions*
(a) $f_\infty(d) > 0, \ \forall 0 \neq d \in \ker A$,
(b) f is convex and $Az \neq 0$ for any z such that $f_\infty(z) \leq 0$ and $f_\infty(-z) > 0$, one has Af proper, lsc, and when $Af(y) \neq +\infty$, the infimum in the definition of $(Af)(y)$ is attained for some x. Moreover, for each y, $(Af)_\infty(y) = Af_\infty(y)$, and Af is convex whenever f is convex.

Proof. Thanks to Theorem 3.5.2, we have only to prove that $\mathcal{H}_0'$ and $\mathcal{H}_2'$ hold. Under the case (a), the hypothesis $\mathcal{H}_0'$ trivially holds, but also $\mathcal{H}_2'$. Indeed, the set X (cf. (3.35)) is empty in this case. Now consider the case (b). First, we show that $\mathcal{H}_0'$ holds. In the contrary case, there would exist $0 \neq d$ with $Ad = 0$ and $f_\infty(d) < 0$, and then by the assumptions in (b), we would have $f_\infty(-d) \leq 0$. But since f is convex, f_∞ is also convex, and since it is positively homogeneous, we obtain $0 = f_\infty(0) \leq f_\infty(d) + f_\infty(-d) < 0$, which is impossible. Now we prove that $\mathcal{H}_2'$ holds as well. Consider a sequence $\{x_k\} \subset X$. Then $f_\infty(\bar{x}) = A\bar{x} = 0$, $\bar{x} \neq 0$, and by the assumption (b), it follows that $f_\infty(-\bar{x}) \leq 0$, which in turns implies by using $(\mathcal{H}_0')$ that $f_\infty(-\bar{x}) = 0$. But since f is assumed convex, this means that $\bar{x}$ is a direction of constancy for f and (3.36) is satisfied by choosing for all k, $z_k = \bar{x}$. $\square$

An interesting application of these results is to the infimal convolution of a finite family of convex functions.

Corollary 3.5.8 *Let $f_1, \ldots, f_p$ be lsc, proper, convex functions on $\mathbb{R}^n$. Assume that $\sum_{i=1}^p z_i \neq 0$ for all vectors z_i such that*

$$\sum_{i=1}^p (f_i)_\infty(z_i) \leq 0, \quad \sum_i^p (f_i)_\infty(-z_i) > 0.$$

Then the infimal convolution $h = f_1 \square \cdots \square f_p$ is a proper, lsc, convex function, and the infimum is attained. Moreover, $h_\infty = (f_1)_\infty \square \cdots \square (f_p)_\infty$.

Proof. Take A as the linear map $A : \mathbb{R}^{np} \to \mathbb{R}$ defined by $\sum_{i=1}^p x_i$, $x_i \in \mathbb{R}^n$, and $f(x_1, \ldots, x_p) = \sum_{i=1}^p f_i(x_i)$ in Corollary 3.5.7. $\square$

3.6   Dual Operations and Subdifferential Calculus

Let $f_i : \mathbb{R}^n \to \mathbb{R} \cup \{+\infty\}$, $i = 1, \ldots, p$, be a collection of proper convex functions. As recalled in Chapter 1, the following relations hold:

$$(f_1 \square \cdots \square f_p)^* = f_1^* + \cdots + f_p^*, \tag{3.38}$$
$$\partial f_1 + \cdots + \partial f_p \subset \partial(f_1 + \cdots + f_p). \tag{3.39}$$

In fact, both relations can be verified in a straightforward fashion, using the definition of the mathematical operations involved. However, to obtain, for example, the expected *dual* formula of (3.38),

$$(f_1 + \cdots + f_p)^* = (f_1^* \square \cdots \square f_p^*),$$

and the reverse inclusion for the subdifferential inclusion (3.39), much more analysis is needed. In fact, one also needs to suppose a *qualification* condition on the domains of the function involved. For the two examples above, such a condition simply asks that the sets $\mathrm{ri}(\mathrm{dom}\, f_i)$, $i = 1, \ldots, p$, have a common point. Similar questions can be asked for other fundamental operations such as the composite function with a linear map. Despite its simplicity, the condition on the nonemptiness of the relative interiors of the domains has, in fact, a deep meaning, which turns out to be strongly connected to the notion of asymptotic functions, and it is our purpose in this section to exhibit these connections.

Dual Operations

We first give a simple rule that will ensure in the convex case that the basic assumptions made in Corollary 3.5.7 and Corollary 3.5.8 are satisfied.

Proposition 3.6.1 *Let* $g : \mathbb{R}^m \to \mathbb{R} \cup \{+\infty\}$ *be a proper convex function and let* A *be a linear map from* $\mathbb{R}^n$ *to* $\mathbb{R}^m$. *If there exists an* $x \in \mathbb{R}^n$ *such that* $Ax \in \mathrm{ri}\,\mathrm{dom}\,g$, *then* $A^*y \neq 0$ *for every* y *satisfying* $(g^*)_\infty(y) \leq 0$, *and* $(g^*)_\infty(y) > 0$.

Proof. Recall that $\mathrm{cl}\,g = g^{**}$, an lsc proper convex function, and since $\mathrm{ri}\,\mathrm{dom}\,g = \mathrm{ri}\,\mathrm{dom}\,\mathrm{cl}\,g$, the condition $Ax \in \mathrm{ri}\,\mathrm{dom}\,g$ is equivalent to $Ax \in \mathrm{ri}\,\mathrm{dom}\,g^{**}$ for some $x \in \mathbb{R}^n$. Set $f := g^*$, $h(y) := f(y) - \langle y, Ax \rangle = f(y) - \langle x, A^*y \rangle$. Then, using Proposition 2.5.8(b), one has

$$Ax \in \mathrm{ri}\,\mathrm{dom}\,g \implies h_\infty(y) > 0, \ \forall\, y,$$

except for those $y \in \mathcal{C}_h = \{v : h_\infty(y) = h_\infty(-y) = 0\}$. Suppose the conclusion of the proposition does not hold. Then there exists y such that $(g^*)_\infty(y) \leq 0$, $(g^*)_\infty(-y) > 0$, and $A^*y = 0$. As a consequence, $h = f = g^*$ and $h_\infty(y) \leq 0$ with $y \notin \mathcal{C}_h$, which is impossible. $\qquad\square$

Remark 3.6.1 In fact, it can be verified that the converse statement is also true in Proposition 3.6.1, although the proof of the converse part does not rely on the use of asymptotic calculus.

Corollary 3.6.1 *Let* $f_i : \mathbb{R}^n \to \mathbb{R} \cup \{+\infty\}$, $i = 1, \ldots, p$, *be a collection of proper convex functions. If* $\cap_{i=1}^p \mathrm{ri}\,\mathrm{dom}\,f_i \neq \emptyset$, *then*

$$\sum_{i=1}^p y_i \neq 0, \ \forall\, y_i \ \text{such that} \ \sum_{i=1}^p (f_i^*)_\infty(y_i) \leq 0, \ \sum_{i=1}^p (f_i^*)_\infty(-y_i) > 0.$$

Proof. This is a direct application of Proposition 3.6.1. Indeed, take $A : \mathbb{R}^n \to \mathbb{R}^{np}$ defined by $Ax = (x, \ldots, x)$ with $x \in \mathbb{R}^n$, and $g(x) = \sum_{i=1}^p f_i(x_i)$ for all $x = (x_1, \ldots, x_p)$ with $x_i \in \mathbb{R}^n$. Then, $A^*y = \sum_{i=1}^p y_i$, $\mathrm{dom}\,g = \cap_{i=1}^p \mathrm{dom}\,f_i$, with g proper convex. A direct calculation shows that $g^*(y) = \sum_{i=1}^p f_i^*(y_i)$, and by Proposition 1.1.9, $\mathrm{ri}\,\mathrm{dom}\,g = \cap_{i=1}^p \mathrm{ri}\,\mathrm{dom}\,f_i$. $\qquad\square$

As already noted in Remark 3.6.1, it can also be shown that the converse statement of the last corollary holds true.

Theorem 3.6.1 *Let* f *be a convex function on* $\mathbb{R}^n$, *let* g *be a convex function on* $\mathbb{R}^m$, *and let* A *be a linear map from* $\mathbb{R}^n$ *to* $\mathbb{R}^m$. *Then:*
(a) $(Af)^* = f^*A^*$.

(b) $((\operatorname{cl} g)A)^* = \operatorname{cl}(A^*g^*)$.
(c) If g is assumed proper, and there exists $x \in \mathbb{R}^n$ such that $Ax \in \operatorname{ri} \operatorname{dom} g$, the closure operation can be omitted in (b), so that $(gA)^*(u) = \inf\{g^*(v) \mid A^*v = u\}$, and the infimum is attained for each u such that there exists $v \in \operatorname{dom} g$ with $A^*v = u$.

Proof. A direct calculation shows that $(Af)^* = f^*A^*$, proving (a), and applying this relation to A^* and g^* we obtain (b). Under the assumption given in (c), Proposition 3.6.1 holds, and therefore we can apply Corollary 3.5.7(b) to the function g^* and the linear map A^*, from which it follows that $A^*g^* = \operatorname{cl}(A^*g^*)$ and for each u such that there exists $v \in \operatorname{dom} g$ with $Av = u$, the infimum in the formula $A^*g^*(u)$ is attained. Now, since $Ax \in \operatorname{ri} \operatorname{dom} g$, it follows from Proposition 2.6.3 that $(\operatorname{cl} g)A = \operatorname{cl}(gA)$. Hence, $((\operatorname{cl} g)A)^* = (gA)^*$, and since we have just proved that $((\operatorname{cl} g)A)^* = \operatorname{cl}(A^*g^*)$, we obtain finally $A^*g^* = (gA)^*$, which completes the proof. □

Proposition 3.6.2 *Let $f_i : \mathbb{R}^n \to \mathbb{R} \cup \{+\infty\}$, $i = 1, \dots, p$, be a collection of proper convex functions such that the sets $\operatorname{ri}(\operatorname{dom} f_i)$, $i = 1, \dots, p$, have a common point. Then*

$$(f_1 + \cdots + f_p)^*(u) = \inf\{f_1^*(x_1) + \cdots + f_p^*(x_p) \mid x_1 + \cdots + x_p = u\},$$

where for each u the infimum is attained whenever it is finite.

Proof. From Proposition 1.2.10 we have $(\operatorname{cl} f_1 + \cdots + \operatorname{cl} f_p)^* = \operatorname{cl}(f_1^* \square \cdots \square f_p^*)$. Since we assume that the sets $\operatorname{ri}(\operatorname{dom} f_i)$, $i = 1, \dots, p$, have a common point, it follows from Proposition 1.2.7 that $\operatorname{cl}(f_1 + \cdots + f_p) = \operatorname{cl} f_1 + \cdots \operatorname{cl} f_p$, so that $(f_1 + \dots + f_p)^* = \operatorname{cl}(f_1^* \square \dots \square f_p^*)$. On the other hand, the same nonemptiness assumption on the relative interiors means that Corollary 3.6.1 is applicable, which in turn implies that we can invoke Corollary 3.5.8 on the functions f_i^*, ensuring that the infimal convolution of the f_i^* is lsc, and the infimum is attained whenever it is finite. □

A useful particular case is the following

Corollary 3.6.2 *For two closed convex sets C_1, C_2 of $\mathbb{R}^n$ satisfying $0 \in \operatorname{ri}(C_1 - C_2)$ one has*

$$\sigma_{C_1 \cap C_2}(d) = (\sigma_{C_1} \square \sigma_{C_2})(d), \ \forall d,$$

with the infimum attained and $\operatorname{dom} \sigma_{C_1 \cap C_2} = \operatorname{dom} \sigma_{C_1} + \operatorname{dom} \sigma_{C_2}$.

Proof. Apply Proposition 3.6.2 with $f_i = \delta_{C_i}$, $i = 1, 2$, recalling that $f_i^* = \sigma_{C_i}$.

□

It is useful to realize that we can relax the relative interior hypothesis in Proposition 3.6.2 when we consider in the infimal convolution a mixture of polyhedral functions with proper convex functions. In fact, we show below that as a consequence of Proposition 3.6.2, the relative interior on the domains of given polyhedral functions can simply be replaced by the domains of these functions.

Corollary 3.6.3 *Let $f_i : \mathbb{R}^n \to \mathbb{R} \cup \{+\infty\}$, $i = 1, \ldots, p$ be a collection of proper convex functions such that $\{f_i : i = 1, \ldots, r\}$ are proper polyhedral and assume that $\cap_{i=1}^r \operatorname{dom} f_i \cap_{i=r+1}^p \operatorname{ri}(\operatorname{dom} f_i) \neq \emptyset$. Then*

$$(f_1 + \cdots + f_p)^*(u) = \inf\{f_1^*(x_1) + \cdots + f_p^*(x_p) \mid x_1 + \cdots + x_p = u\},$$

where for each u the infimum is attained whenever it is finite.

Proof. The proof is based on Proposition 3.6.2 and a simple convex analysis argument. It is enough to prove the result for only two functions, since the general result follows by simple induction. Thus, let f_1 be proper polyhedral and f_2 be proper convex such that $\operatorname{dom} f_1 \cap \operatorname{ri} \operatorname{dom} f_2 \neq \emptyset$. Let $E := \operatorname{aff}(\operatorname{dom} f_2)$. Then the latter intersection property clearly implies that $\operatorname{ri}(E \cap \operatorname{dom} f_1) \cap \operatorname{ri} \operatorname{dom} f_2 \neq \emptyset$. Define $g := f_1 + \delta_E$. Then g is proper convex polyhedral (as the sum of proper polyhedral functions; cf. Chapter 1), with $\operatorname{dom} g = E \cap \operatorname{dom} f_1$, so that one has

$$\operatorname{ri}(\operatorname{dom} g) \cap \operatorname{ri} \operatorname{dom} f_2 \neq \emptyset. \tag{3.40}$$

We also note that by definition of g one has $g + f_2 = f_1 + \delta_E + f_2 = f_1 + f_2$. The rest of the proof consists in applying twice Proposition 3.6.2. Indeed, under (3.40), we obtain $(g + f_2)^* = g^* \square f_2^*$ with the infimum attained in the infimal convolution. But since both δ_E and f_1 are proper polyhedral convex, and thus lsc, on has $(f_1 + \delta_E)^* = \operatorname{cl}(f_1^* \square \delta_E^*)$. Since here one obviously has $\operatorname{dom} f_1 \cap E \neq \emptyset$, then by Corollary 3.5.4, $f_1^* \square \delta_E^*$ is proper polyhedral with the infimum attained in the infimal convolution, so that we have obtained that $g^* = (f_1 + \delta_E)^* = f_1^* \square \delta_E^*$. Thus one has

$$(f_1 + f_2)^* = (g + f_2)^* = f_1^* \square \delta_E^* \square f_2^* = f_1^* \square f_2^*,$$

with the infimum attained throughout in the infimal convolution operations, and in the last equality above we use the fact $\delta_E^* \square f_2^* = (\delta_E + f_2)^* = f_2^*$, which holds, once again by Proposition 3.6.2, since $\operatorname{ri}(\operatorname{aff} \operatorname{dom} f_2) \cap \operatorname{ri} \operatorname{dom} f_2 = \operatorname{aff}(\operatorname{dom} f_2) \cap \operatorname{ri} \operatorname{dom} f_2$ is clearly nonempty. □

Subdifferential Calculus

Theorem 3.6.1 and Proposition 3.6.1 have important implications for computing the subdifferential of a composite function with a linear map and the subdifferential of the sum of a finite collection of convex functions. In turn, these implications lead to important results for the calculus rules concerning normal cones.

Theorem 3.6.2 *Let $g : \mathbb{R}^m \to \mathbb{R} \cup \{+\infty\}$ be a proper convex function, let A be a linear map from $\mathbb{R}^n$ to $\mathbb{R}^m$, and let $h = g \circ A$. Then*

$$\partial h(x) \supset A^* \partial g(Ax), \quad \forall x \in \mathbb{R}^n.$$

Furthermore, if $Az \in \mathrm{ri} \, \mathrm{dom} \, g$ for some $z \in \mathbb{R}^n$, then the inclusion becomes an equality.

Proof. Let $v \in A^* \partial g(Ax)$. Then there exists $y \in \partial g(Ax)$ such that $v = A^* y$, and for every $u \in \mathbb{R}^n$ we have

$$h(u) = (g \circ A)(u) = g(Au) \geq g(Ax) + + \langle y, Au - Ax \rangle = h(x) + \langle v, u - x \rangle,$$

which means that $v \in \partial h(x)$. Conversely, let $v \in \partial h(x)$. Then by Proposition 1.2.18, we have $g(Ax) + (gA)^*(v) = \langle x, v \rangle$, with $(gA)^*(v) \in \mathbb{R}$. Suppose now that there exists some z with $Az \in \mathrm{ri} \, \mathrm{dom} \, g$. Then we can apply Theorem 3.6.1, so that there exists some y satisfying $A^* y = v$, $(gA)^*(v) = g^*(y)$. Thus, the formula $g(Ax) + (gA)^*(v) = \langle x, v \rangle$ reduces to $g(Ax) + g^*(y) = \langle Ax, y \rangle$, and hence by Proposition 1.2.18, one has $y \in \partial g(Ax)$, and therefore $v \in A^* \partial g(Ax)$. $\qquad \square$

The next result is often referred to as the Theorem of Moreau-Rockafellar.

Theorem 3.6.3 *Let $f_i : \mathbb{R}^n \to \mathbb{R} \cup \{+\infty\}$, $i = 1, \ldots, p$, be a collection of proper convex functions such that the sets $\mathrm{ri}(\mathrm{dom} \, f_i)$, $i = 1, \ldots, p$, have a common point. Then*

$$\partial(f_1 + \cdots + f_p) = \partial f_1 + \cdots + \partial f_p.$$

If some of the functions are polyhedral, say $\{f_i \mid i = 1, \ldots, r\}$, then the equality holds under the weaker condition $\cap_{i=1}^{r} \mathrm{dom} \, f_i \cap_{i=r+1}^{p} \mathrm{ri}(\mathrm{dom} \, f_i) \neq \emptyset$.

Proof. By Proposition 1.2.19 we already have the inclusion $\partial f_1 + \cdots + \partial f_p \subset \partial(f_1 + \cdots + f_p)$, and thus it remains to prove the reverse inclusion. To achieve this, we proceed exactly in the same manner as in the proof of Theorem 3.6.2, but here we invoke Proposition 3.6.2 instead of Theorem 3.6.1. Similarly, we leave the details to the reader to verify that the polyhedral case follows using similar arguments to those of Theorem 3.6.2, and with the help of Corollary 3.6.3. $\qquad \square$

We note that in the important particular case of two functions, i.e., when $p = 2$, the equation above holds and takes the simpler form

$$0 \in \mathrm{ri}(\mathrm{dom} \, f_1 - \mathrm{dom} \, f_2) \implies \partial(f_1 + f_2) = \partial f_1 + \partial f_2.$$

Let $C \subset \mathbb{R}^n$ be a nonempty closed set. Then since for each $x \in \mathbb{R}^n$ the normal cone $N_C(x)$ at x is the subdifferential of the indicator function $\delta_C(x)$ of C at x, the latter result leads to the following useful formula for a finite collection of normal cones.

Corollary 3.6.4 *Let $\{C_i\}_{i=1}^m$ be a collection of closed convex sets in $\mathbb{R}^n$ whose relative interiors have a point in common and let $C = \cap_{i=1}^m C_i$. Then*

$$\forall x \in C, \ \ N_C(x) = \sum_{i=1}^m N_{C_i}(x).$$

If some of the sets, say $\{C_i \mid i = 1, \dots, r\}$, are polyhedral, then the equation holds under the weaker assumption $\cap_{i=1}^r (C_i) \cap_{i=r+1}^m \mathrm{ri}\, C_i \neq \emptyset$.

Proof. Apply Theorem 3.6.3 with $f_i = \delta_{C_i}$, recalling that with $C := \cap_{i=1}^m C_i$ one has $\delta_C = \sum_{i=1}^m \delta_{C_i}$. $\qquad\square$

3.7   Additional Results in the Convex Case

Suppose that an ordinary convex program is assumed only to have a finite infimum. Then two natural questions arise. For which class of functions is the optimum attained? For which class of functions is the marginal function of this convex program with (vertical) perturbations lsc? The second question is of key importance in the context of duality theory (see Chapter 5). We shall see that convex asymptotically level stable functions satisfy these two properties as a consequence of the main theorem proven below (Theorem 3.7.1). This theorem also implies two important corollaries. Consider a set described by a finite number of inequalities involving convex als functions. The first consequence says that the image of such a set is closed; the second consequence is a Helly-type theorem in that context and will be established in the next section.

Theorem 3.7.1 *Let $f_i : \mathbb{R}^n \to \mathbb{R} \cup \{+\infty\}$ be als and convex, $i = 1, \dots, m$ with a common effective domain $C := \mathrm{dom}\, f_i$. Suppose there exist sequences $\{\lambda_i^k\}$ converging to λ_i for $i = 1, \dots, m$ such that*

$$\bigcap_{i=1}^m \mathrm{lev}(f_i, \lambda_i^k) \neq \emptyset.$$

Then $\cap_{i=1}^m \mathrm{lev}(f_i, \lambda_i)$ is nonempty.

Proof. The proof is by induction on m. Let $m = 1$. Let $x_k \in \text{lev}(f_1, \lambda_1^k)$ be of minimal norm. If $\{x_k\}$ has a bounded subsequence, there exists at least a cluster point x, and since f_1 is lsc, obviously $f_1(x) \le \lambda_1$. In the opposite case we can suppose without loss of generality that $||x_k|| \to \infty$, $x_k||x_k||^{-1} \to \bar{x}$. Since

$$||x_k||^{-1} f_1 \left(\frac{x_k ||x_k||}{||x_k||} \right) \le ||x_k||^{-1} \lambda_1^k,$$

passing to the limit, we obtain $(f_1)_\infty(\bar{x}) \le 0$. If $(f_1)_\infty(\bar{x}) = 0$, then since f_1 is als, (3.15) is satisfied; i.e., for $\rho > 0$, $x_k - \rho\bar{x} \in \text{lev}(f_1, \lambda_1^k)$ for k sufficiently large. But for k sufficiently large we also have

$$
\begin{aligned}
||x_k - \rho\bar{x}|| &= ||(1 - \rho||x_k||^{-1})x_k + \rho(x_k||x_k||^{-1} - \bar{x})|| \\
&\le (1 - \rho||x_k||^{-1})||x_k|| + \rho||x_k||x_k||^{-1} - \bar{x}|| \\
&\le ||x_k|| + \rho(||(x_k||x_k||^{-1} - \bar{x}|| - 1). \quad (3.41)
\end{aligned}
$$

Since x_k is of minimal norm, it follows that

$$1 \le ||x_k||x_k||^{-1} - \bar{x}||. \quad (3.42)$$

Passing to the limit we get $1 \le 0$, which is impossible. As a consequence it follows that $(f_1)_\infty(\bar{x}) < 0$. Let $x \in \text{dom } f_1$. Since

$$f_1(x + t\bar{x}) \le f_1(x) + t(f_1)_\infty(\bar{x}),$$

it follows that for t sufficiently large $f_1(x + t\bar{x}) < \varepsilon_1$, and the result is proved.

Now we assume by induction that the theorem holds for $m \le l$. Consider the case $m = l + 1$. Let x_k be the solution of smallest norm in $\bigcap_{i=1}^m \text{lev}(f_i, \lambda_i^k)$. If the sequence $\{x_k\}$ has a bounded subsequence, there exists at least a cluster point x of this bounded subsequence, and since the functions f_i are lsc, obviously $f_i(x) \le \lambda_i$ for each i. In the opposite case we can suppose without loss of generality that

$$||x_k|| \to +\infty, \quad x_k||x_k||^{-1} \to \bar{x}.$$

By the same argument given in the first part of the proof we obtain

$$(f_i)_\infty(\bar{x}) \le 0 \quad \forall i = 1, \ldots, m. \quad (3.43)$$

We once again consider two cases.
Case 1: $(f_i)_\infty(\bar{x}) = 0$ for each i. Since f_i is als, it follows that for any $\rho > 0$, we have for k sufficiently large

$$x_k - \rho\bar{x} \in \bigcap_{i=1}^m \text{lev}(f_i, \lambda_i^k).$$

Since x_k is of minimal norm on this set, using (3.41) we obtain again (3.42), and passing to the limit, we get a contradiction.

Case 2: There exists a j such that $(f_j)_\infty(\bar{x}) < 0$. Without loss of generality, let $j = l + 1$. Since $\bigcap_{i=1}^l \text{lev}(f_i, \lambda_i^k)$ is nonempty, the induction hypothesis implies that there exists some $x \in \bigcap_{i=1}^l \text{lev}(f_i, \lambda_i)$. Consider now the point $x + \rho\bar{x}$. By (3.43) one has $\bar{x} \in (\bigcap_{i=1}^l \text{lev}(f_i, \lambda_i))_\infty$, and it follows that

$$f_i(x + \rho\bar{x}) \le \lambda_i \quad \forall i = 1, \ldots, l \quad \forall \rho > 0.$$

Furthermore, since $\text{dom} f_i$ is the same set for each i, it follows that $x \in \text{dom} f_{l+1}$, and we have

$$f_{l+1}(x + \rho\bar{x}) \le f_{l+1}(x) + \rho(f_{l+1})_\infty(\bar{x}).$$

Since $(f_{l+1})_\infty(\bar{x}) < 0$, it follows that for ρ sufficiently large we have $f_{l+1}(x + \rho\bar{x}) < \lambda_{l+1}$, and for such ρ, $x + \rho\bar{x} \in \bigcap_{i=1}^m \text{lev}(f_i, \lambda_i)$. $\square$

Corollary 3.7.1 Let $f_i : \mathbb{R}^n \to \mathbb{R}$ be als and convex for $i = 1, \ldots, m$. Set $C = \bigcap_{i=1}^m \text{lev}(f_i, 0)$, and let A be a linear map defined on $\mathbb{R}^n$. Then if C is nonempty, $A(C)$ is closed.

Proof. Consider a convergent sequence $\{y_k\} \in A(C)$ with $\lim_{k \to \infty} y_k = y$. Then there exists $x_k \in C$ such that $y_k = Ax_k$. Set

$$f_0(x) = \|y - Ax\|^2, \quad \epsilon_k = f_0(x_k).$$

Then $\epsilon_k \to 0^+$ and f_0 is als. Furthermore, for each k the system

$$f_0(x) \le \epsilon_k, \quad f_i(x) \le 0 \quad \forall i = 1, \ldots, m$$

admits a solution x_k. Using Theorem 3.7.1 it follows that the system

$$f_0(x) \le 0, \quad f_i(x) \le 0 \quad \forall i = 1, \ldots, m$$

admits a solution, which is equivalent to saying that $y \in A(C)$. $\square$

We now turn to another application of Theorem 3.7.1 on the marginal function of a convex program. Let $f_i : \mathbb{R}^n \to \mathbb{R} \cup \{+\infty\}$, $i = 0, 1 \ldots, m$, be lsc and proper functions. For $y \in \mathbb{R}^m$ consider the optimization problem

$$(P(y)) \quad h(y) = \inf\{f_0(x) \mid x \in F(y)\},$$

where $F(y) = \{x \mid f_i(x) \le y_i \quad \forall i = 1, \ldots, m\}$. Denote by $S(y)$ the set of optimal solutions of $P(y)$.

Corollary 3.7.2 *Let $f_i : \mathbb{R}^n \to \mathbb{R} \cup \{+\infty\}$ be als and convex for $i = 0, \ldots, m$ with common effective domain $C = \operatorname{dom} f_i$. Suppose there exists y such that $h(y)$ is finite. Then the following properties hold:*
(a) The optimal set $S(y)$ is nonempty.
(b) h is proper and convex.
(c) $\operatorname{dom} h$ is closed and h is lsc.

Proof. (a) Let $\{\nu_k\}$ be a positive sequence converging to 0^+ and consider the set

$$S_k = F(y) \cap \operatorname{lev}(f_0, h(y) + \nu_k).$$

By definition of h, S_k is nonempty, and then by Theorem 3.7.1 it follows that $S(y)$ is nonempty.
(b) Let us prove first that h is lsc at y. Let $\{\varepsilon_i^k\}$, $i = 1, \ldots, m$, be sequences of reals converging to zero and $\varepsilon^k = (\varepsilon_1^k, \ldots, \varepsilon_m^k)^T$. We have to prove that

$$h(y) \le \lim_{k \to \infty} \inf \ h(y + \varepsilon^k). \tag{3.44}$$

Without loss of generality we can suppose that $F(y + \varepsilon^k) \ne \emptyset$ and that

$$\lim_{k \to \infty} \inf \ h(y + \varepsilon^k) = \lim_{k \to \infty} h(y + \varepsilon^k) = \alpha.$$

Suppose that (3.44) is not satisfied, i.e., $\alpha < h(y)$. Then we consider two cases.
Case 1: α is finite. Then $h(y + \varepsilon^k)$ is finite for k sufficiently large, and by part (a) the system of inequalities

$$f_0(x) \le \alpha + (h(y + \varepsilon^k) - \alpha), \quad f_i(x) \le y_i + \varepsilon_i^k, \ i = 1, \ldots, m,$$

is nonempty. Again by Theorem 3.7.1 it follows that the system of inequalities

$$f_0(x) \le \alpha, \quad f_i(x) \le y_i, \ i = 1, \ldots, m,$$

is nonempty. Since $\alpha < h(y)$, this is impossible.
Case 2: $\alpha = -\infty$. Let $\tau < h(y)$. Then for k sufficiently large the system of inequalities

$$f_0(x) \le \tau, \quad f_i(x) \le y_i + \varepsilon_i^k, \ i = 1, \ldots, m,$$

is nonempty. Using Theorem 3.7.1 again, it follows that there exists x_τ satisfying

$$f_0(x_\tau) \le \tau, \quad f_i(x_\tau) \le y_i, \ i = 1, \ldots, m.$$

Passing to the limit as $\tau \to -\infty$, it follows that $h(y) = -\infty$, which yields a contradiction. Since the functions f_i are convex, it follows by Theorem 1.2.2 that h is convex. As a consequence h is proper. Indeed, suppose that there exists $\tilde{y}$ such that $h(\tilde{y}) = -\infty$. Then for $y(\lambda) := y + \lambda(\tilde{y} - y)$, $\lambda \in (0, 1)$, $h(y(\lambda)) = -\infty$, and passing to the limit as $\lambda \to 0$, this contradicts the lower semicontinuity of h at y.

(c) Now let us prove that dom h is closed. This will also obviously imply that h is lsc on the whole space. Suppose the contrary. Then there exists $u \in \mathrm{cl}(\mathrm{dom}\, h) \setminus \mathrm{dom}\, h$. Take $\tilde{y} \in \mathrm{ri}\,\mathrm{dom}\, h$ and set $y(\lambda) = u + \lambda(\tilde{y} - u)$ for $0 < \lambda < 1$. Then $y(\lambda) \in \mathrm{ri}\,\mathrm{dom}\, h$ and $h(y(\lambda))$ is finite. As a consequence the system of inequalities

$$f_i(x) \le y_i(\lambda), \quad i = 1, \ldots, m,$$

is nonempty. Passing to the limit as $\lambda \to 0^+$ it follows from Theorem 3.7.1 that $F(u)$ is nonempty. Then since dom $f_0 = \mathrm{dom}\, f_i$ for each $i = 1, \ldots, m$, it follows that $h(u) < +\infty$, which is impossible.   □

3.8   The Feasibility Problem.

Given a finite or infinite system of convex inequalities, a fundamental question is to characterize whether such a system is empty. This leads to what is usually called alternative theorems or Helly-type theorems. We shall derive here such types of results as a consequence of Theorem 3.7.1 for systems of inequalities involving als convex functions.

Theorem 3.8.1 *Let $f_i : \mathbb{R}^n \longrightarrow \mathbb{R} \cup \{+\infty\}$ be als and convex for $i = 1, \ldots, m$ with common effective domain $C = \mathrm{dom}\, f_i$. Then one and only one of the following alternatives holds:*
(a) There exists a vector x such that $f_i(x) \le 0$ $\forall i = 1, 2 \ldots, m$.
(b) There exist nonnegative real numbers λ_i, not all zero, such that for some $\varepsilon > 0$ one has

$$\sum_{i=1}^{m} \lambda_i f_i(x) \ge \varepsilon, \qquad \forall x. \tag{3.45}$$

Proof. Obviously, (a) and (b) cannot hold simultaneously. Assume that (a) does not hold. We shall prove that (b) holds, and that will establish the theorem. Let $G = \{y \mid \exists x \text{ such that } f_i(x) \le y_i \quad \forall i = 1, \ldots, m\}$. Then G is convex, $0 \notin G$, and by Theorem 3.7.1 G is closed. As a consequence we can separate strongly $\{0\}$ and G by a hyperplane. Thus, for some nonzero $\lambda = (\lambda_1, \ldots, \lambda_m)$ and two real numbers $\alpha \in \mathbb{R}$ and $\epsilon > 0$ we have

$$0 = \langle 0, \lambda \rangle \le \alpha - \epsilon \le \alpha + \epsilon \le \langle \lambda, y \rangle \qquad \forall y \in G. \tag{3.46}$$

Clearly, this implies that for each i, λ_i is nonnegative (if λ_j were negative the last inequality would be violated with $y_j \to \infty$). Furthermore, $\lambda \ne 0$. Since $\alpha + \epsilon \ge 2\epsilon$, setting $y_i = f_i(x)$ in (3.46) for each i ends the proof.   □

Theorem 3.8.2 *Let* $\{f_i\}_{i \in I}$ *be an arbitrary collection of proper, lsc, convex functions defined on* $\mathbb{R}^n$. *Assume that the functions* f_i *have no common asymptotic direction except 0. Then one and only one of the following alternatives holds:*
(a) There exists a vector $x \in \mathbb{R}^n$ *such that* $f_i(x) \leq 0 \quad \forall i \in I$.
(b) There exist nonnegative real numbers λ_i, *only finitely many nonzero, such that for some* $\epsilon > 0$ *one has*

$$\sum_{i \in I} \lambda_i f_i(x) \geq \epsilon, \quad \forall x \in \mathbb{R}^n.$$

If alternative (b) holds, the multipliers λ_i *can be chosen so that at most* $(n+1)$ *of them are nonzero.*

Proof. As in the proof of Theorem 3.8.1, obviously (a) and (b) cannot hold simultaneously. Assume that (a) does not hold, then we have only to prove that (b) holds. Let $f(x) = \sup_{i \in I} f_i(x)$. Then the function f is convex, lsc, and as can be seen later, without loss of generality we can assume that f is proper. As a consequence $f_\infty = \sup_{i \in I} (f_i)_\infty$. Therefore, under the hypothesis of the theorem, it follows that $f_\infty(d) > 0$, $\forall d \neq 0$, so that f is coercive and $\operatorname{argmin} f \neq \emptyset$. Furthermore, since (a) does not hold, it follows that $f^*(0) = -\inf f < 0$. Now from Proposition 1.2.12(b) we have $f^* = \operatorname{cl}(\operatorname{conv}\{f_i^* \mid i \in I\})$. Let us prove that $f^*(0) = \operatorname{conv}\{f_i^* \mid i \in I\}(0)$. From Proposition 1.2.5 it is sufficient to verify that $0 \in \operatorname{ri} \operatorname{dom}(\operatorname{conv}\{f_i^* \mid i \in I\})$. Suppose the contrary holds. Then we can separate the point 0 from $\operatorname{dom}(\operatorname{conv}\{f_i^* \mid i \in I\})$, which contains $\cup_{i \in I} \operatorname{dom} f_i^*$. As a consequence there exists $0 \neq d$ such that $\langle d, x \rangle \leq 0$, $\forall x \in \operatorname{dom} f_i^*$, $\forall i \in I$. But the last inequality means precisely that $(f_i)_\infty(d) \leq 0$, $\forall i \in I$, which is clearly impossible in view of the assumed hypothesis of the Theorem on the lack of asymptotic direction for f_i. Thus we have proved that $f^*(0) = \operatorname{conv}\{f_i^* \mid i \in I\}(0)$. Set $\varepsilon := -f^*(0) > 0$. Then one has $\operatorname{conv}\{f_i^* : i \in I\}(0) = -\varepsilon$ and $(0, -\varepsilon) \in \operatorname{conv}\{\operatorname{epi} f_i^* \mid i \in I\}$. By definition of the convex hull of functions, and invoking Caratheodory's Theorem 1.1.1, there exist y_i and $\lambda_i \geq 0$ with at most $n + 1$ real numbers λ_i that are nonzero such that $\sum_{i \in I} \lambda_i y_i = 0$, $\sum_{i \in I} \lambda_i f_i^*(y_i) = -\varepsilon$. Without loss of generality, we can suppose that the indices corresponding to the nonzero λ_i are just the first m integers ($m \leq n + 1$), and if we set $z_i = \lambda_i y_i$, since $(\lambda_i f_i)^*(z_i) = \lambda_i f_i^*(y_i)$, we then have $\sum_{i=1}^m z_i = 0$, $\sum_{i=1}^m (\lambda_i f_i)^*(z_i) = -\varepsilon$. By definition of the infimal convolution (cf. Definition 1.2.6), this implies that $(\lambda_1 f_1)^* \square \cdots \square (\lambda_m f_m)^*(0) \leq -\varepsilon$. By Proposition 1.2.10, one has $(\sum_{j=1}^m \lambda_j f_j)^*(0) = \operatorname{cl}((\lambda_1 f_1)^* \square \cdots \square (\lambda_m f_m)^*)(0) \leq -\epsilon$, and the latter inequality is equivalent to $\inf_x \sum_{j=1}^m \lambda_j f_j(x) \geq \varepsilon$, proving the desired result.
$\square$

3.9 Notes and References

Existence and stability of optimal solutions in extremum problems and conditions under which such solutions exist and change under perturbation of the data is a topic of fundamental importance in several branches of nonlinear and variational analysis. Existence of optimal solutions under level boundedness is the usual assumption in most optimization problems, and all the results in Section 3.1 are classical and can be found in [119] and [123]. The definition of asymptotically directional constant and weakly coercive functions was given by Auslender [18], and Proposition 3.2.1 takes its arguments in Rockafellar's book [119, Corollary 27.3.1]. The Characterization of weakly coercive convex functions given in Proposition 3.2.2 and the equivalence between (a) and (b) in Theorem 3.2.1 were given by Auslender in [12] and [18]. The equivalence between (b) and (c) in this theorem is new, while the equivalence between (c) and (d) is classical. Proposition 3.2.3 is due to Goberna and Lopez [77], while the rest of the section is from the work of Auslender, Cominetti, and Crouzeix [10]. Asymptotically level stable functions, a terminology coined in this book, were introduced by Auslender in [18], and the results of Section 3.3 are taken from [18]. Existence results (without the level boundedness assumption) for the important special case of quadratic optimization originated with the work of Frank and Wolfe [73], and several extensions can be found in the works of Perold [109]. Other important contributions can be found in Rockafellar [119], Belousov [25], and Baiocchi, Buttazo, Gastaldi and Tomarelli [22] and references therein. For recent extensions and variants in Banach spaces, we refer the reader to the recent work of Penot [108]. The necessary and sufficient condition given in Theorem 3.4.1 is due to [12], but the proof technique used for the sufficiency part is essentially taken from [22]. Likewise, Corollary 3.4.1 is from [22], while Corollaries 3.4.2–3.4.3 are from [18]. For classical results on stability in parametric optimization we refer to the monographs of Bank et al. [23], Belousov [25], and Kummer [84]. For a recent and comprehensive work on the analysis of perturbed optimiztion problems and related second-order optimality conditions, a topic not covered here, we refer the reader to the book of Bonnans and Shapiro [37] and references therein. Most of the stability results given in Section 3.5 are given in Auslender [12], while the results concerning the nonconvex coercive case was first established by Zalinescu [135], and the results concerning the weakly coercive convex case can be found in Rockafellar's book [119]. The material on subdifferential calculus of Section 3.6 is classical and can be found in [119]. The additional results given for the convex case in Section 3.7 and Theorem 3.8.1 are due to Auslender [18], while Theorems 3.7.1–3.8.1 were first derived for quadratic convex functions in [95] and [96], and Theorem 3.8.2 is classical and can be found in [119].

4

Minimizing and Stationary Sequences

Solving an optimization problem usually consists in generating a sequence by some numerical algorithm. The key question is then to show that such a sequence converges to some solution as well as to evaluate the efficiency of the convergent procedure. In general, the coercivity hypothesis on the problem's data is assumed to ensure the asymptotic convergence of the produced sequence. For convex minimization problems, if the produced sequence is stationary, i.e., the sequence of subgradients approaches zero, it is interesting to know for what class of functions we can reach convergence under a weaker assumption than coercivity. This chapter introduces the concept of well-behaved asymptotic functions, which in turn are linked to the problems of error bounds associated with a given subset of a Euclidean space. A general framework is developed around these two themes to characterize asymptotic optimality and error bounds for convex inequality systems.

4.1   Optimality Conditions in Convex Minimization

A basic minimization principle is the well-known Fermat principle, which states that for a continuously differentiable function f, a necessary condition for $\bar{x} \in \mathbb{R}^n$ to be a local minimum of the function f is that the gradient of f at $\bar{x}$ be equal to zero, i.e., $\nabla f(\bar{x}) = 0$. This condition becomes sufficient whenever the function is also assumed convex. This result can be extended to handle more general problems, leading to the *abstract Fermat principle*.

Theorem 4.1.1 *Let $f_0 : \mathbb{R}^n \to \mathbb{R} \cup \{+\infty\}$ be a proper, lsc, convex function, $C \subset \mathbb{R}^n$ a closed convex set, and consider the minimization problem $\inf\{f_0(x) \mid x \in C\}$. Suppose that $\operatorname{ri} \operatorname{dom} f_0 \cap \operatorname{ri} C \neq \emptyset$. Then $\bar{x}$ minimizes f_0 on C if and only if $0 \in \partial f_0(\bar{x}) + N_C(\bar{x})$, or equivalently, if and only if*

$$\exists g \in \partial f_0(\bar{x}) \text{ such that } \langle x - \bar{x}, g \rangle \geq 0, \ \forall x \in C.$$

Proof. Let $f := f_0 + \delta_C$. Then $\bar{x}$ minimizes f on $\mathbb{R}^n$, and from Proposition 1.2.17, this is equivalent to $0 \in \partial f(\bar{x})$. Since we assume that $\operatorname{ri} \operatorname{dom} f_0 \cap \operatorname{ri} C$ is nonempty, using Theorem 3.6.3 this means that $0 \in \partial f_0(\bar{x}) + \partial \delta_C(\bar{x}) = \partial f_0(\bar{x}) + N_C(\bar{x})$. The latter inclusion is equivalent to saying that there exists $g \in \partial f_0(\bar{x})$ such that $-g \in N_C(\bar{x})$, and by definition of the normal cone, the second assertion follows. $\square$

We now specialize the abstract Fermat principle whenever the constraint set C is described by a finite number of convex and affine inequalities to obtain necessary and sufficient optimality conditions, often called in the literature the Karush–Kuhn–Tucker (KKT for short) theorem for convex programs. We thus consider the convex program

$$\text{(P)} \quad \inf\{f_0(x) \mid x \in C\},$$

with $C = \cap_{i=1}^m C_i$, where the closed convex sets C_i are defined by

$$C_i = \{x \mid f_i(x) \leq 0, \ i = 1, \ldots, r, \ \langle a_i, x \rangle \leq \alpha_i, \ i = r+1, \ldots, m\}$$

where the functions $f_i : \mathbb{R}^n \to \mathbb{R} \cup \{+\infty\}$, $i = 0, \ldots, r$, are supposed proper, convex, and lsc and where $a_i \in \mathbb{R}^n$, $\alpha_i \in \mathbb{R}$ are given. Unless otherwise specified, throughout this section we make the following standard regularity assumptions:

(i) $\operatorname{dom} f_0 \subset \operatorname{dom} f_i$, $\operatorname{ri} \operatorname{dom} f_0 \subset \operatorname{ri} \operatorname{dom} f_i$, $i = 1, \ldots, m$.

(ii) $\exists \hat{x} \in \operatorname{ri} \operatorname{dom} f_0$ such that $f_i(\hat{x}) < 0, \ \forall i = 1, \ldots, r$.

Note that (i) is always satisfied whenever the functions f_i are finite valued, while (ii) is a constraint qualification on the nonaffine constraints f_i, usually called the *Slater* condition, which is needed for the application of subdifferential calculus rules.

Theorem 4.1.2 *Consider the convex optimization problem (P) described above and satisfying the hypotheses (i) and (ii). Then $\bar{x} \in C$ minimizes f_0 on C if and only if there exist $g_0 \in \partial f_0(\bar{x})$, $g_i \in \partial f_i(\bar{x})$, $\lambda_i \geq 0 \ \forall i \in I(\bar{x}) := \{i \in [1, r] \mid f_i(\bar{x}) = 0\}$, and $\mu_j \geq 0$ for $j \in J(\bar{x}) := \{j \in [r+1, m] \mid \langle a_j, x_j \rangle = \alpha_j\}$ such that*

$$g_0 + \sum_{i \in I(\bar{x})} \lambda_i g_i + \sum_{j \in J(\bar{x})} a_j \mu_j = 0. \tag{4.1}$$

Proof. We apply Theorem 4.1.1 with $C = \cap_{i=1}^{m} C_i$, recalling that the subdifferential calculus rules given in Theorem 3.6.3 can be applied by hypothesis (i). Thus one has $N_C(\bar{x}) = \sum_{i=1}^{m} N_{C_i}(\bar{x})$, and from Proposition 1.2.21, for $i \geq r+1$ we obtain

$$N_{C_i}(\bar{x}) = \begin{cases} \mathbb{R}_+ a_i & \text{if } \langle a_i, \bar{x} \rangle = \alpha_i, \\ \mathbb{R}^n & \text{otherwise.} \end{cases}$$

For $i = 1, \ldots, r$, from Slater's condition using Proposition 1.2.22 it follows that

$$N_{C_i}(\bar{x}) = \begin{cases} \mathbb{R}_+ \partial f_i(\bar{x}) & \text{if } f_i(\bar{x}) = 0, \\ \{0\} & \text{if } f_i(\bar{x}) < 0. \end{cases}$$

Using these formulas assertion (4.1) follows. □

Approximate Solutions

Let $f : \mathbb{R}^n \to \mathbb{R} \cup \{+\infty\}$ be a proper lsc function with inf $f > -\infty$. In general, as we have seen in Chapter 3, we need additional conditions to ensure that the infimum will be attained, but there are always approximatively ε solutions, i.e., points x_ε for $\varepsilon > 0$ satisfying

$$\inf f \leq f(x_\varepsilon) \leq \inf f + \varepsilon. \tag{4.2}$$

This leads to the notion of ε-subdifferential of f at x.

Definition 4.1.1 *Let $f : \mathbb{R}^n \to \mathbb{R} \cup \{+\infty\}$ be lsc. For any $x \in \text{dom } f$ and any $\varepsilon > 0$ the ε-subdifferential of f at x, denoted by $\partial_\varepsilon f(x)$, is defined by*

$$\partial_\varepsilon f(x) = \{x^* \mid f(y) \geq f(x) + \langle x^*, y - x \rangle - \varepsilon \quad \forall y\}.$$

Obviously, whenever $\varepsilon = 0$, one has $\partial_0 f(x) = \partial f(x)$. In the convex case the ε-subdifferential enjoys many properties. In particular, as we shall see, it is never empty on dom f, and ε-subdifferential calculus rules can be developed. In the convex case we obtain immediately from the definition the following approximate optimality condition.

Theorem 4.1.3 *Let $f : \mathbb{R}^n \to \mathbb{R} \cup \{+\infty\}$ be lsc, proper, convex, and let $\varepsilon > 0$. Then for each $x \in \text{dom } f$, $\partial_\varepsilon f(x)$ is a nonempty closed convex set. Furthermore, if inf $f > -\infty$, then a necessary and sufficient condition for x_ε to satisfy (4.2) is that $0 \in \partial_\varepsilon f(x_\varepsilon)$.*

Proof. From the definition of the ε-subdifferential we get

$$\partial_\varepsilon f(x) = \{x^* \mid f^*(x^*) + f(x) - \langle x^*, x \rangle \leq \varepsilon \},$$

from which it follows that $\partial_\varepsilon f(x)$ is closed and convex. To see that $\partial_\varepsilon f(x)$ is nonempty, we argue by contradiction. Suppose $\partial_\varepsilon f(x) = \emptyset$. Let $g(x^*) = f^*(x^*) - \langle x^*, x \rangle$. Then we have

$$g(x^*) > \varepsilon - f(x) \quad \forall x^*,$$

so that $-g^*(0) \geq \varepsilon - f(x)$. But $g^*(y) = f(x+y)$, which leads to a contradiction. Finally, the second part of the theorem follows immediately from the definition of the ε-subdifferential. $\square$

The next result, called the Ekeland variational principle, is a fundamental tool with respect to several approximating approaches to optimization problems. Note that convexity is no longer needed, except for the last part of the result.

Theorem 4.1.4 *Let* $f : \mathbb{R}^n \to \mathbb{R} \cup \{+\infty\}$ *be an lsc, proper function with* $\inf f > -\infty$. *Let* $\varepsilon > 0$ *and suppose* $x_\varepsilon \in \mathbb{R}^n$ *satisfies* $f(x_\varepsilon) \leq \inf f + \varepsilon$. *Then given* $\lambda > 0$, *there exists a point* $\bar{x} \in \mathrm{dom}\, f$ *such that:*
(a) $f(\bar{x}) \leq f(x_\varepsilon)$.
(b) $\|\bar{x} - x_\varepsilon\| \leq \lambda$.
(c) $f(\bar{x}) < f(x) + \varepsilon\lambda^{-1}\|x - \bar{x}\| \quad \forall x \neq \bar{x}$.
(d) *With* f *assumed convex, there exists* $g \in \partial f(\bar{x})$ *satsisfying* $\|g\| \leq \varepsilon\lambda^{-1}$.

Proof. Let $h(x) := f(x) + \varepsilon\lambda^{-1}\|x - x_\varepsilon\|$. Then $h_\infty(d) \geq f_\infty(d) + \varepsilon\lambda^{-1}\|d\|$, and since $\inf f > -\infty$, one has $f_\infty(d) \geq 0$ and hence $h_\infty(d) > 0 \;\; \forall d \neq 0$, proving that h is a coercive function and $\mathrm{argmin}\, h$ is nonempty. Let $\bar{x} \in \mathrm{argmin}\, h$. Then

$$f(\bar{x}) + \varepsilon\lambda^{-1}\|\bar{x} - x_\varepsilon\| \leq f(x) + \varepsilon\lambda^{-1}\|x - x_\varepsilon\|, \quad \forall x \in \mathbb{R}^n, \qquad (4.3)$$

and by setting $x = x_\varepsilon$ in the last inequality one obtains

$$\varepsilon\lambda^{-1}\|\bar{x} - x_\varepsilon\| \leq f(x_\varepsilon) - f(\bar{x}) \leq f(x_\varepsilon) - \inf f \leq \varepsilon,$$

from which (a) and (b) follow. Part (c) is an immediate consequence of (4.3) and the triangle inequality $\|x - x_\varepsilon\| < \|x - \bar{x}\| + \|\bar{x} - x_\varepsilon\|, \;\; \forall x \neq \bar{x}$. To prove (d), since f is assumed convex, it follows from (c) that $0 \in \partial h(\bar{x})$ with $h(x) = f(x) + \varepsilon\lambda^{-1}\|x - \bar{x}\|$. Since $\mathrm{dom}\, \|\cdot - \bar{x}\| = \mathbb{R}^n$, by Theorem 3.6.3 we thus have $\partial h(\bar{x}) = \partial f(\bar{x}) + \varepsilon\lambda^{-1}\partial(\|\cdot - \bar{x}\|)(\bar{x}) = \partial f(\bar{x}) + \varepsilon\lambda^{-1}\mathbb{B}$, where in the last equality we have used the fact $\partial\|.\|(0) = \mathbb{B}$, with $\mathbb{B}$ the closed unit ball, and (d) is proved. $\square$

As an immediate consequence of Theorem 4.1.4 we obtain that $\bar{x}$ is a kind of "almost critical point" for the differentiable minimization problem $\inf\{f(x) \mid x \in \mathbb{R}^n\}$.

Corollary 4.1.1 *Let f be a proper lsc function on $\mathbb{R}^n$ such that $\inf f > -\infty$ and let $\lambda > 0$. Suppose $f : \mathbb{R}^n \to \mathbb{R}$ is differentiable. Then for each $\varepsilon > 0$, there exists $\bar{x} \in \mathbb{R}^n$ such that $\|\nabla f(\bar{x})\| \leq \varepsilon\lambda^{-1}$.*

Proof. Invoking Theorem 4.1.4, there exists $\bar{x} \in \mathbb{R}^n$ such that

$$f(\bar{x}) \leq f(x) + \varepsilon\lambda^{-1}\|x - \bar{x}\|, \quad \forall x \in \mathbb{R}^n. \tag{4.4}$$

Let $d \in \mathbb{R}^n$ and $t > 0$ be arbitrary. Take $x = \bar{x} + td$ in (4.4), to obtain $t^{-1}(f(\bar{x}+td) - f(\bar{x})) \geq -\varepsilon\lambda^{-1}\|d\|$, and passing to the limit as $t \to 0^+$, since f is assumed differentiable, we get $\langle d, \nabla f(\bar{x})\rangle \geq -\varepsilon\lambda^{-1}\|d\|$ $\forall d \in \mathbb{R}^n$. Since the last inequality is also true when d is replaced by $-d$, we thus obtain $|\langle d, \nabla f(\bar{x})\rangle| \leq \varepsilon\lambda^{-1}\|d\|$, $\forall d \in \mathbb{R}^n$, from which it follows that $\|\nabla f(\bar{x})\| \leq \varepsilon\lambda^{-1}$. $\qquad\square$

As an application we obtain the Bronsted–Rockafellar theorem.

Theorem 4.1.5 *Let $f : \mathbb{R}^n \to \mathbb{R} \cup \{+\infty\}$ be lsc, convex, and proper. Let $x \in \operatorname{dom} f$, $\varepsilon > 0$, $\lambda > 0$. Then for each $g \in \partial_\varepsilon f(x)$ there exist $\bar{x} \in \operatorname{dom} f$ and $\bar{g} \in \partial f(\bar{x})$ such that*

$$\|x - \bar{x}\| \leq \lambda, \quad \|g - \bar{g}\| \leq \varepsilon\lambda^{-1}.$$

Proof. Let $h(u) = f(u) - \langle u, g\rangle$. Then since $g \in \partial_\varepsilon f(x)$, we have

$$h(u) \geq f(x) - \langle x, g\rangle - \varepsilon = h(x) - \varepsilon \quad \forall u,$$

from which it follows that h is lower bounded and that x is an ε-minimizer for h. Using Theorem 4.1.4(b)–(d) it follows that there exist $\bar{x} \in \operatorname{dom} f$ and $\bar{g} \in \partial f(\bar{x})$ such that $\|x - \bar{x}\| \leq \lambda$, $\quad \|g - \bar{g}\| \leq \varepsilon\lambda^{-1}$. $\qquad\square$

As announced in Section 3.1, using Theorem 4.1.4, we shall prove now that a function that is bounded below and satisfies the Palais–Smale condition is level bounded. Recall that for a C^1 function f, a sequence $\{x_k\}$ such that $\{f(x_k)\}$ is bounded and $\lim_{k\to\infty} \nabla f(x_k) = 0$, is said to be a Palais–Smale (PS) sequence for f. Furthermore, f is said to satisfy the Palais–Smale condition if each (PS) sequence is bounded.

Theorem 4.1.6 *Suppose that $f : \mathbb{R}^n \longrightarrow \mathbb{R}$ is a C^1 function bounded below on $\mathbb{R}^n$ that satisfies the Palais–Smale condition. Then f is level bounded.*

Proof. Define the following sets and functions, for a fixed $k \in \mathbb{N}$:

$$X_k := \{x \mid \|x\| \geq k\}, \qquad f_k(x) = f(x) + \delta_{X_k}(x),$$
$$\alpha_k := \inf\{f_k(x) \mid x \in \mathbb{R}^n\}, \qquad \beta_{k+1} = \alpha_{k+1} - \alpha_k.$$

Then the sequence $\{\alpha_k\}$ is a nondecreasing sequence of real numbers, and for each k, $\beta_k \geq 0$. Let $y_k \in X_{k+1}$ be such that $f(y_k) \leq \alpha_{k+1} + k^{-1} =$

$\alpha_k + k^{-1} + \beta_k$. Then since f_k is lsc, and proper with inf $f_k > -\infty$, we can use Ekeland's Theorem 4.1.4, and for $\lambda \in \,]0, 1[$ we can find $x_k \in X_k$ such that

$$f(x_k) \le f(y_k), \qquad \|x_k - y_k\| \le \lambda. \tag{4.5}$$

But since $\|x_k\| \ge \|y_k\| - \lambda > (k+1) - 1 = k$, we have that f_k is differentiable at x_k with $\nabla f(x_k) = \nabla f_k(x_k)$, so that by Theorem 4.1.4(d),

$$\|\nabla f(x_k)\| \le \lambda^{-1}(\beta_k + k^{-1}). \tag{4.6}$$

Now suppose that f is not level bounded, i.e., there exists $v > \inf f$ such that $f(x) \le v$ is unbounded. Then it follows that the sequence $\{\alpha_k\}$ is bounded above by v, so that $\beta_k \to 0_+$, and from (4.5) and (4.6) it follows that $\{x_k\}$ is an unbounded (PS) sequence, which is impossible, since f satisfies the Palais–Smale condition.   □

4.2   Asymptotically Well-Behaved Functions

In this section we consider the problem

$$\text{(P)} \quad \inf f = \inf\{f(x) : \ x \in \mathbb{R}^n\},$$

where $f : \mathbb{R}^n \to \mathbb{R} \cup \{+\infty\}$ is lsc, convex, and proper. We are thus interested in finding a minimizing sequence $\{x_k\}$ for problem (P), that is, a sequence satisfying

$$\lim_{k \to \infty} f(x_k) = \inf f.$$

Usually, the sequence $\{x_k\}$ is produced by some numerical algorithm to solve (P). The standard approach to proving such a limiting result is based on the assumption that the sequence $\{x_k\}$ is bounded or that the objective function is coercive. For example, when $f \in C^1(\mathbb{R}^n)$, if the sequence generated by some algorithm is bounded, then whenever $\nabla f(x_k) \to 0$ one obtains $f(x_k) \to \inf f$ as $k \to \infty$. This is particularly the case when f is assumed coercive. However, these kinds of assumptions on the function f might often be violated, and therefore it is important to know what happens and how to handle more general situations.

Example 4.2.1 Consider the function f defined by

$$f(x_1, x_2) = \begin{cases} \frac{x_1^2}{2x_2} & \text{if } x_2 > 0, \\ 0 & \text{if } (x_1, x_2) = (0, 0), \\ \infty & \text{otherwise.} \end{cases}$$

It can be easily seen that f is the support functional of the closed convex set C, where

$$C = \{x \in \mathbb{R}^2 \mid \frac{x_1^2}{2} + x_2 \le 0\}.$$

Then it follows that f is lsc, convex, and proper. Furthermore, $\inf f = 0$, and the infimum of f is attained at (0,0). Let $x^k = (k, k^2)$. Then

$$f(x^k) = \frac{1}{2}, \quad \nabla f(x^k) = \left(\frac{1}{k}, -\frac{1}{2k^2}\right) \to 0.$$

As a consequence, the sequence $\{x_k\}$ is a stationary sequence but not minimizing, and thus one can say that the function f is not well behaved "at large."

This example indicates that even for the minimization of a proper lsc convex function on a closed convex set, a quite standard model optimization, difficulties arise. This motivates the need for identifying other properties that should be shared by convex functions to guarantee the existence of a minimizing sequence. The following definition introduces the essential notion of asymptotically well-behaved (awb) functions. We first recall the formal definition of stationary and minimizing sequences.

Definition 4.2.1 *Let $f : \mathbb{R}^n \to \mathbb{R} \cup \{+\infty\}$ be a proper, lsc, convex function. A sequence $\{x_k\}$ is said to be stationary if there exists $c_k \in \partial f(x_k)$ with $c_k \to 0$. The sequence is said to be minimizing if $\lim_{k \to \infty} f(x_k) = \inf f$.*

Definition 4.2.2 *A proper, lsc, convex function $f : \mathbb{R}^n \to \mathbb{R} \cup \{+\infty\}$ is said to be asymptotically well behaved (awb) if each stationary sequence is minimizing.*

As we shall see later, the problem to know whether a stationary sequence is minimizing is related to the problem of *global error bound*.

Let $\lambda \ge \inf f$. This problem consists in deriving necessary and sufficient conditions for the existence of a constant $\gamma(\lambda) > 0$ such that

$$\text{dist}(x, \text{lev}(f, \lambda)) \le \gamma(\lambda) f(x)^+ \quad \forall x \in \mathbb{R}^n, \tag{4.7}$$

where $\text{dist}(x, C)$ denotes the usual Euclidean distance of x to C, and $a^+ = \sup(a, 0)$.

An inequality of this kind is called a global error bound for the level set $\text{lev}(f, \lambda)$, see Section 4.3 for more details. Going back to the nonasymptotically well-behaved function given in Example 4.2.1, one can verify that $\text{dist}(x^k, \text{lev}(f, 0)) = \sqrt{k^4 + k^2} \to \infty$, and for $\lambda = 0$ there is no constant $\gamma(\lambda) > 0$ for which (4.7) holds.

Characterization of Asymptotically Well-Behaved Functions

We are now interested in characterizing awb functions in explicit terms of the data and information given on the function f. For that purpose, let us introduce the following constants, which will play a central role. For any $\lambda \geq \inf f$, define

$$l(\lambda) = \inf_{f(x)>\lambda} \frac{f(x) - \lambda}{\operatorname{dist}(x, \operatorname{lev}(f, \lambda))}, \quad r(\lambda) = \inf_{f(x)=\lambda} \inf_{x^* \in \partial f(x)} \|x^*\|,$$

and

$$t(\lambda) = \inf_{f(x)=\lambda} \inf_{0 \neq d \in N_{\operatorname{lev}(f,\lambda)}(x)} f'\left(x; \frac{d}{\|d\|}\right),$$

$$k(\lambda) = \inf_{f(x)=\lambda} \inf_{0 \neq x^* \in \partial f(x)} f'\left(x; \frac{x^*}{\|x^*\|}\right).$$

We begin with two useful technical results needed in our analysis.

Lemma 4.2.1 *Let C be a nonempty closed convex set in $\mathbb{R}^n$ and P_C the usual Euclidean projection on C. Let $x \in C$, $t > 0$, $d \in N_C(x)$. Then*

$$P_C(x + td) = x \quad \forall t > 0 .$$

Proof. Set $x(t) := P_C(x+td)$. Since $x(t) = \arg\min\left\{\frac{1}{2}\|x + td - y\|^2 \mid y \in C\right\}$, it follows from Theorem 4.1.1 that $(x+td-x(t)) \in N_C(x(t))$, which is equivalent to $\langle x(t) - (x+td), u-x(t)\rangle \geq 0 \quad \forall u \in C$. Set $u = x$ in this inequality. Then since $d \in N_C(x)$, it follows that $\|x - x(t)\|^2 \leq t\langle x(t) - x, d\rangle \leq 0$. $\quad\square$

Lemma 4.2.2 *Let $f : \mathbb{R}^n \to \mathbb{R} \cup \{+\infty\}$ be a proper, lsc, and convex function. Let $\lambda \geq \inf f$ and $x \notin \operatorname{lev}(f, \lambda)$ (assumed nonempty when $\lambda = \inf f$) such that $f(x) < \infty$. Let y be the projection of x onto $\operatorname{lev}(f, \lambda)$. Then we have:*
(a) $f(y) = \lambda$, $x - y \in N_{\operatorname{lev}(f,\lambda)}(y)$,
(b) when $\lambda > \inf f$ if $y \in \operatorname{ri}\operatorname{dom} f$ there exists some $\alpha > 0$ such that $\alpha(x - y) \in \partial f(y)$.

Proof. Let us show first that $f(y) = \lambda$. From feasibility we have $f(y) \leq \lambda$. On the other hand, for each $t \in (0, 1)$ the point $x + t(y - x)$ does not belong to $\operatorname{lev}(f, \lambda)$, and therefore

$$\lambda < f(x + t(y - x)) \leq (1 - t)f(x) + tf(y),$$

which after letting $t \to 1$ gives $f(y) \geq \lambda$. From Theorem 4.1.1 one has $y = P_{\operatorname{lev}(f,\lambda)}(x)$ if and only one has $x - y \in N_{\operatorname{lev}(f,\lambda)}(y)$, and (a) is proved. In the case where $\lambda > \inf f$ if $y \in \operatorname{ri}\operatorname{dom} f$, then by Proposition 1.2.22 it follows that $N_{\operatorname{lev}(f,\lambda)}(y) = \mathbb{R}_+\partial f(y)$, and since $x \neq y$, there exists an $\alpha > 0$ such that $\alpha(x - y) \in \partial f(y)$.

$\square$

Theorem 4.2.1 *Let* $f : \mathbb{R}^n \to \mathbb{R} \cup \{+\infty\}$ *be an lsc, proper, and convex function. Then*

$$l(\lambda) = t(\lambda) \quad \forall \lambda > \inf f, \tag{4.8}$$

and (4.8) holds also for $\lambda = \inf f$ *if* $\operatorname{argmin} f \neq \emptyset$. *Furthermore, when* $\lambda > \inf f$ *if for each* x *such that* $f(x) = \lambda$ *we have* $x \in \operatorname{ri dom} f$, *then*

$$k(\lambda) = t(\lambda). \tag{4.9}$$

Proof. We must prove the two inequalities $l \geq t$ and $t \geq l$. The first amounts to saying that for every x such that $f(x) > \lambda$ we have

$$t(\lambda) \leq \frac{f(x) - \lambda}{\operatorname{dist}(x, \operatorname{lev}(f, \lambda))},$$

which follows from Lemma 4.2.2. In fact, it suffices to consider the case $f(x) < \infty$, and then taking the projection y of x onto the set $\operatorname{lev}(f, \lambda)$ we have

$$t(\lambda) \leq f'\left(y; \frac{x - y}{\|x - y\|}\right) \leq \frac{f(x) - f(y)}{\operatorname{dist}(x, \operatorname{lev}(f, \lambda))},$$

from which the result follows, since $f(y) = \lambda$.

For the converse inequality we must show that given x with $f(x) = \lambda$ and given $d \in N_{\operatorname{lev}(f, \lambda)}(x)$, $d \neq 0$, we have

$$l(\lambda) \leq f'(x; d).$$

But from Lemma 4.2.1 we have $P_{\operatorname{lev}(f, \lambda)}(x + td) = x \,\, \forall t > 0$, so that $\operatorname{dist}(x + td, \operatorname{lev}(f, \lambda)) = t\|d\| > 0$. We deduce

$$\frac{f(x + td) - f(x)}{t\|d\|} = \frac{f(x + td) - \lambda}{\operatorname{dist}(x + td, \operatorname{lev}(f, \lambda))} \geq l(\lambda),$$

so that letting $t \to 0^+$ we get the desired conclusion. Now suppose that for each x such that $f(x) = \lambda$ we have $x \in \operatorname{ri dom} f$. Then from Proposition 1.2.22 it follows that $\mathbb{R}_+ \partial f(x) = N_{\operatorname{lev}(f, \lambda)}(x)$, from which (4.9) follows. $\quad\square$

Proposition 4.2.1 *Let* $f : \mathbb{R}^n \to \mathbb{R} \cup \{+\infty\}$ *be an lsc, proper, convex function such that* $f(x) > \inf f \Rightarrow x \in \operatorname{ri dom} f$. *Then:*
(a) $k(\lambda') \geq r(\lambda') \geq l(\lambda) \geq r(\lambda) \,\, \forall \lambda' > \lambda > \inf f$.
(b) For $\lambda > \inf f$ *we have the following characterizations:*

$$r(\lambda) = \inf_{f(x) \geq \lambda} \inf_{x^* \in \partial f(x)} \|x^*\|, \quad l(\lambda) = \inf_{f(x) \geq \lambda} \inf_{0 \neq x^* \in \partial f(x)} f'\left(x; \frac{x^*}{\|x^*\|}\right).$$

Proof. (a) Since for $x^* \in \partial f(x)$ one has $\langle x^*, d \rangle \leq f'(x; d) \; \forall d \in \mathbb{R}^n$, using formula (4.8) it follows easily that $l \geq r$. Thus it suffices to show that $r(\lambda') \geq l(\lambda)$. Let us take x' with $f(x') = \lambda'$ and let x be the projection of x' on $\mathrm{lev}(f, \lambda)$. Then we can use Lemma 4.2.2 and write for some $\alpha > 0$,

$$l(\lambda) = t(\lambda) \leq f'\left(x; \frac{\alpha(x' - x)}{\|\alpha(x' - x)\|}\right) = f'\left(x; \frac{(x' - x)}{\|x' - x\|}\right) \leq \frac{f(x') - f(x)}{\|x' - x\|}.$$

Hence, for every $x^* \in \partial f(x')$ we get

$$l(\lambda) \leq \frac{\langle x^*, x' - x \rangle}{\|x' - x\|} \leq \|x^*\|,$$

and the desired inequality follows.
(b) This is an immediate consequence of the monotonicity of l and r.

□

Theorem 4.2.2 *Let $f : \mathbb{R}^n \to \mathbb{R} \cup \{+\infty\}$ be lsc, proper, convex, and such that $f(x) > \inf f \implies x \in \mathrm{ri}\,\mathrm{dom}\,f$. Then the following statements are equivalent:*
(a) f is asymptotically well behaved.
(b) All stationary sequences $\{x_k\}$ with $f(x_k)$ bounded satisfy $f(x_k) \to \inf f$.
(c) $r(\lambda) > 0$ for all $\lambda > \inf f$.
(d) $t(\lambda) > 0$ for all $\lambda > \inf f$.
(e) $l(\lambda) > 0$ for all $\lambda > \inf f$.

Proof. The implication (a) $\implies$ (b) as well as the equivalence among (c), (d), and (e) are obvious from the definition of an awb function and the previous results, respectively. To prove (b) $\implies$ (c) we observe that otherwise there exists $\lambda > \inf f$ with $r(\lambda) = 0$, so that we can find sequences x_k and $x_k^* \in \partial f(x_k)$ with $x_k^* \to 0$ and $f(x_k) = \lambda > \inf f$, contradicting (b). The implication (c) $\implies$ (a) follows similarly. If (a) does not hold, we can find a stationary sequence that is not minimizing. Extracting a subsequence, we can find $\lambda > \inf f$ and sequences x_k and $x_k^* \in \partial f(x_k)$ such that $f(x_k) \geq \lambda$ and $x_k^* \to 0$. The alternative characterization of r given in Proposition 4.2.1 yields $r(\lambda) = 0$, contradicting (c). □

Next, we provide a simple sufficient condition that guarantees that a function is awb.

Proposition 4.2.2 *Let $f : \mathbb{R}^n \to \mathbb{R} \cup \{+\infty\}$ be an lsc, proper, convex function. Suppose that $0 \notin \mathrm{cl}\,\mathrm{dom}\,f^*$ or equivalently that there exists a vector d such that $f_\infty(d) < 0$. Then f is awb.*

Proof. Since $0 \notin \operatorname{cl} \operatorname{dom} f^*$, there exists $r > 0$ such that $\|c\| \le r \Rightarrow c \notin \operatorname{dom} f^*$. Since

$$\operatorname{dom} \partial f^* \subset \operatorname{dom} f^*, \quad c_n \in \partial f(x_n) \Longleftrightarrow x_n \in \partial f^*(c_n),$$

as a consequence it appears that there cannot exist a stationary sequence.

$\square$

We shall see now that a weakly coercive convex function is awb, but in this case we can say even more.

Proposition 4.2.3 *Let $h : \mathbb{R}^n \to \mathbb{R} \cup \{+\infty\}$ be a convex function with $0 \in \operatorname{ri} \operatorname{dom} h$ and let E be the affine hull of its domain and $h_E = h \circ P_E$, where P_E is the projection operator on E. Then for each sequence $u_k \to 0$ and $u_k^* \in \partial h(u_k)$ we have*

$$\operatorname{dist}(u_k^*, \partial h(0)) = \operatorname{dist}(P_E(u_k^*), \partial h_E(0)) \to 0 \text{ and } h^*(u_k^*) \to -h(0) = \inf h^*.$$

Proof. Since $\partial h(u_k)$ is nonempty, we must have $u_k \in E$, and from Proposition 3.2.3 we get $P_E(u_k^*) \in \partial h_E(u_k)$. From Proposition 1.4.3 the upper semicontinuity of $\partial h_E(\cdot)$ at $0 \in \operatorname{int} \operatorname{dom} h_E$ yields at once

$$\operatorname{dist}(u_k^*, \partial h(0)) = \operatorname{dist}(P_E(u_k^*), \partial h_E(0)) \to 0$$

as well as the boundedness of the sequence $P_E(u_n^*)$. Therefore, since $u_k \in E$, we have

$$h^*(u_k^*) + h(u_k) = \langle P_E(u_k^*), u_k \rangle \to 0,$$

and since h is continuous relative to $\operatorname{ri} \operatorname{dom} h$ and in particular at 0, we conclude that $h^*(u_k^*) \to -h(0)$, as claimed.

$\square$

Corollary 4.2.1 *Let $f : \mathbb{R}^n \to \mathbb{R} \cup \{+\infty\}$ be an lsc, proper, convex function and weakly coercive. Then f is awb. Moreover, if $\{x_k\}$ is a sequence such that $\operatorname{dist}(0, \partial_{\varepsilon_k} f(x_k)) \to 0$ where $\varepsilon_k \to 0^+$, then:*
(a) $f(x_k) \to \inf f$.
(b) $\operatorname{dist}(x_k, \operatorname{argmin} f) \to 0$.

Proof. Since f is weakly coercive, $0 \in \operatorname{ri} \operatorname{dom} f^*$, and when $\varepsilon = 0$ the conclusion of the corollary is an immediate consequence of Proposition 4.2.3.

Let us consider the general case and select $x_k^* \in \partial_{\varepsilon_k} f(x_k)$ with $x_k^* \to 0$. Using Theorem 4.1.5 we may find y_k and x_k such that $\|y_k - x_k\| \le \sqrt{\varepsilon_k}$ and $y_k^* \in \partial f(y_k)$ with $\|y_k^* - x_k^*\| \le \sqrt{\varepsilon_k}$. It follows that $y_k^* \to 0$ and so applying this result (in the case $\varepsilon_k = 0$) to the sequence y_k, we get

$\text{dist}(y_k, \text{argmin } f) \to 0$ and $f(y_k) \to \inf f$. From this it is clear that (a) must hold, since

$$\inf f \le f(x_k) \le f(y_k) - \langle x_k^*, y_k - x_k \rangle + \varepsilon_k,$$

and assertion (b) follows immediately.

$\square$

Dual Characterization of Well-Behaved Functions

We now consider dual characterizations for a given lsc proper convex function to be awb. In order to derive such characterizations we have to compute the conjugate of the positive part of a convex function and also the conjugate of the distance function to a closed convex set. The next results provide the necessary technical tools.

Lemma 4.2.3 *Let $f : \mathbb{R}^n \to \mathbb{R} \cup \{+\infty\}$ be a proper, lsc, convex function such that $r > \inf f > -\infty$ and let $C := \{x \mid f(x) \le r\}$. Then*

$$\sigma_C(d) = \min_{t \ge 0} t(f^*(t^{-1}d) + r) = \inf_{t > 0} t(f^*(t^{-1}d) + r), \ \forall d \ne 0,$$

where $0f^*(\frac{d}{0})$ stands for $(f^*)_\infty(d) = \lim_{t \to 0} t(f^*(t^{-1}d))$.

Proof. Define $g = f - r$ and let $L : (x, \alpha) \to x$ be the projection map. Then one can write $C = L(\text{epi } g \cap \mathbb{R}^n \times \mathbb{R}_-)$, and by Proposition 1.3.3(e) it follows that

$$\sigma_C(d) = \sigma_{\text{epi } g \cap (\mathbb{R}^n \times \mathbb{R}_-)}(d, 0).$$

Since $\inf g < 0$, one has $\text{epi } g \cap \text{int}(\mathbb{R}^n \times \mathbb{R}_-) \ne \emptyset$, and thus applying Corollary 3.6.2 we obtain

$$
\begin{aligned}
\sigma_C(d) &= (\sigma_{\text{epi } g} \square \sigma_{\mathbb{R}^n \times \mathbb{R}_-})(d, 0) \\
&= (\sigma_{\text{epi } g} \square \delta_{\{0\} \times \mathbb{R}_+})(d, 0) \\
&= \min_{s \ge 0} \sigma_{\text{epi } g}(d, -s) \\
&= \min_{s \ge 0} s g^*(s^{-1}d) = \min_{t \ge 0} t(f^*(t^{-1}d) + r),
\end{aligned}
$$

where the first part of the last equality uses Lemma 2.5.2.

$\square$

Lemma 4.2.4 *Let $f : \mathbb{R}^n \to \mathbb{R} \cup \{+\infty\}$ be a lsc, convex, proper function with $\inf f > -\infty$ and let $\lambda \in \mathbb{R}$. Then for all $y \in \mathbb{R}^n$ one has*

$$\left((f - \lambda)^+\right)^*(y) = \inf_{s \in]0,1]} s f^*\left(\frac{y}{s} + \lambda\right) = \min_{s \in [0,1]} s f^*\left(\frac{y}{s} + \lambda\right), \qquad (4.10)$$

where

$$f^+ = \sup(f, 0) \ \text{and} \ 0f^*\left(\frac{z}{0}\right) \ \text{stands for} \ (f^*)_\infty(z).$$

Proof. Since $(f - \lambda)^* = f^* + \lambda$, it is sufficient to prove (4.10) for $\lambda = 0$. For all $y \in \mathbb{R}^n$, using Lemma 2.5.2 one has

$$(f^+)^*(y) = \sigma_{\text{epi } f^+}(y, -1) = \sigma_{\text{epi } f \cap \mathbb{R}^n \times \mathbb{R}_+}(y, -1).$$

Since $\text{int}(\mathbb{R}^n \times \mathbb{R}_+) \cap \text{epi } f \neq \emptyset$, we apply Corollary 3.6.2, and noting that $\sigma_{\mathbb{R}^n \times \mathbb{R}_+}(\cdot) = \delta_{\{0\} \times \mathbb{R}_-}(\cdot)$, we obtain

$$(f^+)^*(y) = \sigma_{\text{epi } f} \square (\delta_{\{0\} \times \mathbb{R}_-})(y, -1) = \min_{t \leq 0} \sigma_{\text{epi } f}(y, -1 - t)$$

$$= \min_{s \in [0,1]} s f^*(\frac{y}{s}) = \inf_{s \in (0,1]} s f^*(\frac{y}{s}).$$

$\square$

Lemma 4.2.5 *Let C be a nonempty closed convex subset of $\mathbb{R}^n$. Then the conjugate* $\text{dist}(\cdot, C)^*$ *is given by the formula*

$$\text{dist}(\cdot, C)^*(y) = \sigma_C(y) \qquad \forall y \in \mathbb{B}, \tag{4.11}$$

where $\mathbb{B} = \{y : \|y\| \leq 1\}$.

Proof. Set $g(u) = \|u\|$. Then we have $\text{dist}(\cdot, C) = (\delta_C \square g)(\cdot)$. Furthermore, it can be easily seen that $g^* = \delta_\mathbb{B}$. Then using Proposition 3.6.1 it follows that $\text{dist}(\cdot, C)^* = \sigma_C + g^*$, which is equivalent to (4.11). $\square$

Theorem 4.2.3 *Let $f : \mathbb{R}^n \to \mathbb{R} \cup \{+\infty\}$ be an lsc, proper, convex function such that $f(x) > \inf f \Rightarrow x \in \text{ri dom } f$, and $\inf f > -\infty$. Then f is awb if and only if for any $\lambda > \inf f$ there exists $\alpha > 0$ such that*

$$((f - \lambda)^+)^*(y) = \sigma_{\text{lev}(f, \lambda)}(y) \qquad \forall y \in \alpha \mathbb{B}. \tag{4.12}$$

Proof. We know by Theorem 4.2.2 that f is awb if and only if $\forall \lambda > \inf f$ there exists $\alpha > 0$ such that

$$f(x) \geq \lambda + \alpha \, \text{dist}(x, \text{lev}(f, \lambda)), \qquad x \notin \text{lev}(f, \lambda),$$

which is equivalent to

$$(f - \lambda)^+ \geq \alpha \, \text{dist}(\cdot, \text{lev}(f, \lambda)).$$

Dually, and using Lemma 4.2.5, this is equivalent to

$$((f - \lambda)^+)^* \leq (\alpha \, \text{dist}(x, \text{lev}(f, \lambda)))^* = \sigma_{\text{lev}(f, \lambda)} + \delta_{\alpha \mathbb{B}}.$$

Using Lemma 4.2.3 and Lemma 4.2.4 this is equivalent to

$$\min_{s \in [0,1]} s f^* \left(\frac{y}{s} + \lambda \right) \leq \min_{s \geq 0} s f^* \left(\frac{y}{s} + \lambda \right) \qquad \forall y \in \alpha \mathbb{B},$$

which obviously implies

$$\min_{s\in[0,1]} sf^*\left(\frac{y}{s}+\lambda\right) = \min_{s\geq 0} sf^*\left(\frac{y}{s}+\lambda\right) \qquad \forall y \in \alpha\mathbb{B},$$

i.e., exactly formula (4.12).

□

As we have seen, the position of the origin relative to the domain of f^* is of main importance. When $0 \notin \mathrm{cl\,dom}\, f^*$ or when $0 \in \mathrm{ri\,dom}\, f^*$, then f is awb. In fact, when 0 belongs to the relative boundary, this requires a particular analysis. The correct notion is the following.

Definition 4.2.3 *Let $f : \mathbb{R}^n \to \mathbb{R}\cup\{+\infty\}$ be an lsc, proper, convex function. The domain of its conjugate $\mathrm{dom}\, f^*$ is said to be locally conical at 0 if there exists $\alpha > 0$ such that $\mathbb{R}_+ \mathrm{dom}\, f^* \cap \alpha\mathbb{B} \subset \mathrm{dom}\, f^*$.*

Theorem 4.2.4 *Let $f : \mathbb{R}^n \to \mathbb{R}\cup\{+\infty\}$ be an lsc, convex, proper function with $\inf f > -\infty$. Assume that $\mathrm{dom}\, f^*$ is locally conical at 0 and that f^* is bounded above on the intersection of a neighborhood of 0 and $\mathrm{dom}\, f^*$. Then f is awb.*

Proof. Let $\alpha > 0$ be such that $\mathbb{R}_+ \mathrm{dom}\, f^* \cap \alpha\mathbb{B} \subset \mathrm{dom}\, f^*$, and $c > f^*(0)$ such that $f^*(y) \leq c \ \forall y \in \mathrm{dom}\, f^* \cap \alpha\mathbb{B}$. Given $\varepsilon \in \,]0,1[$ and $y \in \mathrm{dom}\, f^* \cap \varepsilon\alpha\mathbb{B}$, one has $\varepsilon^{-1}y \in \mathbb{R}_+ \mathrm{dom}\, f^* \cap \alpha\mathbb{B} \subset \mathrm{dom}\, f^* \cap \alpha\mathbb{B}$, so that

$$f^*(y) \leq (1-\varepsilon)f^*(0) + \varepsilon f^*(\varepsilon^{-1}y) \leq (1-\varepsilon)f^*(0) + \varepsilon c,$$

yielding $\limsup_{y\to 0, y\in\mathrm{dom}\, f^*} f^*(y) \leq f^*(0)$.
Since f^* is lsc, we get $\lim_{y\to 0, y\in\mathrm{dom}\, f^*} f^*(y) = f^*(0)$. As a consequence we have for $\delta > 1$,

$$\lim_{y\to 0, y\in\mathrm{dom}\, f^*} \delta f^*(y) - f^*(\delta y) = (\delta - 1)f^*(0). \qquad (4.13)$$

Now let $\{x_k\}$ be a stationary sequence and let us prove that $\{x_k\}$ is minimizing. Let $y_k \in \partial f(x_k)$ with $y_k \to 0$. Then $x_k \in \partial f^*(y_k)$, and we have

$$\langle x_k, (\delta - 1)y_k\rangle + f^*(y_k) \leq f^*(\delta y_k), \quad -f^*(0) \leq f(x_k) = \langle x_k, y_k\rangle - f^*(y_k).$$

Thus we get

$$\begin{aligned}-(\delta - 1)f^*(0) \leq (\delta - 1)f(x_k) &= \langle x_k, (\delta-1)y_k\rangle - (\delta-1)f^*(y_k)\\ &\leq f^*(\delta y_k) - \delta f^*(y_k),\end{aligned}$$

so that

$$0 \leq (\delta - 1)(f(x_k) + f^*(0)) \leq f^*(\delta y_k) - \delta f^*(y_k) + (\delta - 1)f^*(0),$$

and then using (4.13) it follows that $\lim_{k\to\infty} f(x_k) = -f^*(0) = \inf f$. □

4.3   Error Bounds for Convex Inequality Systems

An initial motivation for studying error bounds in mathematical programming arose from a practical consideration in the implementation of numerical methods for solving optimization problems. Furthermore, on the theoretical side, error bounds are playing a key role in the convergence rate analysis of these numerical methods. Given a subset $\mathcal{F}$ of $\mathbb{R}^n$, an error bound is an inequality that bounds the distance from an arbitrary vector to $\mathcal{F}$ in terms of some easily computable *residual* function ρ, that is, a function that satisfies $\rho(x) \geq 0$, $\forall x \in \mathbb{R}^n$, and $\rho(x) = 0$ if and only if $x \in \mathcal{F}$. Thus formally an error bound consists in establishing that for some positive constant θ one has

$$\text{dist}(x, \mathcal{F}) \leq \theta \rho(x), \quad \forall x \in B \subset \mathbb{R}^n,$$

for some (often bounded) subset B of $\mathbb{R}^n$, in which case the error bound is called local, while when $B = \mathbb{R}^n$, the error bound is called a *global* error bound for the set $\mathcal{F}$. In fact, we are also interested in finding global error bounds that are *sharp*, i.e., to identify the best possible value of the constant θ, and this will be the main objective of this section.

The origin of error bounds takes its root in the well-known Hoffman's result for a polyhedral set $\mathcal{F} := \{x \in \mathbb{R}^n : Ax \leq b\}$, with given data $A \in \mathbb{R}^m \times \mathbb{R}^n$, $b \in \mathbb{R}^m$, which asserts that there exists $\theta > 0$ depending only on the data A and such that

$$\text{dist}(x, \mathcal{F}) \leq \theta \|(Ax - b)^+\|, \quad \forall x \in \mathbb{R}^n.$$

Note that in this result the best possible value of θ is not known. As we shall see later, this result will be obtained as a byproduct of our analysis, which includes sharp error bounds for the more general problem described by polyhedral sets and convex systems. Thus, throughout this section we concentrate on error bounds for a set $\mathcal{F}$ described by a finite number of convex and affine inequalities. Let r be a nonnegative integer, and we suppose that for $i = r+1, \ldots, m$ we have $f_i(x) = \langle a_i, x \rangle - \alpha_i$. When $r > 0$ we assume that $f_i : \mathbb{R}^n \to \mathbb{R} \cup \{+\infty\}, i = 1, \ldots, r$ are lsc, proper, convex functions with dom f_i open. We will also use the following notation. Let

$$F(x) = (f_1(x), \ldots, f_m(x))^t, \quad F^+(x) = (f_1^+(x), \ldots, f_m^+(x))^t,$$

with $f_i^+(x) = \max(f_i(x), 0)$, $i = 1, \ldots, m$, and define

$$C_i = \{x : f_i(x) \leq 0\}, \qquad C = D \cap_{i=r+1}^m C_i, \qquad (4.14)$$

with $D = \mathbb{R}^n$ if $r = 0$, and $D = \cap_{i=1}^r C_i$ otherwise. In the following, $||| \cdot |||$ denotes a monotonic norm on $\mathbb{R}^n$, that is to say which satisfies

$$|x_i| \leq |y_i| \quad i = 1, \ldots, m \quad \Longrightarrow \quad |||x||| \leq |||y|||.$$

In particular, the p-norms $\|\cdot\|_p$ with $1 \le p \le +\infty$ are monotonic norms. Our aim is to derive error bounds for the set C, and if possible sharp, i.e., to find a constant $\theta > 0$, the "best" possible, such that

$$\mathrm{dist}(x, C) \le \theta |||F^+(x)||| \qquad \forall x \in \mathbb{R}^n.$$

To this end, we shall assume Slater's condition for C; i.e., if $r > 0$, then there exists $x_0 \in C$ such that $f_i(x_0) < 0$ for $i = 1, \ldots, r$. We know (cf. Corollary 3.6.4, Proposition 1.2.21, and Proposition 1.2.22) that under this condition we have for each $x \in C$,

$$N_C(x) = \left\{ d \mid \exists \lambda_i \ge 0, c_i \in \partial f_i(x),\ i \in I(x) \text{ with } d = \sum_{i \in I(x)} \lambda_i c_i \right\} \quad (4.15)$$

with $I(x) = \{i \mid f_i(x) = 0\}$ and $c_i = a_i$ for $i \ge r+1$. For $y \in C$ we denote by $(f_i^+)'(y; d)$ the directional derivative of f_i^+ at y in the direction d. We define,

$$\begin{aligned}
M(y) &= \{d \in N_C(y) \text{ such that } ||d|| = 1\}, \\
(F_a^+)'(y; d) &= ((f_i^+)'(y; d) \mid i \in I(y)),
\end{aligned}$$

and the following real numbers:

$$l = \inf_{x \notin C} \frac{|||F^+(x)|||}{\mathrm{dist}(x, C)}, \qquad t = \inf_{y \in \mathrm{bd}(C)} \inf_{d \in M(y)} |||(F_a^+)'(y; d)|||.$$

Lemma 4.3.1 *Let C be a nonempty closed convex set in $\mathbb{R}^n$.*
(a) For $x \notin C$ and $y \in \mathrm{bd}(C)$, one has $y = P_C(x)$ if and only if $\frac{x-y}{||x-y||} \in M(y)$.
(b) Let $y \in \mathrm{bd}(C)$ and $d \in M(y)$. Then for each $\theta > 0$ we have

$$x(\theta) = y + \theta d \notin C, \qquad \mathrm{dist}(x(\theta), C) = \theta.$$

Proof. (a) Obviously, if $y = P_C(x)$, then $y \in \mathrm{bd}\, C$. Furthermore, $y = P_C(x)$ minimizes $u \to ||x - u|| + \delta(u|C)$, and then (a) is a direct consequence of the optimality conditions given in Theorem 4.1.1, while (b) is an immediate consequence of Lemma 4.2.1. $\qquad\square$

In order to obtain an error bound we need to introduce some kind of asymptotic constraint qualification for the system of convex inequalities.

Definition 4.3.1 *Let C be defined by the system of convex inequalities given in (4.14).*
(a) Let $\mathcal{S}$ be the set of sequences $\{x_k\}$ in C with $||x_k|| \to \infty$ and for which there exists a nonempty subset I of $\{1, \ldots, m\}$ such that for k sufficiently

large $f_i(x_k) = 0 \; \forall i \in I$. Let $\{x_k\} \subset \mathcal{S}$ and $c_{i,k} \in \partial f_i(x_k) \; \forall k \in \mathbb{N} \; \forall i \in I$. Define the following index sets:

$$I_b := \{i \in I \mid \{c_{i,k}\} \text{ is bounded}\}, \qquad I_{ub} := I \setminus I_b.$$

Then the sequence $\{x_k\}$ is said to satisfy the property (P) if for each subsequence $\{n_k\}$ of $\{k\}$ such that

$$\{c_{i,n_k}\} \text{ converges if } i \in I_b, \quad \left\{\frac{c_{i,n_k}}{c_{i,n_k}}\right\} \text{ converges if } i \in I_{ub},$$

*with limit denoted by $g_i \; \forall i$, one has necessarily $0 \notin \operatorname{conv}\{g_i \mid i \in I\}$.
(b) We say that C satisfies the asymptotic constraint qualification (A.C.Q.) if either $\mathcal{S}$ is empty or each sequence $\{x_k\} \subset \mathcal{S}$ satisfies property (P).*

Note that an obvious situation where the A.C.Q. is automatically satisfied is when C is a compact set.

Theorem 4.3.1 *Let $f_i : \mathbb{R}^n \to \mathbb{R} \cup \{+\infty\}$ be lsc and convex with $\operatorname{dom} f_i$ open for $i = 1, \ldots, r$, and let $f_i(x) = \langle a_i, x \rangle - \alpha_i$ for $r + 1 \le i \le m$. Suppose that Slater's condition holds for the set C defined in (4.14). Then*

$$l := \inf_{x \notin C} \frac{|||F^+(x)|||}{\operatorname{dist}(x, C)} = \inf_{y \in \operatorname{bd}(C)} \inf_{d \in M(y)} |||(F_a^+)'(y; d)||| := t.$$

If in addition A.C.Q. holds, then $t > 0$.

Proof. Let $x \notin C$ and $y = P_C(x)$. Since f_i^+ is convex and since $f_i^+(y) = 0 \; \forall i \in I(y)$ we get $f_i^+(x) \ge (f_i^+)'(y; x - y)$. Since $\operatorname{dist}(x, C) = \|x - y\|$, it follows that

$$\frac{f_i^+(x)}{\operatorname{dist}(x, C)} \ge (f_i^+)'\left(y; \frac{x - y}{\|x - y\|}\right) \qquad \forall i \in I(y).$$

Furthermore, we have

$$(f_i^+)'(y; d) = (f_i'(y; d))^+. \tag{4.16}$$

Since $||| \cdot |||$ is monotonic, it follows that

$$\frac{|||F^+(x)|||}{\operatorname{dist}(x, C)} \ge \left|\left|\left|(F_a^+)'\left(y; \frac{x - y}{\|x - y\|}\right)\right|\right|\right|.$$

Then by Lemma 4.3.1 we obtain

$$\frac{|||F^+(x)|||}{\operatorname{dist}(x, C)} \ge t,$$

which implies that $l \geq t$. Now let $\{y_k\}, \{d_k\}$ be sequences such that $y_k \in \mathrm{bd}\, C, d_k \in M(y_k)$ and

$$t = \lim_{k \to \infty} |||\{(f_i^+)'(y_k; d), i \in I(y_k)\}|||. \tag{4.17}$$

Let $\{\theta_k\}$ be a sequence of positive reals converging to 0 and set $x_k(\theta) = y_k + \theta d_k$ for $\theta > 0$. By Lemma 4.3.1, $x_k(\theta) \notin C$ and $\mathrm{dist}(x_k(\theta), C) = \theta$. Since $f_i^+(y_k) = 0$ for $i \in I(y_k)$, by definition of the directional derivative we have

$$\frac{f_i^+(y_k + \theta d_k)}{\theta} = (f_i^+)'(y_k; d_k) + \varepsilon_k(\theta) \text{ with } \lim_{\theta \to o} \varepsilon_k(\theta) = 0.$$

Let $j(k)$ be such that $\varepsilon_k(\theta_{j(k)}) \to 0$ and set $x_k = x_k(\theta_{j(k)})$. Then $f_i(x_k) < 0$ for $i \notin I(y_k)$ and

$$\frac{f_i^+(x_k)}{\mathrm{dist}(x_k, C)} = (f_i^+)'(y_k; d_k) + \varepsilon_k(\theta_{j(k)}), \qquad i \in I(y_k),$$

and then from (4.17) it follows that

$$\lim_{k \to \infty} \frac{|||F^+(x_k)|||}{\mathrm{dist}(x_k, C)} = t,$$

and we have proved $l = t$. Suppose now that A.C.Q. holds and let us prove that $t > 0$. Let

$$I_0 := \{i \geq r + 1 \mid \langle a_i, x \rangle = \alpha_i \quad \forall x \in C\}, \quad I_1 := \{i \geq r + 1 \mid i \notin I_0\}.$$

Observing that

$$C = \{x \mid f_i(x) \leq 0\ i = 1, \ldots, r, \quad \langle a_i, x \rangle \leq \alpha_i\ i \in I_1, \ \langle a_i, x \rangle = \alpha_i\ i \in I_0\},$$

without loss of generality (deleting the other equations if necessary), we can suppose that the vectors $\{a_i \mid i \in I_0\}$ are linearly independent. We first prove that for each $y \in C$, and each subset $I \subset I(y)$ we cannot have

$$0 = \sum_{i \in I} \lambda_i c_i, \quad \lambda_i > 0, \quad c_i \in \partial f_i(y) \ \forall i \in I. \tag{4.18}$$

Three cases are possible:
(i) $I = I_0$. In this case (4.18) is impossible, since the vectors $\{a_i \mid i \in I_0\}$ are linearly independent.
(ii) $I = I_0 \cup I_1$ with I_1 nonempty. Then for each $i \in I_1$ there exists $x_i \in C$ such that $\langle a_i, x_i \rangle < \alpha_i$. Let $|I_1|$ be the number of elements of I_1 and let $\bar{x} = |I_1|^{-1} \sum_{i \in I_1} x_i$. Then $\bar{x} \in C$, and we have

$$\langle a_i, \bar{x} \rangle < \alpha_i \qquad \forall i \in I_1.$$

As a consequence, we obtain from (4.18)

$$0 = \sum_{i \in I_0 \cup I_1} \lambda_i \langle c_i, \bar{x} - y \rangle \leq \sum_{i \in I_0 \cup I_1} \lambda_i (\langle a_i, \bar{x} \rangle - \alpha_i) < 0,$$

which is impossible.

(iii) There exists $i \leq r$,with $i \in I$. Then using Slater's condition with (4.18) we obtain

$$0 = \sum_{i \in I} \lambda_i \langle c_i, x_0 - y \rangle \leq \sum_{i \in I} \lambda_i f_i(x^0) < 0,$$

which again is impossible.

Suppose now that $t = 0$. Then there exist sequences $\{y_k\} \in \mathrm{bd}\, C, \{d_k\} \in M(y_k)$ such that $\lim_{k \to \infty} (f_i^+)'(y_k; d_k) = 0$ for $i \in I(y_k)$. Without loss of generality we can suppose that $I(y_k) = I$ for each $k \in \mathbb{N}$. Then using (4.16) we get

$$\limsup_{k \to \infty} f_i{}'(y_k; d_k) \leq 0 \qquad \forall i \in I. \tag{4.19}$$

Since $d_k \in M(y_k)$, there exist $\lambda_i^k \geq 0$, $c_{i,k} \in \partial f_i(y_k)$, $i \in I$, such that $d_k = \sum_{i \in I} \lambda_i^k c_{i,k}$.

Without loss of generality we can suppose that

$$\lim_{k \to \infty} \|c_{i,k}\| = \infty, \quad \lim_{k \to \infty} \frac{c_{i,k}}{\|c_{i,k}\|} = g_i \text{ for } i \in I_{ub}, \lim_{k \to \infty} c_{i,k} = g_i \text{ for } i \in I_b.$$

Let $t_k = \{\lambda_i^k, \ i \in I_b, \ \lambda_i^k \|c_{i,k}\|, \ i \in I_{ub}\}$ and set

$$\bar{\lambda}_i^k = \frac{\lambda_i^k}{\|t_k\|_1} \text{ for } i \in I_b, \qquad \bar{\lambda}_i^k = \frac{\lambda_i^k \|c_{i,k}\|}{\|t_k\|_1} \text{ for } i \in I_{ub}.$$

Since $\bar{\lambda}_i{}^k \geq 0$, $\sum_{i \in I} \bar{\lambda}_i^k = 1$, it follows that $\{\bar{\lambda}_i^k\}$ is bounded, and again without loss of generality we can suppose that the sequence $\{\bar{\lambda}_i^k\}$ converges to $\bar{\lambda}_i \geq 0$ with $\sum_{i \in I} \bar{\lambda}_i = 1$. Now two cases are possible, depending on the boundedness of the sequence $\{y_k\}$:

(i) Suppose that the sequence $\{y_k\}$ is bounded. Then since $\mathrm{dom}\, f_i$ is open, the sequences $\{c_{i,k}\}, i \in I$, are bounded, and there exist subsequences $\{y_{k_l}\}$, $\{c_{i,k_l} : \ i \in I\}$, $\{d_{k_l}\}$ converging respectively to $y \in C$, $c_i \in \partial f_i(y), i \in I \subset I(y)$, and d with $\|d\| = 1$. Furthermore, I_{ub} is empty, and we cannot have $\lim_{l \to \infty} \|t_{k_l}\| = \infty$. Indeed, we have

$$\frac{d_{k_l}}{\|t_{k_l}\|} = \sum_{i \in I_b} \bar{\lambda}_i^{k_l} c_{i,k_l} \to \sum_{i \in I_b} \bar{\lambda}_i g_i,$$

so that

$$0 = \sum_{i \in I_b} \bar{\lambda}_i c_i \text{ with } c_i \in \partial f_i(y), \qquad \sum_{i \in I_b} \bar{\lambda}_i = 1, \ \bar{\lambda}_i \geq 0,$$

which is impossible from (4.18). Then we can suppose without loss of generality that the sequences $\{\lambda_i^{k_l}\}, i \in I$, are bounded and converge to some $\lambda_i \geq 0$ such that

$$d = \sum_{i \in I} \lambda_i c_i, \qquad \|d\| = 1. \tag{4.20}$$

From (4.19) we have

$$\limsup_{k \to \infty} \langle c_{i,k}, d_k \rangle \leq 0 \qquad \forall i \in I.$$

Passing to the limit we obtain $\langle c_i, d \rangle \leq 0$, $\forall i \in I$, and from (4.20) we obtain $\|d\| = 0$, which is impossible.

(ii) Suppose now that the sequence $\{y_k\}$ is unbounded. Then we can write

$$\frac{d_k}{\|t_k\|} = \sum_{i \in I} \bar{\lambda}_i^k d_{i,k}, \text{ with } d_{i,k} = c_{i,k} \text{ if } i \in I_b \text{ and } d_{i,k} = \frac{c_{i,k}}{\|c_{i,k}\|} \text{ otherwise.} \tag{4.21}$$

Let us prove that the sequence $\{t_k\}$ is bounded. In the contrary case, without loss of generality we can suppose that $\lim_{k \to \infty} \|t_k\| = \infty$, and passing to the limit in (4.21) we obtain $0 = \sum_{i \in I} \bar{\lambda}_i g_i$, which is impossible from A.C.Q.

Now from (4.19) we have

$$\limsup_{k \to \infty} \langle c_{i,k}, d_k \rangle \leq 0 \quad \forall i \in I_b, \qquad \limsup_{k \to \infty} \left\langle \frac{c_{i,k}}{\|c_{i,k}\|}, d_k \right\rangle \leq 0 \quad \forall i \in I_{ub},$$

from which it follows that

$$\langle g_i, d \rangle \leq 0 \ \forall i \in I, \tag{4.22}$$

where d is any cluster point of the sequence $\{d_k\}$ and satisfies $\|d\| = 1$. Since the sequence $\{\|t_k\|_1\}$ is bounded, we can suppose without loss of generality that it converges to some $\alpha \geq 0$. It can be easily seen that $\alpha \neq 0$; otherwise, by the definition of d_k, passing to the limit we would obtain a vector $d = 0$, with $\|d\| = 1$, which is impossible. Then from (4.21), passing to the limit we obtain $\alpha^{-1} d = \sum_{i \in I} \bar{\lambda}_i g_i$. Using (4.22), it follows that $d = 0$, which is impossible. $\qquad \square$

As a corollary we obtain the following result, which concerns two important cases, and in particular allows for refining Hoffman's error bound, which is now shown to be sharp.

Corollary 4.3.1 *Let $f_i : \mathbb{R}^n \to \mathbb{R} \cup \{+\infty\}$ be lsc and convex with dom f_i open for $i = 1, \ldots, p$, and let $C = \{x \mid f_i(x) \leq 0, \ i = 1, \ldots, p\}$. If either one of the following conditions is satisfied:*
(a) C is bounded and Slater's condition holds,
(b) C is a polyhedral set, i.e., $f_i(x) = \langle a_i, x \rangle - \alpha_i \quad \forall i$,
then $l = t$ and $t > 0$; i.e., we have a sharp error bound.

Proof. Part (a) is exactly given by Theorem 4.3.1. For part (b), when C is a polyhedral set, as remarked in the proof of Theorem 4.3.1, we can suppose that the vectors $\{a_i \mid i \in I_0\}$ are linearly independent. In this case A.C.Q. is equivalent to saying that for each sequence $\{x_k\}$ with $\|x_k\| \to \infty$ one has $0 \notin \text{conv}\{a_i \mid i \in I\}$. But as seen in the proof of Theorem 4.3.1, this cannot occur. $\square$

As a final result for such kinds of bounds, we give now a simpler sufficient condition for the above set to satisfy a global error bound by invoking Theorem 4.2.2. However, the error bound will not be sharp in that case.

Proposition 4.3.1 *Let $f_i : \mathbb{R}^n \to \mathbb{R} \cup \{+\infty\}$ be lsc and convex with $\text{dom } f_i$ open for $i = 1, \ldots, p$, and let $C = \{x \mid f_i(x) \leq 0, \ i = 1, \ldots, p\}$. Suppose that Slater's condition holds and there exists a scalar $\delta > 0$ such that*

$$\forall \bar{x} \in \text{bd } C, \quad \exists \bar{z} \in \text{int } C \ \text{ such that } \ \frac{\|\bar{x} - \bar{z}\|}{\min_{1 \leq i \leq p}\{-f_i(\bar{z})\}} \leq \delta.$$

Then there exists $\theta > 0$ such that

$$\text{dist}(x, C) \leq \theta \max_{1 \leq i \leq p} f_i^+(x) \quad \forall x \in \mathbb{R}^n.$$

Proof. Let $f(x) = \max_{1 \leq i \leq p} f_i(x)$. Then one can write

$$C = \{x \mid f(x) \leq 0\}, \quad \text{int } C = \{x \mid f(x) < 0\}, \quad \text{bd } C = \{x \mid f(x) = 0\}.$$

We claim that our assumption implies that $0 \notin \text{cl}(\partial f(f^{-1}(0)))$, and thus the desired error bound follows from Theorem 4.2.2. Assume that the claim is false. Then there exist sequences $\{x_k\}$ and $\{c_k\}$ such that

$$c_k \in \partial f(x_k), \quad f(x_k) = 0, \quad \lim_{k \to \infty} c_k = 0.$$

By the assumptions there exists $z_k \in \text{int } C$ such that

$$\|x_k - z_k\| \leq \delta \min_{1 \leq i \leq p}\{-f_i(z_k)\},$$

or equivalently, $-f(z_k) \geq \delta^{-1}\|x_k - z_k\|$. Since $c_k \in \partial f(x_k)$, we have

$$f(z_k) = f(z_k) - f(x_k) \geq \langle c_k, z_k - x_k \rangle \geq -\|c_k\| \|z_k - x_k\|,$$

which implies that $\|c_k\| \geq \delta^{-1} \ \forall k$, a contradiction. $\square$

Definition 4.3.2 *Let C be defined by (4.14). Then C is said to be H-metrically regular if there exist $M > 0$, $\gamma \in \,]0, 1]$ such that*

$$\text{dist}(x, C) \leq M r(x) \quad \forall x \in \mathbb{R}^n$$

with

$$r(x) = \max_{1 \le i \le m} \max \left\{ (f_i^+(x), f_i^{+\gamma}(x)) \right\}.$$

The H-metric regularity stipulates that error bounds of Holder type hold for C. When $\gamma = 1$, this means that a usual error bound with $M > 0$ exists for C. In this case we have Lipschitz type error bounds. We shall say that C is L-metrically regular.

This concept is of particular importance, since as we shall see in the next section it ensures in the constrained case that asymptotically feasible stationary sequences or approximate Karush–Kuhn–Tucker sequences are minimizing sequences. Furthermore, this concept is particularly useful because not only does it enlarge the concept of L-metric regularity, but in addition it ensures in the quadratic case that C is H-metrically regular with $\gamma = \frac{1}{2}$.

4.4   Stationary Sequences in Constrained Minimization

In this section we are back to the convex constrained optimization problem

$$(P) \qquad m = \inf\{f_0(x) | x \in C\}$$

with

$$C = \{x \in \mathbb{R}^n \mid f_i(x) \le 0 \quad i = 1, \ldots, m\},$$

where for $i = 0, \ldots, m$, the functions $f_i : \mathbb{R}^n \to \mathbb{R} \cup \{+\infty\}$ are lsc convex proper functions. We begin with some basic definitions.

Definition 4.4.1 *Let $f = f_0 + \delta_C$. A sequence $\{x_k\}$ is said to be:*
(a) asymptotically residual feasible(arf) if

$$\limsup_{k \to \infty} f_i(x_k) \le 0 \text{ for each } i = 1, \ldots, m;$$

(b) asymptotically feasible (af) if $\lim_{k \to \infty} \text{dist}(x_k, C) = 0$;
(c) asymptotically residual feasible stationary (arfs) if $\{x_k\}$ is arf and if the projected sequence $\{\bar{x}_k\}$ ($\bar{x}_k = P_C(x_k)$) is stationary, i.e., if there exists $a_k \in \partial f(\bar{x}_k)$ with $a_k \to 0$.

As a consequence of the concept of H-metric regularity, it follows that an arf sequence is af when C is H-metrically regular.
We recall that a function $F : \mathbb{R}^n \to (\mathbb{R} \cup \{+\infty\})^m$ is uniformly continuous near a given sequence $\{x_k\}$ if for every $\varepsilon > 0$ there exists $\delta > 0$ such that for each k we have

$$\|y - x_k\| \le \delta \Rightarrow \|F(y) - F(x_k)\| \le \varepsilon.$$

The next theorem gives conditions under which a sequence $\{x_k\}$ is asymptotically feasible and minimizing with respect to problem (P).

Theorem 4.4.1 *Let $f_i : \mathbb{R}^n \to \mathbb{R} \cup \{+\infty\}$, $i = 0, \ldots, m$, be a collection of proper lsc convex functions with $\operatorname{dom} f_0$ open and $\operatorname{dom} f_0 \cap C \neq \emptyset$. Suppose that C and $L(\lambda) := \{x \in C \mid f_0(x) \leq \lambda\}$ with $\lambda > m$ are H-metrically regular. Let $\{x_k\}$ be an arfs sequence. Then if f_0 is uniformly continuous near $\{x_k\}$, the sequence $\{x_k\}$ is an af minimizing sequence; i.e.,*

$$\lim_{k\to\infty} \operatorname{dist}(x_k, C) = 0 \quad and \quad \lim_{k\to\infty} f_0(x_k) = m. \tag{4.23}$$

Proof. Since C is H-metrically regular and $\{x_k\}$ is an arf sequence, it follows that $\{x_k\}$ satisfies $\lim_{k\to\infty} \operatorname{dist}(x_k, C) = 0$ and we have only to prove $\lim_{k\to\infty} f_0(x_k) = m$. Let $\bar{x}_k = P_C(x_k)$. Then since $\{x_k\}$ is an arfs sequence, it follows by Definition 4.3.1(c) that there exists $a_k \in \partial f(\bar{x}_k)$ with $a_k \to 0$. The sequence $\{\bar{x}_k\}$ is feasible, and we want to prove first that it is a minimizing sequence. Suppose the contrary. Then there exists $\lambda > m$ such that $\liminf_{k\to\infty} f_0(\bar{x}_k) > \lambda$. Let $y_k = P_{L(\lambda)}(\bar{x}_k)$. Such an y_k exists, since $L(\lambda)$ is nonempty. Since $\operatorname{dom} f_0$ is open and since $\bar{x}_k \in C$, we have $f_0(y_k) = \lambda$. Since f_0 is convex, then

$$\lambda = f_0(y_k) \geq f_0(\bar{x}_k) + \langle c_k, y_k - \bar{x}_k \rangle \quad \forall c_k \in \partial f_0(\bar{x}_k). \tag{4.24}$$

Now by Theorem 4.1.1 it follows that there exist $c_k \in \partial f_0(\bar{x}_k)$, $b_k \in N_C(\bar{x}_k)$ such that $a_k = c_k + b_k$. But since $b_k \in N_C(\bar{x}_k)$ and $y_k \in C$, it follows that $\langle y_k - \bar{x}_k, c_k - a_k \rangle \geq 0$, and then from (4.24) we get

$$f_0(\bar{x}_k) - \lambda \leq -\langle a_k, y_k - \bar{x}_k \rangle.$$

Using the Cauchy–Schwarz inequality and recalling that $L(\lambda)$ is assumed H-metrically regular, it follows that there exist constants $M > 0$, $\gamma \in]0, 1]$ such that

$$f_0(\bar{x}_k) - \lambda \leq M \|a_k\| \max\{f_0(\bar{x}_k) - \lambda, \ (f_0(\bar{x}_k) - \lambda)^\gamma\}.$$

Dividing by $f_0(\bar{x}_k) - \lambda$, which is positive, we obtain

$$1 \leq M \|a_k\| \max\{1, (f_0(\bar{x}_k) - \lambda)^{\gamma-1}\}.$$

Since $a_k \to 0$ and $\liminf_{k\to\infty} f_0(\bar{x}_k) > \lambda$, we obtain a contradiction from the above expression by passing to the limit when $k \to \infty$. Then we have

$$\bar{x}_k \in C, \quad \lim_{k\to\infty} f_0(\bar{x}_k) = m.$$

Now since $\|x_k - \bar{x}_k\| = \operatorname{dist}(x_k, C) \to 0$ and since f_0 is uniformly continuous near x_k, it follows from the definition that $\liminf_{k\to\infty} f_0(x_k) = m$.   □

Remark 4.4.1 If the functions f_i are quadratic and convex, then it can be proved that C and $L(\lambda)$ are H-metrically regular. If f_0 is linear, then f_0 is uniformly continuous.

In many algorithms, in fact, one often computes an arf sequence $\{x_k\}$ and a sequence of multipliers $\mu^k = (\mu_1^k, \ldots, \mu_m^k)$ that satisfies approximatively the Karush–Kuhn–Tucker conditions, i.e.,

$$\mu_i^k \geq 0, \quad \lim_{k\to\infty} \mu_i^k f_i(x_k) = 0 \ \forall i = 1, \ldots, m, \qquad (4.25)$$

$$\lim_{k\to\infty} \nabla f_0(x_k) + \sum_{i=1}^{m} \mu_i^k \nabla f_i(x_k) = 0. \qquad (4.26)$$

In general, with Slater's condition the algorithms generate a sequence $\{\mu^k\}$ that is bounded. Such typical situations arise in particular with augmented Lagrangian methods. The following theorem tells us when we can expect that such a sequence is minimizing.

Theorem 4.4.2 *Let $f_i : \mathbb{R}^n \to \mathbb{R} \cup \{+\infty\}$ be lsc convex proper functions with $\mathrm{dom}\, f_i$ open and f_i differentiable on their domains, for each $i = 0, \ldots, m$. We suppose that C and $L(\lambda) := \{x \in C : f_0(x) \leq \lambda\}$ with $\lambda > m$, are H-metrically regular. Let $\{x_k\}$ be an arf sequence and $\{\mu^k\}$ a bounded sequence of multipliers satisfying (4.25)–(4.26). Then if for each $i = 0, \ldots, m$, f_i, ∇f_i are uniformly continuous near $\{x_k\}$, then $\{x_k\}$ is an af minimizing sequence.*

Proof. The proof is very similar to the proof of Theorem 4.4.1. Let $\{x_k\}$ be an arf sequence. Since C is H-metrically regular, it follows that $\{x_k\}$ is arf and we have only to prove that $\lim_{k\to\infty} f_0(x_k) = m$. Let $\bar{x}_k = P_C(x_k)$. Then since f_i, ∇f_i are uniformly continuous near $\{\bar{x}_k\}$ and since $\{\mu_i^k\}$ is bounded satisfying (4.25)–(4.26) we have

$$\varepsilon_k := \sum_{i=1}^{m} \mu_i^k f_i(\bar{x}_k) \to 0, \quad a_k := \nabla f_0(\bar{x}_k) + \sum_{i=1}^{m} \mu_i^k \nabla f_i(\bar{x}_k) \to 0.$$

The sequence $\{\bar{x}_k\}$ is feasible, and we want to prove first that it is a minimizing sequence. Suppose the contrary, then there exists $\lambda > m$ such that $\liminf_{k\to\infty} f_0(\bar{x}_k) \geq \lambda$. Let $y_k = P_{L(\lambda)}(\bar{x}_k)$. Such a y_k exists, since $L(\lambda)$ is nonempty. Since $\mathrm{dom}\, f_0$ is open and since $\bar{x}_k \in C$, we have $f_0(y_k) = \lambda$, and by convexity it follows that

$$\lambda = f_0(y_k) \geq f_0(\bar{x}_k) + \langle \nabla f_0(\bar{x}_k), y_k - x_k \rangle, \qquad (4.27)$$

and since $y_k \in C$, we also have

$$0 \geq f_i(y_k) \geq f_i(\bar{x}_k) + \langle \nabla f_i(\bar{x}_k), y_k - x_k \rangle.$$

Multiplying this last inequality by μ_i^k and summing up for $i = 1, \ldots, m$, we obtain

$$-\varepsilon_k \geq \sum_{i=1}^m \mu_i^k \langle \nabla f_i(\bar{x}_k), y_k - x_k \rangle.$$

Adding this last inequality to (4.27) we obtain $f_0(\bar{x}_k) - \lambda \leq \varepsilon_k + \langle a_k, y_k - x_k \rangle$, and the proof can be completed the same way as the proof of Theorem 4.4.1.
$\square$

Note that all the assumptions of Theorem 4.4.2 are satisfied for linear programming. For more general situations the assumptions requiring f_i for all $i = 0, \ldots, m$, uniformly continuous near $\{x_k\}$, are usually not easy to verify, and one has to consider stronger assumptions, as in the following theorem, which corresponds to the coercive case and will be proved in Chapter 6, in the more general framework of variational inequalities.

Theorem 4.4.3 *Let $f_i : \mathbb{R}^n \to \mathbb{R} \cup \{+\infty\}$ be lsc convex proper functions with $\mathrm{dom}\, f_i$, $i = 1, \ldots, m$, open. Suppose that the optimal set of (P) is nonempty and compact and that $C \subset \mathrm{dom}\, \partial f_0$. Let $\{u^k\}$ be a bounded sequence in $\mathbb{R}_+^m$ and consider a sequence $\{x_k, g_k, g_{i,k}, i = 1, \ldots, m\}$ with $g_k \in \partial f_0(x_k)$, $g_{i,k} \in \partial f_i(x_k)$, $i = 1, \ldots, m$, such that*

$$\lim_{k \to \infty} \left\{ g_k + \sum_{i=1}^m u_i^k g_{i,k} \right\} = 0,$$

$$\limsup_{k \to \infty} f_i(x_k) \leq 0, \ \forall i = 1, \ldots, m,$$

$$\lim_{k \to \infty} u_i^k f_i(x_k) = 0, \ \forall i = 1, \ldots, m.$$

Then the sequence $\{x_k\}$ is bounded, and each limit point of this sequence is an optimal solution of (P).

4.5   Notes and References

Results given in Section 4.1 are standard except Theorem 4.1.6, whose proof is due to Corvallec [50]. Well-behaved functions were introduced by Auslender and Crouzeix [8], from which all the results presented here have been taken, except for Proposition 4.2.3 and Corollary 4.2.1, which were given in Auslender, Cominetti and Crouzeix [10]. Dual characterizations of well-behaved functions were first given by Angleraud [1]. An easier characterization was given by Cominetti [49], and more recently Aze and Michel [21] have completed the results of [49]. All the results and material given here are from [21]. For relationships between the notions of well-behaved functions and well-posedness, a notion not discussed here, we refer to the work of Penot [107]. In the last two decades, the study of error bounds has grown significantly. An important part of the literature is devoted to linear systems, starting with the classical result of Hoffman [79] and later on with the work of Walkups and Wets [130], Robinson [111], and Mangasarian [97]. Many results on linear systems providing new insights and further analysis can be found in Bergthaller and Singer [32], Burke and Tseng [45], Guler, Hoffman, and Rothblum [78], Klatte and Thiere [81], Li [88], Luo and Tseng [94], and Renegar [110]. Results on error bounds for convex inequality systems with compact convex sets under Slater's condition are due to Robinson [111], [112]. Mangasarian [98] introduced an asymptotic constraint qualification (ACQ) and derived a global error bound for convex systems with differentiable functions under (ACQ) and assuming Slater's condition. Auslender and Crouzeix then gave a sharp error bound for the case of nondifferentiable convex functions by varying ACQ; see Theorem 4.3.1. Luo and Luo [92] avoid having to assume any type of constraint qualifications by restricting their analysis to systems described by quadratic convex functions. To obtain computable error bounds, we refer, for example, to Mangasarian [99], from which Proposition 4.3.1 comes, and later on to Bertsekas [35] and Deng [62]. Several papers have also been devoted to global error bounds, not of Liptshiz type, but of Holder type; see, for example, the work of Luo and Tseng [94] which studies systems of analytic functions, and the work of Wang and Pang on convex quadratic functions but without Slater's assumption. For further reading and references we refer the reader to Lewis and Pang [86] and to Pang [106]. The final section of this chapter summarizes some ideas and results developed by Chou, Ng, and Pang [48], and all the results described here come out from this work, except Theorem 4.4.3.

5

Duality in Optimization Problems

Duality theory plays a fundamental role in the analysis of optimization and variational problems. It not only provides a powerful theoretical tool in the analysis of these problems, but also paves the way to designng new algorithms for solving them. The basic idea behind a duality framework is greatly similar to the general mathematical line of thinking, namely to transform a hard problem into an easy or at least easier one to be analyzed and to be solved. A key player in any duality framework is the Legendre–Fenchel conjugate transform. Often, duality is associated with convex problems, yet it turns out that duality theory also has a fundamental impact even on the analysis of nonconvex problems. This chapter gives the elements of duality theory for optimization problems. Starting with a very general and abstract scheme for duality based on perturbation functionals, we derive the basic conditions leading to strong duality results, and characterization of optimal solutions. This covers well-known Fenchel and Lagrangian duality schemes, as well as minimax theorems for convex–concave functionals within a unified approach that emphasizes the importance of asymptotic functions.

5.1 Perturbational-Conjugate Duality

This section introduces a general abstract duality framework that includes all the basic ingredients necessary to derive a viable duality theory for optimization problems. The general idea of duality can be described as follows.

Given an initial problem, called the primal, construct a new problem based on the *same data*, which is called the dual and is hopefully easier to analyze and to solve. The key question is then to find the relationships between the primal and dual problems, in particular, to know when these two problems are equivalent in the sense of having the same optimal values and in being able to characterize the existence of the corresponding optimal solutions as well as to recover optimal solutions of one problem from the optimal solutions of the other.

We first outline results that do not assume convexity, and then establish stronger duality relations for the convex case.

A General Abstract Dual Scheme

Let $f : \mathbb{R}^n \to \mathbb{R} \cup \{+\infty\}$ be a proper function and consider the abstract minimization problem

$$(\mathrm{P}) \quad V_\mathrm{P} := \inf\{f(x) \mid x \in \mathbb{R}^n\}.$$

Problem (P) will be called the primal problem. Recall that since f is extended valued, the above formulation includes problems with a set of constraints $C \subset \mathbb{R}^n$ by simply redefining the function f with $f + \delta_C$. Note that we also assume that f is proper in order to exclude the degenerate case $f \equiv +\infty$. The solution set (possibly empty) of problem (P) is denoted by S_P, and when $\inf f$ is finite one has

$$S_\mathrm{P} = \left\{ \bar{x} \in \mathbb{R}^n \mid f(\bar{x}) = \inf_x f(x) \right\}.$$

We do not assume convexity for the moment, so that the above formalism encompasses a wide range of optimization problems, and the results described below hold in particular for nonconvex problems.

A powerful concept for developing and studying a duality theory is through the use of *perturbations*. More precisely, we embed problem (P) in a family of perturbed problems depending on a parameter $u \in \mathbb{R}^m$ by defining a function $\Phi : \mathbb{R}^n \times \mathbb{R}^m \to \mathbb{R} \cup \{+\infty\}$ such that

$$f(x) = \Phi(x, 0).$$

For each $u \in \mathbb{R}^m$, we then consider the perturbed problem of (P):

$$(\mathrm{P}_u) \quad \inf\{\Phi(x, u) \mid x \in \mathbb{R}^n\}.$$

Clearly, for $u = 0$ problem (P_0) coincides with the original primal problem (P).

Definition 5.1.1 *For any proper function* $\Phi : \mathbb{R}^n \times \mathbb{R}^m \to \mathbb{R} \cup \{+\infty\}$*, the function*

$$\varphi(u) := \inf\{\Phi(x, u) \mid x \in \mathbb{R}^n\}$$

is called the perturbation or marginal function associated with (P)*.*

The marginal function plays a fundamental role in any duality framework. We start with an elementary result on the conjugate functions associated with the marginal function that paves the way to defining a dual problem.

Lemma 5.1.1 *For any proper function* $\Phi : \mathbb{R}^n \times \mathbb{R}^m \rightarrow \mathbb{R} \cup \{+\infty\}$ *one has:*
(a) $V_P = \varphi(0) = \inf_x f(x)$.
(b) $\varphi^*(y) = \Phi^*(0, y), \; \forall y \in \mathbb{R}^m$.
(c) $\varphi^{**}(0) = \sup\{-\Phi^*(0, y) \mid y \in \mathbb{R}^m\}$.

Proof. Part (a) is just the definition of φ at $u = 0$. Using the definition of the conjugate function (cf. Chapter 1, Definition 1.2.5), one has

$$
\begin{aligned}
\varphi^*(y) &= \sup_u \{\langle u, y \rangle - \varphi(u)\} = \sup_u \{\langle u, y \rangle - \inf_x \Phi(x, u)\}, \\
&= \sup_x \sup_u \{\langle x, 0 \rangle + \langle u, y \rangle - \Phi(x, u)\}, \\
&= \sup_{(x,u)} \{\langle (x, u), (0, y) \rangle - \Phi(x, u)\} = \Phi^*(0, y),
\end{aligned}
$$

proving (b). By definition of the biconjugate $\varphi^{**}(z) = \sup_y \{\langle z, y \rangle - \varphi^*(y)\}$, and then using (b), one obtains $\varphi^{**}(0) = \sup_y \{-\varphi^*(y)\} = \sup_y \{-\Phi^*(0, y)\}$, which is exactly (c).   □

This simple and general result encompasses the basic ingredients of duality and leads one naturally to define a dual problem associated with (P). Indeed, one always has the following relation between a function and its biconjugate (cf. Chapter 1):

$$\varphi(u) \geq \varphi^{**}(u), \quad \forall u \in \mathbb{R}^m.$$

Thus, in particular, for $u = 0$ one obtains with the help of Lemma 5.1.1

$$V_P = \inf_x f(x) = \varphi(0) \geq \varphi^{**}(0) = \sup\{-\Phi^*(0, y) \mid y \in \mathbb{R}^m\}.$$

Define the *dual objective* function by

$$h(y) := -\Phi^*(0, y), \quad \forall y \in \mathbb{R}^m.$$

Then the latter inequality leads to the definition of a *dual* problem (D) associated with (P) through the perturbation function Φ as

$$\text{(D)} \quad V_D := \sup\{h(y) \mid y \in \mathbb{R}^m\}.$$

The above scheme, which is free from any assumptions on the problem's data, already uncovers two fundamental facts for the pair of primal–dual problems (P)–(D):

- Weak duality holds: $\varphi(0) = V_P = \inf\{f(x) \mid x \in \mathbb{R}^n\} \geq \sup\{h(y) \mid y \in \mathbb{R}^m\} = V_D = \varphi^{**}(0)$.

- The dual objective function $y \to h(y) = -\Phi^*(0, y)$ is a concave function, so that problem (D) is *always* a convex problem (as the maximization of a concave function).

As we can already see, the duality results rely on the properties of the marginal function at the point zero. In a similar way, we can define a perturbation function associated with the dual problem (D). Set $h(y) = -\Phi^*(0, y)$, and for each $v \in \mathbb{R}^n$, consider the perturbed problem of (D):

$$(D_v) \quad \sup\{-\Phi^*(v, y) \mid y \in \mathbb{R}^m\},$$

which for $v = 0$ coincides with the dual problem (D). The marginal function associated with (D) can then be defined by

$$\psi(v) = \inf_y \Phi^*(v, y),$$

and hence $-\psi(0) = \sup_y h(y) = V_D$.

Thus, we have a systematic way to define a pair of primal–dual problems through the perturbation framework. Moreover, starting, for example, with the dual problem (D), applying the perturbation framework we can construct the *bidual* of (D) as

$$(DD) \quad \inf\{\Phi^{**}(x, 0) \mid x \in \mathbb{R}^n\},$$

where Φ^{**} is the biconjugate of Φ. Once again one can see the fundamental role played by the point $u = 0$. Indeed, complete symmetry between the pair (P)–(D) will be ensured whenever (DD) coincides with (P), namely whenever $\Phi(\cdot, 0) = \Phi^{**}(\cdot, 0)$. As we shall see later, this occurs precisely when Φ is a proper lsc and *convex* function.

Note that repeating the dualization process does not help much, since the perturbed problem associated with (DD) is given by

$$(DD_v) \quad \inf\{\Phi^{**}(x, v) \mid x \in \mathbb{R}^n\}.$$

Thus, the dual problem of (DD) becomes

$$(DDD) \quad \sup\{-\Phi^{***}(0, y) \mid y \in \mathbb{R}^m\}.$$

However, since for any function Φ we always have $\Phi^{***} = \Phi^*$, then problem (DDD) is nothing else but the dual problem (D).

We summarize the elementary properties, relationships, and notation of this general dual framework in the next result.

Proposition 5.1.1 *(Weak Duality) Let* $\Phi : \mathbb{R}^n \times \mathbb{R}^m \to \mathbb{R} \cup \{+\infty\}$ *be proper and consider the pair of primal–dual problems (P)–(D) defined through the function* Φ *via*

$$\text{(P) } V_P = \varphi(0) = \inf \Phi(x,0) = \inf_x f(x);$$

$$\text{(D) } V_D = -\psi(0) = \sup_y -\Phi^*(0,y) = \sup_y h(y).$$

Then:
(a) $-\infty \le \sup_y h(y) = V_D \le V_P = \inf_x f(x) \le +\infty$.
(b) The dual function $h : \mathbb{R}^m \to \mathbb{R} \cup \{-\infty\}$ *is concave.*
(c) $0 \in \operatorname{dom} \varphi \Longrightarrow \sup_y h(y) \le \inf_x f(x) < \infty$.
(d) $0 \in \operatorname{dom} \psi \Longrightarrow -\infty < \sup_y h(y) \le \inf_x f(x)$.
Moreover, under (c) and (d) one has $-\infty < \sup_y h(y) \le \inf_x f(x) < +\infty$.

Proof. The proof of (a) and (b) have already been observed from the fact $\varphi(0) \le \varphi^{**}(0)$, while the concavity of $h(y) = -\Phi^*(0,y)$ follows since the conjugate $\Phi^*(0,y)$ is always convex. We prove only (c), since (d) follows by a similar argument. Since $0 \in \operatorname{dom} \varphi$, then there exists $\bar{x}$ such that $f(\bar{x}) = \Phi(\bar{x},0) < \infty$, and therefore by (a) one has $\sup_y h(y) \le \inf_x f(x) \le \Phi(\bar{x},0) < \infty$. When both (c) and (d) hold, then clearly the last statement is obtained. $\square$

In the next result we establish a sufficient condition that guarantees that a pair $(\bar{x}, \bar{y})$ is a solution of the bidual and dual problems respectively. Once again, this result is very general, since it does not assume convexity and thus can be applied to any optimization problems.

Theorem 5.1.1 *Let* $\Phi : \mathbb{R}^n \times \mathbb{R}^m \to \mathbb{R} \cup \{+\infty\}$ *be such that* $\operatorname{conv} \Phi$ *is proper, and let (P)–(D)–(DD) be the corresponding triple of primal dual-bidual problems associated with* Φ. *If* $\bar{x} \in \mathbb{R}^n$ *and* $\bar{y} \in \mathbb{R}^m$ *satisfy one of the following equivalent conditions:*
(a) $\Phi^{**}(\bar{x},0) + \Phi^*(0,\bar{y}) = 0$,
(b) $(\bar{x},0) \in \partial\Phi^*(0,\bar{y})$,
(c) $(0,\bar{y}) \in \partial\Phi^{**}(\bar{x},0)$
then $\bar{x}$ *solves (DD) and* $\bar{y}$ *solves (D).*

Proof. First, we note that since $\operatorname{conv} \Phi$ is proper, then both Φ^* and Φ^{**} are proper, lsc, and convex functions. Therefore, the equivalence of the three conditions follows from Proposition 1.2.18. It is thus enough to prove the result whenever (a) holds. By definition of the bidual (DD), one has $V_{DD} \le \Phi^{**}(x,0)$, $\forall x \in \mathbb{R}^n$, while by definition of the dual problem (D), one has $V_D \ge -\Phi^*(0,y)$, $\forall y \in \mathbb{R}^m$. Combining these two inequalities, and using the weak duality relation for the pair of problems (D)–(DD), one thus obtains

$$0 \le V_{DD} - V_D \le \Phi^{**}(x,0) + \Phi^*(0,y), \quad \forall (x,y) \in \mathbb{R}^n \times \mathbb{R}^m.$$

Therefore, assuming that (a) is satisfied, from the above inequality one must have $V_{DD} = V_D$ and $V_{DD} = \Phi^{**}(\bar{x}, 0), V_D = -\Phi^*(0, \bar{y})$, proving that $\bar{x}$ is a solution of (DD) and $\bar{y}$ is a solution of (D).   □

The inequality in the weak duality relation for the pair of problems (P)-(D) can be strict, in which case the quantity

$$\Gamma_{PD} := \inf_x f(x) - \sup_y h(y) = V_P - V_D > 0$$

is called the *duality gap* between the pair of problems (P)–(D).

When $\Gamma_{PD} = 0$ one says that we have a pair of problems with *zero duality gap*. The role of the point $u = 0$ in the marginal function is fundamental. Indeed, the key point is to guarantee the relation $\varphi^{**}(0) = \varphi(0)$, which in turn implies a *zero duality gap*, i.e., $\Gamma_{PD} = 0$, between the pair of problems (P)–(D). In the next result we establish the first basic duality theorem that eliminates the duality gap.

Theorem 5.1.2 *(Zero duality gap) Consider the pair of primal–dual problems (P)–(D) with marginal functions $\varphi : \mathbb{R}^m \to \overline{\mathbb{R}}$ and $\psi : \mathbb{R}^m \to \overline{\mathbb{R}}$, respectively, which are assumed to be convex. The following statements are equivalent:*
(a) $\varphi(0)$ is finite and φ is lsc at 0.
(b) $\psi(0)$ is finite and $-\psi$ is lsc at 0.
(c) $V_P = V_D \in \mathbb{R}$.

Proof. We prove only the equivalence (a)–(c), since the equivalence (b)–(c) is analogous. Recall (cf. Proposition 1.2.3 and Theorem 1.2.5) that for any point $\bar{u}$ where a convex function $\varphi : \mathbb{R}^m \to \overline{\mathbb{R}}$ is finite one has $\varphi(\bar{u}) = \varphi^{**}(\bar{u})$ if and only if φ is lsc at $\bar{u}$. Therefore, since by Proposition 5.1.1(a) one has $V_P = V_D$ if and only if $\varphi(0) = \varphi^{**}(0)$, the equivalence between (a) and (c) follows.   □

It is important to notice that the above result is still quite general since we have supposed only the convexity of the marginal function, which is guaranteed by Proposition 1.2.2, whenever Φ is assumed convex. Yet, we do not assume that the primal problem (P) is convex, nor any convexity properties of the perturbation function Φ. Clearly, the convexity of the marginal function φ does not necessarily imply that (P) is a convex problem, but it is possible to have situations where (P) is a nonconvex optimization problem, yet with a convex (and lsc at zero) marginal function φ.

Example 5.1.1 *(Minimum eigenvalue)* Let Q be a symmetric and positive semidefinite $n \times n$ matrix. The minimal eigenvalue of the matrix Q is obtained by solving

$$\lambda_n(Q) = \min \left\{ x^T Q x \mid \|x\| = 1, \ x \in \mathbb{R}^n \right\}.$$

Eventhough the objective function is convex, this problem is clearly non-convex due to the nonconvex equality constraint. Define the perturbation function

$$\Phi(x, u) = \begin{cases} x^T Q x & \text{if } \|x\| = 1 - u, \\ +\infty & \text{otherwise.} \end{cases}$$

Then the marginal function associated with Φ is

$$\varphi(u) = \inf_x \Phi(x, u) \quad = \quad \inf\{x^T Q x \mid \|x\| = 1 - u\}$$

$$= \begin{cases} \lambda_n(Q)(1 - u)^2 & \text{if } u \leq 1, \\ +\infty & \text{otherwise.} \end{cases}$$

Since $\lambda_n(Q) \geq 0$, then it follows that φ is convex, and it is easy to see that it is also lsc at zero. A direct computation shows that $\varphi^{**}(0) = \sup\{-\Phi^*(0, y) \mid y \in \mathbb{R}\} = \sup\{y \mid Q - yI \succeq 0, \ y \in \mathbb{R}\} = \lambda_n(Q)$, where I denotes the $n \times n$ identity matrix, and the latter expression is a dual representation of $\lambda_n(Q)$.

For optimization problems with convex marginal functions there exists an important and useful interplay between subdifferentials of their conjugates and duality results. More precisely, the solution set of the dual problem (D) (possibly empty) is given by $S_D = \operatorname{argmax} h = \partial \varphi^{**}(0)$.

Theorem 5.1.3 *Consider problem (P) and suppose that the marginal function φ associated with (P) is convex. The following statements are equivalent:*

(a) $\partial \varphi(0) \neq \emptyset$.
(b) $\varphi(0)$ is finite and lsc at 0, and the set of optimal dual solutions is nonempty.

Proof. We first show the implication (a) $\Longrightarrow$ (b). We use Proposition 1.2.20. One has $\partial \varphi(0) \neq \emptyset \Longrightarrow \varphi^{**}(0) = \varphi(0)$ and hence also $\partial \varphi^{**}(0) = \partial \varphi(0)$. Therefore, φ is lsc at 0, $\varphi(0)$ is finite, $V_P = V_D$, and the optimal set of dual solutions, which by definition is $\partial \varphi^{**}(0)$, is nonempty. For the reverse implication, we note that the lower semicontinuity of φ at 0 means that $\varphi^{**}(0) = \varphi(0)$, and thus with the argument just given above, one has $\partial \varphi(0) = \partial \varphi^{**}(0)$. The latter set being the set of optimal dual solutions, assumed nonempty in (b), thus proves the statement (a). $\square$

We now give the conditions on the marginal function that will imply strong duality results, namely a zero duality gap and the existence of dual optimal solutions. We will also be more precise about the structure of the dual optimal set. Recall from Chapter 3 (Proposition 3.2.3) that with $E := \operatorname{aff}(\operatorname{dom} \varphi)$, we denote by φ_E the function restricted to the affine hull E.

Theorem 5.1.4 *(Strong duality, dual attainment) Consider problem (P) and suppose that the marginal function φ associated with (P) is convex and*

$\varphi(0) \in \mathbb{R}$. *Then:*
(a) $0 \in \operatorname{int dom} \varphi \implies \varphi$ *is continuous at* 0, *and the set of optimal dual solutions* $S_{\mathrm{D}} = \operatorname{argmax} h = \partial\varphi(0) = (\partial\varphi^{**}(0))$ *is a nonempty compact set.*
(b) $0 \in \operatorname{ri}(\operatorname{dom}\varphi) \implies S_{\mathrm{D}} = \operatorname{argmax} h = \partial\varphi_E(0) + E^{\perp}$ *with* $\partial\varphi_E(0)$ *nonempty and compact and* $E^{\perp}$ *the orthogonal complement to* E.
Under (a) or (b) one thus has $V_{\mathrm{P}} = V_{\mathrm{D}}$ *with the dual value* V_{D} *attained.*

Proof. (a) Since φ is assumed convex, then whenever $0 \in \operatorname{int dom} \varphi$, since $\varphi(0) \in \mathbb{R}$, one has φ proper. Moreover, invoking Theorem 1.2.3, one has φ continuous at 0. We can thus apply Proposition 1.2.16 to φ to conclude that $\partial\varphi(0)$ is a nonempty compact set, and therefore by Theorem 5.1.3 it follows that the solution set of the dual problem (D) given by $S_{\mathrm{D}} = \operatorname{argmax} h = \partial\varphi^{**}(0) = \partial\varphi(0)$ is a nonempty compact set. To prove (b) we first note that under the given hypothesis, the convex function φ is proper, so that $\operatorname{cl}\varphi = \varphi^{**}$, and thus by Proposition 1.2.5 it follows that $0 \in \operatorname{ri}(\operatorname{dom}\varphi^{**}) = \operatorname{ri}(\operatorname{dom}\varphi)$ and $\operatorname{aff}(\operatorname{dom}\varphi^{**}) = \operatorname{aff}(\operatorname{dom}\varphi) := E$. The dual problem (D) is defined by $\sup_y -\varphi^*(y)$ with φ^* proper, lsc, and convex, and so by Theorem 3.2.1, $0 \in \operatorname{ri}(\operatorname{dom}\varphi^{**})$ implies that φ^* is weakly coercive, and hence by Corollary 3.2.2 it follows that $S_{\mathrm{D}} = \operatorname{argmax} h = \partial\varphi_E(0) + E^{\perp}$. The last assertion is then immediate from Theorem 5.1.2.   □

It is interesting to note that in Theorem 5.1.4 one can establish the converse statements.

Corollary 5.1.1 *Consider problem (P) and suppose that the marginal function* φ *associated with (P) is convex. Then:*
(a) $S_{\mathrm{D}} = \partial\varphi(0)$ *is a nonempty compact set* $\implies 0 \in \operatorname{int dom} \varphi$.
(b) $S_{\mathrm{D}} = \operatorname{argmax} h = \partial\varphi_E(0) + E^{\perp}$ *with* $\partial\varphi_E(0)$ *nonempty and compact* $\implies 0 \in \operatorname{ri}(\operatorname{dom}\varphi)$.
Under (a) or (b) one thus has $\varphi(0) \in \mathbb{R}$, φ *is lsc at* 0 *and* $V_{\mathrm{P}} = V_{\mathrm{D}}$.

Proof: Assumption (a) is using Proposition 2.1.2 to say that $(\partial\varphi(0))_\infty = \{0\}$, which in turn is by Proposition 2.5.4 equivalent to saying that $0 \in \operatorname{int dom} \varphi$. Statement (b) follows similarly by noting that in that case, $\partial\varphi_E(0)$ nonempty and compact is equivalent to $0 \in \operatorname{int dom} \varphi_E$. Then, since by Proposition 3.2.3(b) one has $\operatorname{int dom} \varphi_E = \operatorname{ri dom} \varphi + E^{\perp}$, the desired conclusion follows. The last statement is immediate from Theorem 5.1.3.   □

To derive a strong duality result for the primal problem similar to Theorem 5.1.4, i.e., to obtain primal attainment, we need to preserve symmetry in order to have a convex bidual problem. Thus, we now make the blanket assumption

$$\Phi : \mathbb{R}^n \times \mathbb{R}^m \to \mathbb{R} \cup \{+\infty\} \quad \text{is a proper, lsc, and convex function.}$$

Associated with this Φ, note that f is convex and lsc, while h is concave and usc, so that the pair of problems (P)–(D) then becomes respectively a pair of *convex* and *concave* optimization problems. This situation leads to the strongest possible duality results for primal–dual convex–concave problems.

Theorem 5.1.5 *(Strong duality, primal attainment) Let $\Phi : \mathbb{R}^n \times \mathbb{R}^m \to \mathbb{R} \cup \{+\infty\}$ be a proper, lsc, and convex function and let*

$$\psi(v) = \inf\{\Phi^*(v, y) \mid y \in \mathbb{R}^m\}$$

be the marginal function associated with the concave dual problem and $\psi(0) \in \mathbb{R}$. Then:
(a) $0 \in \mathrm{int\,dom}\,\psi \implies \psi$ is continuous at 0 and the set of optimal primal solutions $S_\mathrm{P} = \mathrm{argmin}\, f = \partial(\psi(0))$ is a nonempty compact set.
(b) $0 \in \mathrm{ri}(\mathrm{dom}\,\psi) \implies S_\mathrm{P} = \mathrm{argmin}\, f = \partial(\psi_M(0)) + M^\perp$ with $\partial(\psi_M(0))$ nonempty and compact and $M^\perp$ the orthogonal complement to $M = \mathrm{aff}(\mathrm{dom}\,\psi)$. Under (a) or (b) one has $V_\mathrm{P} = V_\mathrm{D}$, with V_P attained.

Proof. Apply Theorem 5.1.4, by substituting (D) for (P), with the perturbation function Φ^*. Since Φ is proper, convex, and lsc, then $\Phi = \Phi^{**}$, and the dual functional $\Phi^{**}(\cdot, 0)$ satisfies $\Phi^{**}(x, 0) = \Phi(x, 0) = f(x)$, from which we obtain (a) and (b). □

As in the dual case, we can state a converse of Theorem 5.1.5 in the same manner as stated in Corollary 5.1.1, but in terms of the marginal function ψ. The details are left to the reader. However, it is interesting to note that in this case, since under our blanket assumption the perturbation function Φ is lsc on $\mathbb{R}^n \times \mathbb{R}^m$, we can in fact arrive at the stronger conclusion that the primal marginal function φ is lsc on the whole space $\mathbb{R}^m$. Indeed, for example, the assumption $0 \in \mathrm{int\,dom}\,\psi$, which is equivalent to the fact that the optimal set of the primal problem (P), namely $\mathrm{argmin}\, f = \partial\psi(0)$, is nonempty and compact, simply means that $f_\infty(d) > 0$, $\forall d \neq 0$, since $\Phi_\infty(d, 0) = f_\infty(d)$. We can then invoke Corollary 3.5.6 to conclude that φ is lsc on $\mathbb{R}^m$.

We end this section by the characterization of optimal solutions through primal-dual relations for the convex case.

Corollary 5.1.2 *(Optimality conditions) Suppose that Φ is lsc, proper, and convex. Then $\bar{x}$ solves (P), $\bar{y}$ solves (D), and $V_\mathrm{P} = V_\mathrm{D}$ if and only if $(\bar{x}, 0) \in \partial\Phi^*(0, \bar{y})$, the latter inclusion being equivalent to $(0, \bar{y}) \in \partial\Phi(\bar{x}, 0)$.*

Proof. The optimality conditions follow immediately from Theorem 5.1.1, noting that under the stated assumptions one has here $\Phi^{**} = \Phi$. □

In the following sections we present particular instances of this duality framework for optimization problems exhibiting special structures.

5.2   Fenchel Duality

A very general and powerful duality scheme, which encompasses a broad class of optimization problems, is the Fenchel duality scheme.

Let $f : \mathbb{R}^n \to \mathbb{R} \cup \{+\infty\}$ and $g : \mathbb{R}^m \to \mathbb{R} \cup \{+\infty\}$ be proper, convex functions and let $A : \mathbb{R}^n \to \mathbb{R}^m$ be a linear map. Consider a convex primal optimization problem given in the form

$$(P) \quad p := \inf\{f(x) + g(Ax) \mid x \in \mathbb{R}^n\}.$$

The dual problem (D) associated with (P) can be easily constructed through the perturbation function on $\mathbb{R}^n \times \mathbb{R}^m$ defined by

$$\Phi(x, u) := f(x) + g(Ax + u). \tag{5.1}$$

From the assumptions on the problem data, the function Φ is clearly a proper, convex function on $\mathbb{R}^n \times \mathbb{R}^m$. To construct the dual problem, we compute its conjugate:

$$
\begin{aligned}
\Phi^*(v, y) &= \sup_{x,u}\{\langle x, v \rangle + \langle u, y \rangle - f(x) - g(Ax + u)\} \\
&= \sup_{x,w}\{\langle x, v \rangle + \langle w, y \rangle - \langle Ax, y \rangle - f(x) - g(w)\} \\
&= \sup_{x}\{\langle x, -A^T y + v \rangle - f(x)\} + \sup_{w}\{\langle w, y \rangle - g(w)\} \\
&= f^*\left(-A^T y + v\right) + g^*(y).
\end{aligned}
$$

Therefore, since the dual objective for problem (P) is by definition $h(y) = -\Phi^*(0, y)$, we obtain the dual problem (D) associated with (P):

$$(D) \quad p^* := \sup\left\{-f^*(A^T y) - g^*(y) \mid y \in \mathbb{R}^m\right\}.$$

Note that the construction of the dual itself is of course general and does not require convexity. Moreover, from the Fenchel inequality we immediately obtain the weak duality inequality $p \geq p^*$. We are interested in establishing strong duality results for the pair (P)–(D). Once again, to achieve this goal we will have to establish the lower semicontinuity at 0 of the marginal function $\varphi : \mathbb{R}^m \to [-\infty, +\infty]$, which by Definition 5.1.1, using (5.1), is given here by

$$\varphi(u) := \inf\{f(x) + g(Ax + u) \mid x \in \mathbb{R}^n\}. \tag{5.2}$$

The proof of the Fenchel duality theorem will be obtained as a direct application of the duality results established for the abstract pair of optimization problems considered in Section 5.1. First, we need to collect some elementary facts on the perturbed function defined in (5.2).

Proposition 5.2.1 *Let f and g be proper, lsc, convex functions on $\mathbb{R}^n$ and $\mathbb{R}^m$, respectively, and let $\varphi(u) = \inf_x \Phi(x, u)$, and $\psi(v) := \inf_y \Phi^*(v, y)$ with Φ defined in (5.1). Then:*
(a) φ and ψ are convex.
(b) $\operatorname{dom} \varphi = \operatorname{dom} g - A \operatorname{dom} f$ and $\operatorname{dom} \psi = \operatorname{dom} f^* + A^T \operatorname{dom} g^*$, and both domains are nonempty convex sets.

Proof. We verify only the statement for the function φ, since the statements for ψ are obtained similarly. Since the function $(x, u) \to f(x) + g(Ax + u)$ is convex, assertion (a) follows from Proposition 1.2.2. To obtain (b), simply write the definition of the domain of φ to obtain

$$\begin{aligned} \operatorname{dom} \varphi &= \{u \mid \exists (x, w) \text{ such that } f(x) < \infty, g(w) < \infty, \ u = w - Ax\} \\ &= \operatorname{dom} g - A \operatorname{dom} f. \end{aligned}$$

$\square$

Consequently, we have the following relations:

$$\begin{aligned} p < +\infty &\iff 0 \in \operatorname{dom} g - A \operatorname{dom} f, \\ p^* > -\infty &\iff 0 \in \operatorname{dom} f^* + A^T \operatorname{dom} g^*. \end{aligned}$$

We first give the standard Fenchel duality theorem, and then we will refine the result.

Theorem 5.2.1 *Let $f : \mathbb{R}^n \to \mathbb{R} \cup \{+\infty\}$ and $g : \mathbb{R}^n \to \mathbb{R} \cup \{+\infty\}$ be proper lsc convex functions, let $A : \mathbb{R}^n \to \mathbb{R}^m$ be a linear map, and let the marginal function associated with (P) $\varphi : \mathbb{R}^m \to [-\infty, +\infty]$ be*

$$\varphi(u) := \inf\{f(x) + g(Ax + u) \mid x \in \mathbb{R}^n\}.$$

The Fenchel primal and dual problems are defined by

$$p = \inf\{f(x) + g(Ax) \mid x \in \mathbb{R}^n\}; \quad p^* = \sup\{-f^*(A^T y) - g^*(-y) \mid y \in \mathbb{R}^m\},$$

with optimal primal and dual solution sets S_P and S_D respectively. The following statements hold:
(a) $p = \varphi(0) \in \mathbb{R}$, $0 \in \operatorname{int}(\operatorname{dom} g - A \operatorname{dom} f)$ if and only if S_D is nonempty and compact.
(b) $p^* = \psi(0) \in \mathbb{R}$, $0 \in \operatorname{int}(\operatorname{dom} f^* + A^T \operatorname{dom} g^*)$ if and only if S_P is nonempty and compact.
(c) Under condition (a) or (b) we have $p = p^*$. Moreover, the optimal solutions (x, y) of the pair of primal-dual problems are characterized through the primal–dual optimality conditions

$$x \in \partial f^*(-A^T y), \ Ax \in \partial g^*(y) \iff y \in \partial g(Ax), -A^T y \in \partial f(x). \quad (5.3)$$

Proof. We first prove only (a), since (b) is just the dual statement, which follows in a similar way. In fact, invoking Theorem 5.1.4(a) applied to the marginal function φ, together with the computations made in Proposition 5.2.1 on the domain of φ, the result follows at once. The first part of (c) is clear whenever (a) or (b) holds true. To obtain the optimality conditions, we note that by Theorem 4.1.2, one has that x is a solution of the primal problem if and only if $0 \in \partial[f(x)+g(Ax)]$, which under the stated assumptions is equivalent to $0 \in \partial f(x)+\partial g(Ax) = \partial f(x)+A^T\partial g(Ax)$. The latter inclusion can be written as $0 \in \partial f(x) + A^Ty$, $y \in \partial g(Ax)$. Recalling (cf. Proposition 1.2.18 (a)) that for any proper lsc convex function c one has $(\partial c)^{-1} = \partial c^*$, this inclusion is thus equivalent to $x \in \partial f^*(-A^Ty)$, $Ax \in \partial g^*(y)$, showing that y solves the dual problem, and completing the proof of (c).   □

Fenchel's duality theorem can be refined, by replacing interiors with relative interiors, in the conditions on the domains of the functions as given in Theorem 5.2.1. This is, in fact, a direct application of Theorem 5.1.4(b) and Theorem 5.1.5(b) which also give more information on the structure of the optimal primal–dual sets.

Corollary 5.2.1 *For the Fenchel primal–dual pair under the assumptions of Theorem 5.2.1 the following statements hold:*
(a) $p = \varphi(0) \in \mathbb{R}$, $0 \in \mathrm{ri}(\mathrm{dom}\, g - A\,\mathrm{dom}\, f)$ if and only if S_D is the sum of a compact set and a linear space.
(b) $p^* = \psi(0) \in \mathbb{R}$, $0 \in \mathrm{ri}(\mathrm{dom}\, f^* + A^T\,\mathrm{dom}\, g^*)$ if and only if S_P is the sum of compact set and a linear space.
(c) Under condition (a) or (b) one has $p = p^*$. Moreover, the optimal solutions (x,y) of the pair of primal–dual problems are characterized through the optimality conditions (5.3).

A nice application of the Fenchel duality theorem is to the case of linearly constrained problems.

Corollary 5.2.2 *(Linear constraints) Let $f : \mathbb{R}^n \to \mathbb{R} \cup \{+\infty\}$ and $g : \mathbb{R}^n \to \mathbb{R} \cup \{+\infty\}$ be proper, lsc, and convex functions, let $A : \mathbb{R}^n \to \mathbb{R}^m$ be a linear map and $b \in \mathbb{R}^m$. If $b \in \mathrm{ri}(A\,\mathrm{dom}\, f)$, then*

$$p = \inf\{f(x) \mid Ax = b, x \in \mathbb{R}^n\} = p^* = \sup\{\langle b, y \rangle - f^*(A^Ty) \mid y \in \mathbb{R}^m\},$$

the supremum being attained whenever it is finite.

Proof. Apply Theorem 5.2.1 with g the indicator of the point $\{b\}$.

□

Example 5.2.1 An important special case that arises in applications is the case of linearly decomposable problems given in primal form as

$$p = \inf \left\{ \sum_{i=1}^{p} f_i(x_i) \mid \sum_{i=1}^{p} A_i x_i = b, \ x_i \in \mathbb{R}^{n_i}, i = 1, \ldots, p \right\},$$

where the functions f_i are proper, lsc, and convex on $\mathbb{R}^{n_i}$, A_i are $m \times n_i$ matrices, and $b \in \mathbb{R}^m$. Then the dual problem is

$$p^* = \sup \left\{ - \sum_{i=1}^{p} f_i^*(A_i^T y) + b^T y \mid y \in \mathbb{R}^m \right\}.$$

Then in this case with $g = \delta_{\{b\}}$ the conditions in Fenchel's theorem read

$$0 \in \mathrm{ri}(\mathrm{dom}\, g - A\, \mathrm{dom}\, f) \iff \exists z_i \in \mathrm{ri}\, \mathrm{dom}\, f_i : \sum_{i=1}^{p} A_i z_i = b,$$

$$0 \in \mathrm{ri}(\mathrm{dom}\, f^* + A^T \mathrm{dom}\, g^*) \iff \exists y \in \mathbb{R}^m : A_i^T y \in \mathrm{ri}\, \mathrm{dom}\, f_i^*.$$

5.3 Lagrangian Duality

Fenchel duality is an elegant and powerful scheme to construct a dual problem via the conjugate of the functions involved and is particularly useful to prove existence (minima/maximum attainments) results of the associated primal-dual problems by simply checking conditions on the domains of the relevant functions and their conjugates. However, for optimization problems exhibiting more special structures, it is not always appropriate or possible to get explicit and/or useful expressions of the conjugates. For the standard convex programming problem, described by explicit convex inequalities, it turns out that it is often appealing and easier to develop a duality theory via the *Lagrangian duality* framework. The Lagrangian framework makes more transparent the optimality conditions (cf. Chapter 4) and also leads one naturally to consider more general classes of minimax problems, see Section 5.6. The two schemes are not only complementary but are in fact equivalent.

Consider the convex program

$$\text{(P)} \quad V_P := \inf\{f_0(x) \mid f_i(x) \le 0, \ x \in \mathbb{R}^n\},$$

where the functions $f_i : \mathbb{R}^n \to \mathbb{R} \cup \{+\infty\}$, $i = 0, \ldots, m$, are supposed proper, convex, and lsc and for clarity of exposition not affine. The results for the mixed case, i.e., involving affine constraints, will be outlined at the end of the section. Unless otherwise specified, throughout this section we make the following standard regularity assumptions already used in Chapter 4:

Assumption (R). $\operatorname{dom} f_0 \subset \operatorname{dom} f_i$, $\operatorname{ri} \operatorname{dom} f_0 \subset \operatorname{ri} \operatorname{dom} f_i$, $i = 1, \ldots, m$.

Note that the assumption on the domains can be always enforced for any optimization problem by appropriately redefining the objective function when necessary. We will also use the vector notation $F(x) := (f_1(x), \ldots, f_m(x))^T$, and denote the constraint set of problem (P) by $C := \{x \,|\, F(x) \leq 0\}$. Recall that a feasible point of (P) is a point $x \in \cap_{i=1}^m \operatorname{dom} f_i$ satisfying the constraints $F(x) \leq 0$. The function $L : \mathbb{R}^n \times \mathbb{R}^m_+ \to \overline{\mathbb{R}}$ defined by

$$
L(x, y) = \begin{cases} f_0(x) + \sum_{i=1}^m y_i f_i(x) & \text{if } x \in \operatorname{dom} f_0, y \in \mathbb{R}^m_+, \\ -\infty & \text{if } x \in \operatorname{dom} f_0, y \notin \mathbb{R}^m_+, \\ +\infty & \text{if } x \notin \operatorname{dom} f_0, \end{cases}
$$

is called the *Lagrangian* associated with (P). This definition is in fact built into the primal optimization problem, since one clearly has:

$$
\sup_{y \in \mathbb{R}^m_+} L(x, y) = \begin{cases} f_0(x) & \text{if } x \in C, \\ +\infty & \text{otherwise.} \end{cases}
$$

Thus we can write the optimization problem (P) as

$$
\text{(P)} \quad V_{\mathrm{P}} = \inf_{x \in \mathbb{R}^n} \sup_{y \in \mathbb{R}^m_+} L(x, y),
$$

where $V_{\mathrm{P}} \in [-\infty, +\infty]$, denotes the optimal value of problem (P). With this formulation, by reversing the "inf-sup" operation, one can naturally associate with (P) another problem:

$$
\text{(D)} \quad V_{\mathrm{D}} = \sup_{y \in \mathbb{R}^m_+} \inf_{x \in \mathbb{R}^n} L(x, y).
$$

Define the *dual objective* function as

$$
h(y) := \begin{cases} \inf\{L(x, y) \,|\, x \in \operatorname{dom} f_0\} & \text{if } y \in \mathbb{R}^m_+, \\ -\infty & \text{otherwise.} \end{cases}
$$

Then the *dual problem* associated with (P) is

$$
\text{(D)} \quad V_{\mathrm{D}} = \sup\{h(y) \,|\, y \in \mathbb{R}^m\} = \sup\{h(y) \,|\, y \in \mathbb{R}^m_+\},
$$

where $V_{\mathrm{D}} \in [-\infty, +\infty]$ is the optimal dual value.

From the above construction, we immediately obtain that for any primal–dual feasible problem, the weak duality relation $V_{\mathrm{P}} \geq V_{\mathrm{D}}$ holds. We notice once again that this duality framework remains valid without any convexity assumptions on the problem's data f_i, and the resulting dual objective function is always concave.

As in the previous section, our main interest is to find conditions under which there is no duality gap, i.e., to ensure that $V_P = V_D$. We know, from our previous analysis, that to achieve this goal we need to define a perturbation function and to study the corresponding marginal function associated with (P). Let $\Phi : \mathbb{R}^n \times \mathbb{R}^m \to \mathbb{R} \cup \{+\infty\}$ be defined by

$$\Phi(x, u) = \begin{cases} f_0(x) & \text{if } F(x) + u \leq 0, \\ +\infty & \text{otherwise,} \end{cases}$$

and let $\varphi : \mathbb{R}^m \to [-\infty, +\infty]$ be the corresponding marginal function defined by

$$\varphi(u) = \inf\{f_0(x) \mid F(x) + u \leq 0, \ x \in \mathbb{R}^n\}. \tag{5.4}$$

Clearly, under assumption (**R**), the function Φ is proper, lsc, and convex with φ convex.

The first result shows that the Lagrangian dual defined above can also be developed through the abstract perturbation framework developed in Section 5.1. Note that this result remains general and no convexity is assumed.

Proposition 5.3.1 *Let $f_i : \mathbb{R}^n \to \mathbb{R} \cup \{+\infty\}$, $i = 0, \ldots, m$, be arbitrary functions. Then:*
(a) The optimal primal value V_P is equal to $\varphi(0)$.
(b)

$$\varphi^*(y) := \begin{cases} -h(y) & \text{if } y \in \mathbb{R}^m_+, \\ +\infty & \text{otherwise.} \end{cases}$$

(c) $h(y) = -\varphi^*(y) \ \forall y \in \mathbb{R}^m$.
(d) The optimal dual value V_D is equal to $\varphi^{**}(0)$.

Proof. The result in (a) is just by definition of V_P. To prove (b) using the definition of $\varphi(u)$ and its conjugate one obtains

$$\begin{aligned}
\varphi^*(y) &= \sup\{\langle y, u \rangle - \varphi(u) \mid u \in \mathbb{R}^m\} \\
&= \sup\{\langle y, u \rangle - \inf_x\{f_0(x) \mid F(x) + u \leq 0\} \mid u \in \mathbb{R}^m\} \\
&= \sup\{\langle y, u \rangle - f_0(x) \mid F(x) + u = -s, x \in \text{dom } f_0, s \in \mathbb{R}^m_+, u \in \mathbb{R}^m\} \\
&= \sup\{\langle y, -F(x) - s \rangle - f_0(x) \mid x \in \text{dom } f_0, s \in \mathbb{R}^m_+\} \\
&= -\inf\{f_0(x) + \langle y, F(x) \rangle \mid x \in \text{dom } f_0\} + \sup\{\langle -y, s \rangle \mid s \in \mathbb{R}^m_+\}, \\
&= -h(y) + \delta_{\mathbb{R}^m_+}(y),
\end{aligned}$$

which proves (b) and (c). Using (b) the last statement (d) then follows from the relations: $V_D = \sup_{y \in \mathbb{R}^m_+} h(y) = \sup_{y \in \mathbb{R}^m_+} -\varphi^*(y) = \varphi^{**}(0)$.

$\square$

From this proposition it is clear once again that the key to obtaining a zero duality gap is through the marginal function and its biconjugate at the point zero. To ensure that the marginal function φ is lsc at 0, we need to make a further assumption on the problem's data, known as a *constraint qualification*.

Assumption (S) $\exists \bar{x} \in \operatorname{ri} \operatorname{dom} f_0$ such that $F(\bar{x}) < 0$.

This constraint qualification is known as Slater's condition (cf. Chapter 4) on the convex problem (P). The next result explains the relations between Slater's condition and the associated marginal function and allows for producing strong duality results in the context of the Lagrangian dual scheme.

Proposition 5.3.2 *Consider the convex program (P) and suppose that V_P is finite. Then:*
*(a) Assumption (**S**) $\Longrightarrow$ $0 \in \operatorname{int} \operatorname{dom} \varphi$.*
*(b) Assumption (**R**) $\Longrightarrow$ $0 \in \operatorname{ri} \operatorname{dom} \varphi$. Moreover, $\operatorname{ri} \operatorname{dom} \varphi = \operatorname{int} \operatorname{dom} \varphi$ and assumption (**S**) holds.*
Thus, in both cases (a) and (b) one has $V_P = V_D$ and the dual optimal set of problem (P) is nonempty and compact.

Proof. The last assertion is an immediate consequence of the general strong duality result Theorem 5.1.4, and thus it remains to establish (a) and (b). First notice that since $V_P = \varphi(0)$ is finite, then $0 \in \operatorname{dom} \varphi$. Moreover, if assumption (S) holds, then it follows that $\exists \varepsilon > 0$ such that with $\|u\|_\infty = \max_{1 \leq i \leq m} |u_i| < \varepsilon$ one has $f_i(\bar{x}) + u_i \leq 0$, $i = 1, \ldots, m$. Therefore, for $\|u\|_\infty$ sufficiently small, $u \in \operatorname{dom} \varphi$, which implies $0 \in \operatorname{int} \operatorname{dom} \varphi$, and (a) is proved. Now to prove (b) let L be the projection map $L : (x, u) \to x$. Then by definition of the marginal function we have $\operatorname{dom} \varphi = L(\operatorname{dom} \Phi)$, where $\operatorname{dom} \Phi = \{(x, u) \in \mathbb{R}^n \times \mathbb{R}^m \mid x \in \operatorname{dom} f_0, F(x) + u \leq 0\}$, and thus by Proposition 1.1.6(c) that $\operatorname{ri} \operatorname{dom} \varphi = L(\operatorname{ri} \operatorname{dom} \Phi)$. Under (b), thanks to assumption (L), we can apply Proposition 1.1.7 and Proposition 1.2.8 to conclude that $\operatorname{ri} \operatorname{dom} \Phi = \{(x, u) \in \mathbb{R}^n \times \mathbb{R}^m \mid x \in \operatorname{ri} \operatorname{dom} f_0, F(x) + u < 0\}$. Therefore, with $0 \in \operatorname{ri} \operatorname{dom} \varphi$ it follows that Slater's condition holds. Moreover, since by (a), Slater's condition implies $0 \in \operatorname{int} \operatorname{dom} \varphi$, one also has $\operatorname{int} \operatorname{dom} \varphi = \operatorname{ri} \operatorname{dom} \varphi$.

$\square$

Let $f = f_0 + \delta_C$ be the objective function of (P) and recall that the marginal function associated with the dual problem (D) is given by $\psi(v) = \inf_y \Phi^*(v, y)$. Let $C_f = \{d \mid (f_i)_\infty(d) = (f_i)_\infty(-d) = 0, \ i = 0, \ldots, m\}$ be the constancy space of f (see Definition 2.5.2).

Proposition 5.3.3 *Consider the convex program (P) with $\psi(0) \in \mathbb{R}$ and satisfying assumption (L). Then, $M := \mathrm{aff}(\mathrm{dom}\,\psi) = \mathcal{C}_f^{\perp}$. Consider the following statements:*
(A) Coercive case:
(a) $0 \in \mathrm{int}\,\mathrm{dom}\,\psi$.
(b) The optimal set of problem (P) is nonempty and compact.
(c) The functions f_i, $i = 0, \ldots, m$, have no common direction of recession.
(B) Weakly coercive case:
(a) $0 \in \mathrm{ri}\,\mathrm{dom}\,\psi$.
(b) $(f_0)_{\infty}(d) > 0$, $\forall 0 \neq d \in M^{\perp}$ and satisfying $(f_i)_{\infty}(d) \leq 0$, $i = 0, \ldots, m$.
(c) The optimal set of problem (P) is given by $\partial\psi_M(0) + M^{\perp}$ with $\partial\psi_M(0)$ nonempty and compact.
In both cases (A) and (B) the statements (a), (b), and (c) are equivalent. Moreover, under case (A) or (B) one has φ lsc on the whole space $\mathbb{R}^m$ and $V_P = V_D$.

Proof. Since $\psi^*(x) = \Phi^{**}(x, 0) = f(x)$, it follows that $M = \mathcal{C}_f^{\perp}$. Then (A) and (B) are direct consequences of Corollary 3.1.2 and Proposition 3.2.5. Furthermore, by Theorem 5.1.5, one has $V_P = V_D$. Finally, a direct computation shows that

$$\Phi_{\infty}(d, 0) = \begin{cases} (f_0)_{\infty}(d) & \text{if } (f_i)_{\infty}(d) \leq 0, \\ +\infty & \text{otherwise,} \end{cases}$$

and by Corollary 3.5.6, it follows that φ is lsc on the whole space $\mathbb{R}^m$. □

The Lagrangian duality result carries over for the convex program (P) with linear constraints. Consider the convex problem (P), but now with linear constraints, i.e.,

$$(\text{PL}) \quad \inf\{f_0(x) \mid Ax = b, F(x) \leq 0\},$$

where A is a $p \times n$ matrix and $b \in \mathbb{R}^m$. The corresponding perturbation function in that case takes the form $\Phi : \mathbb{R}^n \times \mathbb{R}^m \times \mathbb{R}^p \to \mathbb{R} \cup \{+\infty\}$ and is defined by

$$\Phi(x, u, w) = \begin{cases} f_0(x) & \text{if } Ax + w = b, F(x) \leq u, \\ +\infty & \text{otherwise,} \end{cases}$$

and the marginal function is now $\varphi(u, w) = \inf_x \Phi(x, u, w)$. From here it is easy to construct the dual problem

$$(\text{DL}) \quad \sup\{h(y, z) \mid y \in \mathbb{R}_+^m \times \mathbb{R}^p\},$$

where the dual objective h is now defined by

$$h(y, z) = \begin{cases} \langle b, z \rangle + \inf_{x \in \mathrm{dom}\,f_0}\{L(x, y) - \langle z, Ax \rangle\} & \text{if } y \in \mathbb{R}_+^m \times \mathbb{R}^p \\ +\infty & \text{otherwise.} \end{cases}$$

It is then easy to derive strong duality results for the pair of problems (PL)–(DL) via the results developed above, under appropriate conditions. For example, the Slater assumption (S) for problem (PL) takes the form

$$\exists\, \bar{x} \in \mathrm{ri}\,\mathrm{dom}\, f_0 \mid A\bar{x} = b,\ F(\bar{x}) < 0.$$

Example 5.3.1 *(Fenchel duality and Lagrangians)* The Fenchel and Lagrangian schemes are essentially equivalent, in the sense that each scheme can be recovered from the other. The Fenchel formulation of the primal problem

$$p = \inf_{x}\{f(x) + g(Ax)\}$$

can be written equivalently as

$$p = \inf_{x,z}\{f(x) + g(z)\,|\,Ax = z\}.$$

For any $y \in \mathbb{R}^m$, we define the Lagrangian

$$L(x, z, y) = f(x) + g(z) - \langle z - Ax, y\rangle.$$

The dual problem is then

$$p^* = \sup_{y}\inf_{x,z}\{f(x) + g(z) - \langle z - Ax, y\rangle\},$$

and an easy computation shows that $p^* = \sup_{y\in\mathbb{R}^m}\{-f^*(-A^T y) - g^*(-y)\}$, recovering the Fenchel dual problem. Conversely, we can start with a general Lagrangian of the form

$$l(x, y) = f(x) - g^*(y) - \langle y, Ax\rangle$$

to recover the Fenchel primal–dual by considering the problems $\inf_x \sup_y l(x, y)$ and its dual $\sup_y \inf_x l(x, y)$, where the functions f, g are lsc proper convex.

5.4   Zero Duality Gap for Special Convex Programs

We consider again problems (P) and (D) as described in Section 5.3. As seen in all the analysis above, a constraint qualification (Slater's condition) is needed to obtain a zero duality gap, and this condition is equivalent to the primal or the dual problem being coercive or weakly coercive. What happens for problems for which such conditions are not assumed and for which we suppose only that the infimum is finite? We now give some interesting results for two classes of convex optimization problems where a zero duality gap can be obtained without constraint qualifications.

The class of Asymptotic Level Stable Convex Programs

The Class of als functions has been introduced in Section 3.3.

Theorem 5.4.1 *Let $f_i : \mathbb{R}^n \to \mathbb{R} \cup \{+\infty\}$ be als and convex for $i = 0, \ldots, m$ with common effective domain $C = \operatorname{dom} f_i$. Suppose that V_{P} is finite. Then the optimal set S_{P} of (P) is nonempty, $V_{\mathrm{P}} = V_{\mathrm{D}}$, and the marginal function φ is lsc on $\mathbb{R}^m$.*

Proof. The fact that S_{P} is nonempty and φ is lsc on $\mathbb{R}^m$ was proved in Corollary 3.7.2. Furthermore, since φ is lsc at 0, it follows from Theorem 5.1.2 that $V_{\mathrm{P}} = V_{\mathrm{D}}$.   □

The Class of Weakly Analytic Convex Programs

We have seen that the key element to proving a duality result for a given pair of optimization problems relies on proving the lower semicontinuity of the associated perturbation functional at the point 0. We exhibit now another class of convex programs for which there is no duality gap and without assuming any constraint qualification.

Definition 5.4.1 *A real-valued convex function $f : \mathbb{R}^n \to \mathbb{R}$ is called weakly analytic if when the function f is constant on a nonempty open interval, then f is constant on the whole line containing this interval.*

Examples of weakly convex analytic functions are the *faithfully convex* functions, which can be defined as follows. Let $s : \mathbb{R}^p \to \mathbb{R}$ be a given strictly convex function, A a linear transformation from $\mathbb{R}^n \to \mathbb{R}^p$, and $b \in \mathbb{R}^p, l \in \mathbb{R}^n, \beta \in \mathbb{R}$. Then the class of faithfully convex functions can be represented as

$$f(x) := s(Ax + b) + \langle l, x \rangle + \beta.$$

Typical examples include the following functions and any of their combinations:

$$s(v) := \log \left(\sum_{k=1}^{p} e^{v_k} \right) \qquad \text{(geometric programs)},$$

$$s(v) := \|v\|^2 \qquad \text{(quadratic programs)},$$

$$s(v) := \sum_{k=1}^{p} p_k^{-1} \|v\|^{p_k} \qquad (l_p \text{ programs } 1 < p_k < +\infty).$$

Let us remark also that the class of weakly analytic functions does not contain the class of als functions (for example, take $\max\{0, x\}$). Conversely, the exponential function is faithfully convex but is not als. So the approach between these two classes is different, and we shall see that for weakly analytic functions we obtain zero duality gap but we do not achieve attainment in the infimum of the primal problem.

In all the rest of this section we make the following standing assumption.

Assumption A. For $i = 0, \ldots, m$, the functions $f_i : \mathbb{R}^n \to \mathbb{R}$ are convex real-valued and:
(i) For any $1 \le i \le r$ with $r \ge 1$, the functions f_i are weakly analytic.
(ii) There exists a feasible point $x_0 \in C$ such that $f_i(x_0) < 0$, $\forall i \in (r, m]$.
(iii) $\varphi(0)$ is finite.

To derive the desired duality result, the main task is to prove that the marginal function φ is lsc at zero. We begin with a series of technical results toward achieving this task. Define the perturbed feasible set by

$$C(u) := \{x \in \mathbb{R}^n \,|\, f_i(x) \le u_i, \ i = 1, \ldots, m\}.$$

We decompose the index set as

$$
\begin{aligned}
I_1 &= \{i \,|\, f_i(x) = 0, \ \forall x \in C(0)\}, \\
I_2 &= \{i \,|\, \exists x(i) \in C(0) \text{ with } f_i(x(i)) < 0\},
\end{aligned}
$$

so that $I_1 \cup I_2 = \{1, \ldots, m\}$. We will also use the following subsets:

$$
\begin{aligned}
C_1(u) &= \{x \,|\, f_i(x) \le u_i, \ \forall i \in I_1\}, \\
C_2(u) &= \{x \,|\, f_i(x) < u_i, \ \forall i \in I_2\}.
\end{aligned}
$$

Without loss of generality, we can suppose in the rest of the analysis that the sets I_1 and I_2 are both nonempty; otherwise, we set $C_1(u) = \mathbb{R}^n$ or $C_2(u) = \mathbb{R}^n$.

Lemma 5.4.1 *Suppose assumption A holds. Then:*
(a) $C_1(0) = \{x \,|\, f_i(x) = 0, \ \forall i \in I_1\}$.
(b) The set $C_1(0)$ is an affine space.
(c) Let L be the subspace such that $C_1(0) = L + x$, with $x \in C_1(0)$. Then

$$(C_1(0))_\infty = L = \{d \,|\, (f_i)_\infty(d) = (f_i)_\infty(-d) = 0, \ \forall i \in I_1\}.$$

Proof. Suppose (a) does not hold. Then there exists $x \in C_1(0)$ such that for some $j \in I_1$ one has $f_j(x) < 0$. Take any $v \in C_1(0) \cap C_2(0)$ and $t \in (0, 1)$. Then $z(t) = tx + (1 - t)v \in C_1(0)$ and for t sufficiently small we have $f_i(z(t)) < 0$ for any $i \in I_2 \cup \{j\}$, which contradicts the definition of I_2. The statement in (b) is an immediate consequence of the representation given in (a) and the definition of weakly analytic functions. To prove (c), note that $(C_1(0))_\infty = \{d \,|\, (f_i)_\infty(d) \le 0, \ \forall i \in I_1\}$, and therefore by the definition of weakly analytic functions, it follows that $(C_1(0))_\infty = \{d \,|\, (f_i)_\infty(d) = (f_i)_\infty(-d) = 0, \ \forall i \in I_1\}$. Furthermore, since $C_1(0) = L + x$, with $x \in C_1(0)$, it follows that $L = (C_1(0))_\infty$. $\qquad \square$

Lemma 5.4.2 *Let $\{x_k\} \subset C_1(u^k)$ with $u^k \to 0$. Then there exist $v_k \to 0$ and $w_k \in C_1(0)$ such that $x_k = v_k + w_k$.*

Proof. Any point x_k can be written as $x_k = r_k + s_k$, with $r_k = P_L(x_k)$ and $s_k = P_{L^\perp}(x_k)$, where P_L denotes the projection operator on L. Since $x_k \in C_1(u^k)$, we have by Lemma 5.4.1, $f_i(s_k) = f_i(x_k) \le u_i^k$, $\forall i \in I_1$. Moreover, the sequence $\{s_k\}$ is bounded. Indeed, if not, then we can suppose without loss of generality that $\|s_k\| \to +\infty$, with $\|s_k\|^{-1} s_k \to d \in L^\perp$, $d \ne 0$. Using the last inequality we have $\|s_k\|^{-1} f_i(\|s_k\|^{-1} s_k \|s_k\|) \le \|s_k\|^{-1} u_i^k$ for all $i \in I_1$, and then passing to the limit, this implies that $(f_i)_\infty(d) \le 0$ for all $i \in I_1$, so that $d \in (C_1(0))_\infty$ and from Lemma 5.4.1 that $d \in L$, which clearly contradicts our starting hypothesis. Let s be any limit point of the bounded sequence $\{s_k\}$. Then passing to the limit in $f_i(s_k) = f_i(x_k) \le u_i^k$, $\forall i \in I_1$, it follows that $f_i(s) \le 0$, $\forall i \in I_1$ and $s \in C_1(0)$. Now let S be the set of limit points of the sequence $\{s_k\}$. Then, S is a nonempty compact set in $C_1(0)$. Furthermore, let $\bar{s}_k \in \operatorname{argmin}\{\|x - s_k\| \mid x \in S\}$ and set $v_k = s_k - \bar{s}_k$, $w_k = r_k + \bar{s}_k$. Then it follows that $v_k \to 0$ and $w_k \in C_1(0)$, which proves the desired result. $\qquad\square$

Lemma 5.4.3 *Let* $G(v) := v + C_1(0)$ *and let* $g(v) := \inf\{f_0(x) \mid x \in C_2(0) \cap G(v)\}$. *Then* $g(0) = \varphi(0)$, *and the function* g *is continuous at* 0.

Proof. First note that the set-valued map $v \to G(v) \cap C_2(0)$ is lsc at 0. Since f_0 is continuous, then it follows from Proposition 1.4.2 that g is upper semicontinuous at 0. Let $\{v_k\}$ be a sequence converging to 0. We have to prove that $g(0) \le \liminf_{k\to\infty} g(v_k)$. Take $x_k \in G(v_k) \cap C_2(0)$ such that $\liminf_{k\to\infty} f_0(x_k) = \liminf_{k\to\infty} g(v_k)$. Since g is upper semicontinuous at 0, there exist $a_k \in G(-v_k) \cap C_2(0)$ such that

$$\limsup_{k\to\infty} f_0(a_k) = \limsup_{k\to\infty} g(-v_k) \le g(0).$$

Set $z_k := 2^{-1}(x_k + a_k)$. Then $z_k \in G(0) \cap C_2(0)$, and therefore $g(0) \le f_0(z_k) \le 2^{-1}(f_0(x_k) + f_0(a_k))$. The last inequality can be rewritten as $g(0) \le f_0(x_k) - (g(0) - f_0(a_k))$, and passing to the limit together with the two previous limit arguments proves the statement. $\qquad\square$

Theorem 5.4.2 *Suppose assumption A holds. Then the function* φ *is lsc at the point* 0. *Consequently, there is no duality gap for the pair (P)–(D), where (D) is the usual dual problem associated with (P).*

Proof. Let $\varepsilon > 0$. Then there exist $x \in C_1(0) \cap C_2(0)$ such that $\varphi(0) - f_0(x) \ge -\varepsilon$. Let $x_k \in C(u^k)$ with $u^k \to 0$. From Lemma 5.4.2, there exist $v_k \to 0$, $w_k \in C_1(0)$ such that $x_k = v_k + w_k$. Therefore, $z_k := 2^{-1}(x_k + x) \in G(2^{-1} v_k)$, and for k sufficiently large, since $f_i(x) < 0$, $f_i(x_k) \le u_i^k$, $\forall i \in I_2$, it follows that $z_k \in C_2(0)$. Invoking Lemma 5.4.3, it thus follows that for k sufficiently large we have

$$f_0(z_k) \ge g\left(\frac{v_k}{2}\right) \ge g(0) - \varepsilon = \varphi(0) - \varepsilon.$$

Since f_0 is convex, the latter inequality implies that

$$f_0(x_k) \geq 2\varphi(0) - f_0(x) - 2\varepsilon,$$

and then using $\varphi(0) - f_0(x) \geq -\varepsilon$, we obtain $f_0(x_k) \geq \varphi(0) - 3\varepsilon$. Passing to the limit in the last inequality, it follows that $\liminf_{k \to \infty} \varphi(u^k) \geq \varphi(0) - 3\varepsilon$, and since $\varepsilon > 0$ was arbitrary, this shows that φ is lsc at 0. $\square$

5.5   Duality and Asymptotic Functions

In Section 2.8 we have seen that most optimization problems can be represented via composite models with asymptotic functions. Let $H : \mathbb{R}^m \to \mathbb{R} \cup \{+\infty\}$ be a generating asymptotic kernel (cf. Definition 2.8.1). Then the composite model is

$$(\text{CM}) \quad v = \inf\{\phi(x) \mid x \in \mathbb{R}^n\}$$

with

$$\phi(x) = \begin{cases} f_0(x) + H_\infty(f_1(x), \ldots, f_m(x)) & \text{if } x \in \bigcap_{i=1}^m \text{ dom } f_i, \\ +\infty & \text{otherwise.} \end{cases}$$

Recall that the corresponding approximate model is given by the problem

$$(\text{CM})_r \quad v_r = \inf\{\phi_r(x) \mid x \in \mathbb{R}^n\}$$

with

$$\phi_r(x) = \begin{cases} f_0(x) + H_r(f_1(x), \ldots, f_m(x)) & \text{if } x \in \bigcap_{i=1}^m \text{ dom } f_i, \\ +\infty & \text{otherwise,} \end{cases}$$

and $H_r(y) = rH(r^{-1}y)$, for $r > 0$ and any $y \in \mathbb{R}^m$. We are interested in deriving duality results for these classes of problems. We assume throughout this section that the problem (CM) is convex, namely that H is isotone, and each function f_i, $i = 0, \ldots, m$, is lsc, proper, and convex and that there exists $x_0 \in \text{dom } f_0$ such that $F(x_0) \in \text{dom } H_\infty$, where $F(x) := (f_1(x), \ldots, f_m(x))$. When $0 \notin \text{dom } H$ we suppose in addition that $F(x_0) \in \text{ri dom } H_\infty$. Then by Lemma 2.8.2, ϕ and ϕ_r for any $r > 0$ are lsc proper convex functions and (CM) and (CM)$_r$ are convex problems.

We associate then a dual problem with (CM) by considering the perturbation function

$$\Phi(x, u) = \begin{cases} f_0(x) + H_\infty(F(x) + u) & \text{if } x \in \bigcap_{i=0}^m \text{ dom } f_i, \\ +\infty & \text{otherwise.} \end{cases}$$

Clearly, Φ is proper, lsc, and convex. Thus, thanks to the duality scheme of Section 5.1, the dual problem via this perturbation function is

$$\sup\{-\Phi^*(0, y) \mid y \in \mathbb{R}^m\},$$

where Φ^* denotes the Fenchel conjugate of Φ, which may be computed as

$$\begin{aligned}
\Phi^*(0, y) &= \sup_{x,u}\{\langle x, 0\rangle + \langle u, y\rangle - f_0(x) - H_\infty(F(x) + u)\} \\
&= \sup_x\{-f_0(x)\} + \sup_u\{\langle u + F(x), y\rangle - \langle y, F(x)\rangle - H_\infty(F(x) + u)\}.
\end{aligned}$$

Since $H_\infty(\cdot) = \delta^*_{\mathrm{cl\,dom}\,H^*}(\cdot)$, it follows that

$$\Phi^*(0, y) = \sup_x\ \{-f_0(x) - (y, F(x)) + \delta_{\mathrm{cl\,dom}\,H^*}(y)\},$$

and thus the dual problem of (CM) can be written as

$$\text{(DCM)} \qquad w = \sup\{h(y) \mid y \in \mathbb{R}^m\} \qquad\qquad (5.5)$$

with

$$h(y) = \begin{cases} \inf_x\{f_0(x) + \langle y, F(x)\rangle\} & \text{if } y \in \mathrm{cl\,dom}\,H^*, \\ -\infty & \text{otherwise.} \end{cases}$$

Applying Theorem 5.1.4 to the pair of primal–dual problems (CM)–(DCM) we thus immediately obtain the following duality result.

Theorem 5.5.1 *Consider the convex problem (CM) with associated dual (DCM). Suppose that there exists $x_0 \in \mathrm{dom}\,f_0$ such that $F(x_0) \in \mathrm{int\,dom}\,H_\infty$. Then, if $v = \inf_x \phi(x)$ is finite, one has $v = w$, and the solution set of the dual problem is nonempty and compact.*

Example 5.5.1 *(Duality for standard convex programs)* We have seen in Section 2.8 that the standard convex programming problem

$$\inf\{f_0(x) \mid f_i(x) \leq 0,\ i = 1, \ldots, m\}$$

can be written as $\inf_x\{f_0(x) + H_\infty(F(x))\}$ with $H_\infty = \delta_{\mathbb{R}^m}$, and since $\delta_{\mathbb{R}^m_-} = \delta^*_{\mathbb{R}^m_+}$, then $\mathrm{dom}\,H^* = \mathrm{cl\,dom}\,H^* = \mathbb{R}^m_+$, and the usual dual problem is recovered via the formula (5.5) for $h(y)$. Furthermore, Theorem 5.5.1 can be applied, and since here $\mathrm{int\,dom}\,H_\infty = \mathbb{R}^m_-$, the required assumption thereto, $F(x_0) \in \mathrm{int\,dom}\,H_\infty$, translates to the standard Slater condition $\exists x_0 \in \mathrm{dom}\,f_0 : F(x_0) < 0$.

Similarly, we can associate a dual problem with the approximate problem $(\text{CM})_r$ by considering the perturbation function

$$\Phi_r(x, u) = \begin{cases} f_0(x) + rH(r^{-1}(F(x) + u)) & \text{if } x \in \bigcap_{i=0}^m \mathrm{dom}\,f_i, \\ +\infty & \text{otherwise.} \end{cases}$$

The function Φ_r is lsc, proper, and convex, and the dual problem via this perturbation function is

$$\sup\{-\Phi_r^*(0, y) \mid y \in \mathbb{R}^m\}.$$

An easy computation shows that

$$\Phi_r^*(0, y) = \sup_x \{-f_0(x) - \langle y, F(x)\rangle - rH^*(y) + \delta_{\text{cl dom } H^*}(y)\},$$

and thus the dual problem of $(\text{CM})_r$ can be written as

$$(\text{DCM})_r \qquad w_r = \sup\{h_r(y) \mid y \in \mathbb{R}^m\}, \qquad (5.6)$$

with

$$h_r(y) = \begin{cases} \inf_x\{f_0(x) + \langle y, F(x)\rangle\} - rH^*(y) & \text{if } y \in \text{cl dom } H^*, \\ -\infty & \text{otherwise.} \end{cases}$$

Once again, a dual result similar to Theorem 5.5.1 can then be stated for the pair of primal–dual problems $(\text{CM})_r$ and $(\text{DCM})_r$. Note that the dual problem $(\text{DCM})_r$ has another interesting interpretation; namely, it can be viewed as a *viscosity* regularization method, and the duality results can be used to analyze the convergence of the corresponding trajectories $\{y_r\}$ as $r \to 0^+$, solutions of $(\text{DCM})_r$, to the solution of the original dual problem. To illustrate this situation, we give one such a result.

Proposition 5.5.1 *Under the assumptions of Theorem 5.5.1 and assuming that $0 \in \text{cl dom } H$, then for every $r > 0$ one has $v_r = w_r$, and the set of dual optimal solutions of $(DCM)_r$ is nonempty and compact. Moreover, if $v_r \to v$, each sequence $\{y_r\}$ optimal solution of $(DCM)_r$ is bounded, and as $r \to 0^+$, every limit point of the sequence $\{y_r\}$ is a dual optimal solution of (DCM).*

Proof. Under the stated hypothesis, from Theorem 5.5.1 the set of dual optimal solutions for (DCM) is nonempty and compact. Setting $p := -h$, where h is the dual objective of (DCM), so that p is a proper lsc convex function, this is equivalent to $p_\infty(y) > 0 \quad \forall y \neq 0$. Furthermore, since $h_r = -p - rH^*$, one has

$$\begin{aligned} t_\infty^r(y) &:= (-h_r)_\infty(y) := p_\infty(y) + r(H)_\infty^*(y), \\ (H)_\infty^* &= \delta_{\text{cl dom } H}^*. \end{aligned}$$

Then under the assumption $0 \in \text{cl dom } H$ we deduce

$$t_\infty^r(y) \geq p_\infty(y) > 0 \quad \forall y \neq 0,$$

and it follows that the set of dual optimal solutions of the approximate dual problem $(\text{DCM})_r$ is a nonempty compact set and that $v_r = w_r$.

Now let us prove that the sequence $\{y_r\}_{r>0}$ is bounded when $r \to 0^+$. Suppose the contrary, that is, if $\{y_r\}_{r>0}$ is not bounded, then we can find $r_k \to 0^+$ such that

$$\|y_{r_k}\| \to +\infty, \quad \frac{y_{r_k}}{\|y_{r_k}\|} \to \bar{y} \neq 0.$$

Since $v_r \to v$ and $v = w$, one has $w_r \to w$, and then for $\varepsilon > 0$ we have for k sufficiently large,

$$\frac{p(y_{r_k})}{\|y_{r_k}\|} + r_k \frac{H^*(y_{r_k})}{\|y_{r_k}\|} \leq \frac{w + \varepsilon}{\|y_{r_k}\|}, \quad \frac{H^*(y_{r_k})}{\|y_{r_k}\|} \geq (H^*)_\infty(\bar{y}) - \varepsilon,$$

and then passing to the limit we obtain

$$p_\infty(\bar{y}) \leq 0, \ \bar{y} \neq 0,$$

in contradiction to the compactness of the set of dual solutions of (DCM), i.e., with $p_\infty(y) > 0$, $\forall y \neq 0$. Finally, let $\varepsilon > 0$. Then for r sufficiently small we have

$$p(y_r) + rH^*(y_r) \leq w + \varepsilon.$$

Then if $\bar{y}$ is a limit point of $\{y_r\}$, since H^* and p are lsc functions, passing to the limit we deduce $p(\bar{y}) \leq w + \varepsilon$, and then with $\varepsilon \to 0^+$ if follows that $\bar{y}$ is an optimal dual solution of (DCM). $\quad\square$

Note that for convex constrained problems, coercivity of the primal problem ensures the hypothesis $v_r \to v$ made in Proposition 5.5.1. In the more general case, it is necessary to add some technical assumptions that are satisfied in most problems of interest; see the Notes and References for details.

Example 5.5.2 Duality for Semidefinite Programming Consider the semidefinite optimization problem introduced in Example 2.8.2,

$$\text{(SDP)} \quad \inf c^T x \text{ subject to } B(x) \preceq 0,$$

with $B(x) = B_0 + \sum_{i=1}^m x_i B_i$. The problem data are the vector $c \in \mathbb{R}^m$ and the $(m + 1)$ symmetric matrices $B_0, \ldots, B_m$ of order $n \times n$. We have seen that (SDP) can be written as the composite model

$$\text{(SDP)} \quad \inf_x \left\{ c^T x + H_\infty(B(x)) \right\},$$

and the corresponding approximate (SDP) is then given by

$$\text{(SDP)}_r \quad \inf_x \{ c^T x + H_r(B(x)) \},$$

with $H_r(y) = r^{-1}H(ry)$. Then, using (5.5), it is easy to see that the dual problem of (SDP) is given by

$$\text{(DSDP)}\quad w = \sup\{\text{tr}\, B_0.Z \mid -\text{tr}\, B_i.Z = c_i,\ i = 1,\ldots,m,\ Z \succeq 0\}.$$

Similarly, it follows from (5.6) that the dual of the corresponding approximate problem (SDP)_r is given by

$$\text{(DSDP)}_r\ w_r = \sup\{\text{tr}\, B_0.Z - rH^*(Z) \mid -\text{tr}\, B_iZ = c_i,\ i = 1,\ldots,m,\ Z \succeq 0\}.$$

Under the corresponding Slater assumption for problem (SDP), i.e., that there exists x_0 such $B(x_0) \prec 0$, the duality results established above hold true for the semidefinite programs. An interesting example widely used in Barrier type algorithms for the function H is the choice H_4 given in Section 2.8. Its conjugate, computed via Theorem 2.7.1(a), is given by

$$H_4^*(D) = \begin{cases} -n - \log \det D & \text{for } D \succ 0, \\ +\infty & \text{otherwise,} \end{cases}$$

where $\det D$ stands for the determinant of the matrix D.

5.6   Lagrangians and Minimax Theory

A very general and convenient way to represent and analyze optimization problems as well as many other problems arising in variational analysis (cf. Chapter 6) is via the concept of convex–concave functionals defined on a product space.

Lagrangians

The perturbation functional Φ associated with an abstract optimization problem

$$\text{(P)}\quad \inf\{f(x) \mid x \in \mathbb{R}^n\}$$

leads naturally to the definition of a corresponding Lagrangian.

Definition 5.6.1 *Let $\Phi : \mathbb{R}^n \times \mathbb{R}^m \to \mathbb{R} \cup \{+\infty\}$ be a proper perturbation function for problem (P), such that $\Phi(x, u)$ is lsc and convex in u. The Lagrangian $l : \mathbb{R}^n \times \mathbb{R}^m \to \overline{\mathbb{R}}$ of problem (P) corresponding to Φ is defined by*

$$l(x, y) = \inf\{\Phi(x, u) - \langle u, y \rangle \mid u \in \mathbb{R}^m\}.$$

Similarly, if $l(x, y)$ is the Lagrangian corresponding to problem (P), then the associated perturbation function is given by

$$\Phi(x, u) = \sup\{l(x, y) + \langle u, y \rangle \mid y \in \mathbb{R}^m\}.$$

Note that for each $x \in \mathbb{R}^n$, the function $y \to -l(x,y)$ is just the conjugate of the function $u \to \Phi(x,u)$. To preserve symmetry, namely to ensure that the conjugate of $y \to -l(x,y)$ is exactly $\Phi(x,u)$, one needs the extra condition made in the definition of $\Phi(x,u)$ in the variable u, which guarantees that $\Phi(x,\cdot)$ coincides with its biconjugate.

Lagrangian functions lead naturally to the concept of saddle points and functions.

Definition 5.6.2 Let $l : \mathbb{R}^n \times \mathbb{R}^m \to \overline{\mathbb{R}}$. Then $(\bar{x}, \bar{y})$ is called a saddle point of l if
$$l(\bar{x}, y) \leq l(\bar{x}, \bar{y}) \leq l(x, \bar{y}), \quad \forall (x,y) \in \mathbb{R}^n \times \mathbb{R}^m.$$
or equivalently if $\inf_x l(x,y) = l(\bar{x}, \bar{y}) = \sup_y l(x,y)$.

The set of all saddle points is denoted by $\arg\min\max_{x,y} l$, and the common value $l(\bar{x}, \bar{y})$ is called the saddle value of l.

In terms of the general Lagrangian l given in Definition 5.6.1 one can define a primal problem

$$(\text{P}) \quad \inf_{x \in \mathbb{R}^n} f(x); \quad \text{with } f(x) = \Phi(x,0) = \sup_{y \in \mathbb{R}^m} l(x,y).$$

Likewise, since

$$-\Phi^*(v,y) = \inf_{x,u}\{-\langle v,x\rangle - \langle y,u\rangle + \Phi(x,u)\} = \inf_x\{l(x,y) - \langle v,x\rangle\},$$

the dual problem (D) is defined by

$$(\text{D}) \quad \sup_{y \in \mathbb{R}^m} h(y); \quad \text{with } h(y) = -\Phi^*(0,y) = \inf_{x \in \mathbb{R}^n} l(x,y).$$

We thus have immediately the minimax inequality

$$\inf_x \sup_y l(x,y) \geq \sup_y \inf_x l(x,y),$$

which corresponds to weak duality. Furthermore, in terms of saddle functions the equality corresponds to the existence of a saddle point, namely $(\bar{x}, \bar{y}) \in \arg\min\max l(x,y)$ if and only if

$$\bar{x} \in \operatorname{argmin}_x f(x); \quad \bar{y} \in \operatorname{argmax}_y h(y); \quad \text{and} \quad \inf_x \sup_y l(x,y) = \sup_y \inf_x l(x,y).$$

Note that from the above development, the set of saddle points is nonempty whenever $\inf f = \sup h$ and always given as the product set

$$\operatorname{argmin}\max l = \operatorname{argmin} f \times \operatorname{argmax} h.$$

We summarize this in the following proposition.

Proposition 5.6.1 *Let l be a Lagrangian associated with a perturbation function $\Phi : \mathbb{R}^n \times \mathbb{R}^m \to \mathbb{R} \cup \{+\infty\}$. Then:*
(a) $\inf_x f(x) = \inf_x \sup_y l(x, y) \geq \sup_y \inf_x l(x, y) = \sup_y h(y)$.
(b) $(\bar{x}, \bar{y})$ is a saddle point of l if and only if $\bar{x}$ is an optimal solution of (P), $\bar{y}$ is an optimal solution of (D), and $\inf_x \sup_y l(x, y) = \sup_y \inf_x l(x, y)$.

As already noted, from Definition 5.6.1, since for each $x \in \mathbb{R}^n$, the function $y \to -l(x, y)$ is the conjugate of $\Phi(x, \cdot)$, then $-l(x, y)$ is lsc and convex in y.

Lagrangian functions defined on a jointly convex perturbation Φ enjoy further convexity properties.

Proposition 5.6.2 *Let $\Phi : \mathbb{R}^n \times \mathbb{R}^m \to \mathbb{R} \cup \{+\infty\}$ be a given proper perturbation such that $\Phi(x, u)$ is lsc in u and with associated Lagrangian $l(x, y)$. Then the function $x \to l(x, y)$ is convex if and only if $\Phi(x, u)$ is jointly convex in (x, u). In the latter case, one also has*

$$(v, y) \in \partial\Phi(x, u) \iff v \in \partial_x l(x, y), u \in \partial_y[-l](x, y).$$

Proof. For any $y \in \mathbb{R}^m$, under the given assumption on Φ, the function $(x, u) \to G_y(x, u) := \Phi(x, u) - \langle y, u \rangle$ is convex in (x, u), and therefore since by definition, one has $l(x, y) = \inf_u G_y(x, u)$, the convexity of $x \to l(x, y)$ follows from Proposition 1.2.2. Conversely, the convexity of $x \to l(x, y)$ implies the convexity of $(x, u) \to H_y(x, u) := l(x, y) + \langle u, y \rangle$, and since by Definition 5.6.1, the function $\Phi(x, u)$ is the pointwise supremum of the family of convex functions H_y for $y \in \mathbb{R}^m$, then Φ is convex in (x, u). Using the subgradient inequality for the convex function Φ one has $(\bar{v}, \bar{y}) \in \partial\Phi(\bar{x}, \bar{u})$ if and only if

$$\Phi(x, u) \geq \Phi(\bar{x}, \bar{u}) + \langle \bar{v}, x - \bar{x} \rangle + \langle \bar{y}, u - \bar{u} \rangle, \ \forall(x, u).$$

Take $x = \bar{x}$ in the inequality above. This implies $y \in \partial_u\Phi(\bar{x}, \bar{u})$, which is equivalent to $\bar{u} \in \partial_y[-l](\bar{x}, \bar{y})$ (recall that by definition $-l(\bar{x}, \cdot)$ is the conjugate of $\Phi(\bar{x}, \cdot)$). Furthermore, from the same inequality one also has

$$\inf_u \{\Phi(x, u) - \langle \bar{y}, u \rangle \geq \Phi(\bar{x}, \bar{u}) - \langle \bar{y}, \bar{u} \rangle + \langle \bar{v}, x - \bar{x} \rangle, \ \forall x,$$

which can be rewritten in terms of l as $l(x, \bar{y}) \geq l(\bar{x}, \bar{y}) + \langle \bar{v}, x - \bar{x} \rangle, \ \forall x$, and the latter means that $\bar{v} \in \partial_x l(\bar{x}, \bar{y})$. $\quad\square$

An interesting example for the definition of the Lagrangian above is well illustrated by considering the classical convex programs already discussed in Section 5.3,

$$\text{(P)} \quad \inf\{f_0(x) \mid f_i(x) \leq 0, \ i = 1, \ldots, m \ x \in \mathbb{R}^n\},$$

with the perturbation function given for $u \in \mathbb{R}^m$ by

$$\Phi(x, u) = \begin{cases} f_0(x) & \text{if } f_i(x) + u_i \leq 0, \ i = 1, \ldots, m, \\ +\infty & \text{otherwise.} \end{cases}$$

Let $F(x) = (f_1(x), \ldots, f_m(x))^T$. From Definition 5.6.1, one thus has

$$l(x, y) = \inf_{F(x)+u \leq 0} \{f_0(x) - \langle u, y \rangle\}.$$

Since $\mathrm{dom}\, f_0 \subset \cap_{i=1}^m \mathrm{dom}\, f_i$, then for a fixed $x \in \mathrm{dom}\, f_0$, clearly the above infimum is attained for $u = -F(x)$ whenever $y \geq 0$, and it takes the value $-\infty$ if $y \notin \mathbb{R}_+^m$, while for $x \notin \mathrm{dom}\, f_0$, one clearly has $l(x, y) = +\infty$. Summarizing the above computation, one has then obtained

$$l(x, y) = \begin{cases} f_0(x) + \langle y, F(x) \rangle & \text{if } x \in \mathrm{dom}\, f_0, y \in \mathbb{R}_+^m, \\ -\infty & \text{if } x \in \mathrm{dom}\, f_0, y \notin \mathbb{R}_+^m, \\ +\infty & \text{if } x \notin \mathrm{dom}\, f_0. \end{cases}$$

One thus recognizes the usual Lagrangian associated with the convex program (P) as defined in Section 5.3.

Minimax Theory for Convex–Concave Problems

The theory of dual optimization problems can in fact be cast within the more general convex–concave minimax theory. Our objective here is to demonstrate once more the power of the abstract duality framework, especially of Theorem 5.1.4 and Theorem 5.1.5, in the context of minimax problems and to derive the corresponding minimax theorems. For that purpose, we first need to define a viable convex–concave functional, which will cover every reasonable minimax problem in applications.

Definition 5.6.3 *Let C and D be nonempty convex sets of $\mathbb{R}^n \times \mathbb{R}^m$ and let $K : \mathbb{R}^n \times \mathbb{R}^m \to \overline{\mathbb{R}}$. We say that K is a convex–concave closed function if the following two conditions hold:*
(a) $K(x, y)$ is finite-valued on $C \times D$, and

$$K(x, y) = \begin{cases} -\infty & \text{if } x \in C, \ y \notin D, \\ +\infty & \text{if } x \notin C. \end{cases}$$

(b) For each $(x, y) \in C \times D$, the functions $K(\cdot, y)$ and $-K(x, \cdot)$ are lsc, proper, convex functions.

A typical example of a closed convex-concave function is provided by the Lagrangian $L(x, y)$ associated with the convex program as defined in Section 5.3 with $C = \mathrm{dom}\, f_0$ and $D = \mathbb{R}_+^m$.
Associated with K we define

$$f(x) = \begin{cases} \sup_{y \in D} K(x, y) & \text{if } x \in C, \\ +\infty & \text{if } x \notin C, \end{cases}$$

and

$$h(y) = \begin{cases} \inf_{x \in C} K(x,y) & \text{if } y \in D, \\ -\infty & \text{if } y \notin D. \end{cases}$$

Then the functions f and $-h$ are lsc convex functions. For each $x \in C$, one can take the supremum of $K(x, \cdot)$ over the set D followed by the infimum over the set C to obtain a *primal* convex minimization problem:

$$\text{(P)} \quad V_P = \inf_{x \in C} \sup_{y \in D} K(x,y) = \inf_{x \in \mathbb{R}^n} \sup_{y \in \mathbb{R}^m} K(x,y) = \inf_{x \in \mathbb{R}^n} f(x).$$

Similarly, by reversing the construction above, one can construct a *dual* concave maximization problem:

$$\text{(D)} \quad V_D = \sup_{y \in D} \inf_{x \in C} K(x,y) = \sup_{y \in \mathbb{R}^m} \inf_{x \in \mathbb{R}^n} K(x,y) = \sup_{y \in \mathbb{R}^m} h(y).$$

The primal–dual terminology used here will be further justified below. Obviously, by construction of the above pair of problems (P)–(D) one always has the weak duality relation

$$V_P = \inf_{x \in C} \sup_{y \in D} K(x,y) \geq \sup_{y \in D} \inf_{x \in C} K(x,y) = V_D,$$

but in general, we do not have the equality $V_P = V_D$. In fact, when this equality holds, following Definition 5.6.2, we will say that there exists a saddle point for K, namely $(\bar{x}, \bar{y})$, is a saddle point of K if $(\bar{x}, \bar{y}) \in C \times D$ and

$$K(\bar{x}, y) \leq K(\bar{x}, \bar{y}) \leq K(x, \bar{y}), \quad \forall (x,y) \in C \times D.$$

We then call $K(\bar{x}, \bar{y}) = V_P = V_D$ the saddle value of K. The following proposition summarizes the above concepts.

Proposition 5.6.3 *For any closed convex–concave functional $K : \mathbb{R}^n \times \mathbb{R}^m \to \overline{\mathbb{R}}$, one has that $(\bar{x}, \bar{y})$ is a saddle point of K if and only if $V_P = V_D$, with $\bar{x}$ an optimal solution of (P) and $\bar{y}$ an optimal solution of (D).*

We now turn to the question of finding the appropriate conditions that will lead us to establish minimax theorems, namely strong duality results. These results will be obtained as a direct application of the abstract duality framework of Section 5.1 through the use of appropriate perturbation functions and the corresponding marginal functions associated with problems (P)–(D).

Given a closed convex–concave function K, let $\Phi : \mathbb{R}^n \times \mathbb{R}^m \to \mathbb{R} \cup \{+\infty\}$ be the bi-function defined by

$$\Phi(x,u) = \sup\{\langle u, y \rangle + K(x,y) \mid y \in D\},$$

and let $\varphi : \mathbb{R}^m \to \overline{\mathbb{R}}$ defined by

$$\varphi(u) = \inf_x \Phi(x,u) = \inf_{x \in C} \sup_{y \in D} \{\langle u, y \rangle + K(x,y)\}$$

be the marginal function associated with (P) through Φ. Clearly, one has $\varphi(0) = \inf_{x \in C} \sup_{y \in D} K(x, y) = V_P$.

It will be convenient in the rest of the analysis to use the following notation. For any $x \in C$, let $t_x : \mathbb{R}^m \to \mathbb{R} \cup \{+\infty\}$ be defined by

$$t_x(y) := \begin{cases} -K(x, y) & \text{if } y \in D, \\ +\infty & \text{otherwise.} \end{cases} \tag{5.7}$$

Since K is closed and convex–concave, it follows from Definition 5.6.3 that for any $x \in C$, the function t_x is lsc convex on $\mathbb{R}^m$. The next proposition gives all the useful and necessary facts for the derivation of the main minimax theorems.

Proposition 5.6.4 *Let K be a closed convex-concave function on $\mathbb{R}^n \times \mathbb{R}^m$ and Φ the bi-function associated with K with marginal function $\varphi : \mathbb{R}^m \to \overline{\mathbb{R}}$ defined by*

$$\varphi(u) = \inf_{x \in C} \Phi(x, u) = \inf_{x \in C} \sup_{y \in D} \{\langle u, y \rangle + K(x, y)\}.$$

Then:
(a) Φ is proper, lsc and convex.
(b) $\operatorname{dom} \varphi = \cup_{x \in C} \operatorname{dom} -K(x, \cdot)^*$.
(c) $\varphi^{**}(0) = \sup_{y \in D} \inf_{x \in C} K(x, y) = \sup_y h(y) = V_D$.
(d) $\sigma_{\operatorname{dom} \varphi}(d) = \sup_{x \in C} -K(x, \cdot)_\infty(d) = \sup_{x \in \operatorname{ri} C} -K(x, \cdot)_\infty(d), \forall d \in \mathbb{R}^m$.
(e) Problem (D) is the dual of (P) via the perturbation function Φ.

Proof. Clearly, by the definition of φ with $u = 0$ one has

$$\varphi(0) = \inf_x \sup_y K(x, y) = \inf_x f(x) = V_P.$$

The function $(x, u) \to \langle u, y \rangle + K(x, y)$ is convex for any $y \in D$, and thus as a pointwise supremum, the function Φ is convex and lsc. Furthermore, since for $x \in C$ one has $\Phi(x, \cdot) = (-K(x, \cdot))^*$, it follows that $\Phi(x, \cdot)$ is proper, which in turns implies that Φ is proper, proving (a). Since K is closed and convex–concave, then for any $x \in C$, the function t_x defined in (5.7) is proper, lsc, and convex on $\mathbb{R}^m$ with conjugate given by

$$t_x^*(u) = \sup_{y \in \mathbb{R}^m} \{\langle u, y \rangle - t_x(y)\} = \sup_{y \in D} \{\langle u, y \rangle + K(x, y)\}.$$

Therefore, we can write $\varphi(u) = \inf_{x \in C} t_x^*(u)$, and hence one obtains $\operatorname{dom} \varphi = \cup_{x \in C} \operatorname{dom} t_x^* = \cup_{x \in C} \operatorname{dom} -K(x, \cdot)^*$, proving (b). Using Proposition 1.2.12, the conjugate of φ is easily seen to be $\varphi^*(y) = \sup_{x \in C} t_x^{**}(y)$, but since t_x is lsc, then $t_x^{**} = t_x$, and thus $\varphi^*(y) = \sup_{x \in C} t_x(y)$. Now, since the dual (D) of (P) via the perturbation function Φ consists in maximizing $-\varphi^*(y)$

over $y \in \mathbb{R}^m$, it follows from the last formula and the definition of t_x that $-\varphi^*(y) = h(y)$, proving (e). We can now compute the biconjugate of φ,

$$
\begin{aligned}
\varphi^{**}(u) &= \sup_y \{\langle u, y \rangle - \varphi^*(y)\} = \sup_y \{\langle u, y \rangle - \sup_{x \in C} t_x(y)\}, \\
&= \sup_y \inf_{x \in C} \{\langle u, y \rangle - t_x(y)\} = \sup_{y \in D} \inf_{x \in C} \{\langle u, y \rangle + K(x, y)\},
\end{aligned}
$$

and with $u = 0$ this proves (c). To prove the first equality in (d), we apply the support functions calculus rules given in Proposition 1.3.3(d) followed by Theorem 2.5.4(b), which expressed the support function of the domain of a conjugate of an lsc convex function in terms of its asymptotic function. Thus, with $\operatorname{dom} \varphi = \cup_{x \in C} \operatorname{dom} t_x^*$ proven in (b) this gives

$$
\begin{aligned}
\sigma_{\operatorname{dom} \varphi}(d) = \sigma_{\cup_{x \in C} \operatorname{dom} t_x^*}(d) &= \sup_{x \in C} \sigma_{\operatorname{dom} t_x^*}(d) \\
&= \sup_{x \in C} (t_x)_\infty(d) = \sup_{x \in C} -K(x, \cdot)_\infty(d).
\end{aligned}
$$

To prove the second equality in (d), recall that by Proposition 1.3.2(e) for any convex set S one has $\sigma_S = \sigma_{\operatorname{ri} S}$, and hence since φ is convex, $\sigma_{\operatorname{dom} \varphi} = \sigma_{\operatorname{ri} \operatorname{dom} \varphi}$. Since by (b), $\operatorname{dom} \varphi = \{u \mid \exists x \text{ with } (u, x) \in \operatorname{dom} \Phi\}$, invoking Proposition 1.1.9 on relative interiors one obtains

$$
\operatorname{ri} \operatorname{dom} \varphi = \{u \mid \exists x \in \operatorname{ri} C \text{ with } u \in \operatorname{ri} \operatorname{dom} \Phi(x, \cdot)\} = \cup_{x \in \operatorname{ri} C} \operatorname{ri} \operatorname{dom} t_x^*.
$$

Therefore, applying the same chain of arguments given at the beginning of the proof of (d) we obtain

$$
\begin{aligned}
\sigma_{\operatorname{dom} \varphi}(d) &= \sigma_{\operatorname{ri} \operatorname{dom} \varphi}(d) = \sup_{x \in \operatorname{ri} C} \sigma_{\operatorname{ri} \operatorname{dom} t_x^*}(d) \\
&= \sup_{x \in \operatorname{ri} C} \sigma_{\operatorname{dom} t_x^*} = \sup_{x \in \operatorname{ri} C} (t_x)_\infty(d) = \sup_{x \in \operatorname{ri} C} -K(x, \cdot)_\infty(d),
\end{aligned}
$$

which completes the proof. $\qquad \square$

A similar result can be established for the corresponding dual problem (D). For this it suffices to interchange the roles of (P) and (D). Indeed, consider the convex problem

$$
(D) \quad -V_D = \inf_{y \in D} \sup_{x \in C} -K(x, y)
$$

with associated bi-function and marginal function defined by

$$
\Psi(y, v) = \sup_{x \in C} \{\langle v, x \rangle - K(x, y)\}; \quad \psi(v) = \inf_{y \in D} \Psi(y, v).
$$

Clearly, it holds that $-\psi(0) = \sup_{y \in D} \inf_{x \in C} K(x, y)$, and therefore as an immediate consequence of Proposition 5.6.4 one obtains the following.

Proposition 5.6.5 *Let K be a closed convex–concave function on $\mathbb{R}^n \times \mathbb{R}^m$ and Ψ the bi-function associated to $-K$, with marginal function $\psi : \mathbb{R}^n \to \overline{\mathbb{R}}$. Then:*
(a) The function Ψ is proper, lsc, and convex.
(b) $\operatorname{dom}\psi = \cup_{y \in D} \operatorname{dom} K(\cdot, y)^*$.
(c) $-\psi^{**}(0) = \inf_{x \in C} \sup_{y \in D} K(x, y) = \inf_x f(x) = V_{\mathrm{P}}$.
(d) $\sigma_{\operatorname{dom}\psi}(d) = \sup_{y \in D} K(\cdot, y)_\infty(d) = \sup_{y \in \operatorname{ri} D} K(\cdot, y)_\infty(d), \ \forall d \in \mathbb{R}^n$.
(e) Problem (P) is the dual of (D) via the perturbation function Ψ.

We are now ready to give the minimax theorems.

Theorem 5.6.1 *Let K be a closed convex–concave function on $\mathbb{R}^n \times \mathbb{R}^m$ and consider the primal and dual problems (P) and (D) associated with K.*
(a) If V_{P} is finite and $0 \in \cup_{x \in \operatorname{ri} C} \operatorname{ri} \operatorname{dom} -K(x, \cdot)^*$, then the optimal set of the dual problem (D) is a nonempty set equal to the sum of a compact set with a linear space, and one has

$$\max_{y \in \mathbb{R}^m} \inf_{x \in \mathbb{R}^n} K(x, y) = \inf_{x \in \mathbb{R}^n} \sup_{y \in \mathbb{R}^m} K(x, y).$$

Furthermore, if we replace the relative interior operation on $\operatorname{dom} -K(x, \cdot)^*$ by interior, then the linear space reduces to the singleton $\{0\}$.
(b) Dually, if V_{D} is finite and $0 \in \cup_{y \in \operatorname{ri} D} \operatorname{ri} \operatorname{dom} K(\cdot, y)^*$, then the optimal set of the primal problem (P) is a nonempty set equal to the sum of a compact set with a linear space, and one has

$$\sup_{y \in \mathbb{R}^m} \inf_{x \in \mathbb{R}^n} K(x, y) = \min_{x \in \mathbb{R}^n} \sup_{y \in \mathbb{R}^m} K(x, y).$$

Furthermore, if we replace the relative interior operation on $\operatorname{dom} K(\cdot, y)^*$ by interior, then the linear space reduces to the singleton $\{0\}$.

Proof. We prove only (a), since (b) follows in the same manner. The results follow as an immediate application of Theorem 5.1.4. Indeed, from Proposition 5.6.4(d) one has

$$\operatorname{ri} \operatorname{dom}\varphi = \cup_{x \in \operatorname{ri} C} \operatorname{ri} \operatorname{dom} -K(x, \cdot)^*.$$

Therefore, the hypothesis (a) of the theorem means exactly that $\varphi(0)$ is finite and $0 \in \operatorname{ri} \operatorname{dom}\varphi$. $\qquad\square$

Several important consequences for minimax problems can be extracted from the above analysis. The first one is a translation of the hypothesis that guarantees the existence of a saddle point and saddle value in terms of asymptotic functions.

Corollary 5.6.1 *Let K be a closed convex–concave function on $\mathbb{R}^n \times \mathbb{R}^m$. Then the set of saddle points of K is a nonempty and compact set if*

*and only if the optimal sets of the primal and dual problems (P)–(D) are
nonempty and compact. The latter is equivalent to*

$$\sup_{x \in C} -K(x, \cdot)_\infty(v) > 0, \ \forall v \neq 0; \quad \sup_{y \in D} K(x, \cdot)_\infty(w) > 0, \ \forall w \neq 0. \qquad (5.8)$$

Proof. By definition of f and h, the conditions (5.8) mean that f and $-h$
are coercive. This is equivalent to saying that the optimal sets of problems
(P)-(D) are nonempty and compact, and with Proposition 5.6.3 the result
is proved.   □

As an immediate consequence, we obtain the following two classical minimax theorems.

Corollary 5.6.2 *Let K be a closed convex-concave function on $\mathbb{R}^n \times \mathbb{R}^m$
and suppose that the effective domain of K given by $C \times D$ is compact.
Then one has $V_P = V_D$ and the set of saddle points of K is nonempty and
compact.*

Corollary 5.6.3 *Let K be a closed convex–concave function on $\mathbb{R}^n \times \mathbb{R}^m$
and suppose that there exists some $(x, y) \in C \times D$ such that the functions
$K(\cdot, y)$ and $-K(x, \cdot)$ are coercive. Then, one has $V_P = V_D$ and the set of
saddle points of K is nonempty and compact.*

5.7   Duality and Stationary Sequences

We are back to the duality scheme of Section 5.1 and with the same notation and definitions for the pair of primal–dual optimization problems.
We suppose for all this section that the perturbation function Φ is jointly
convex, proper, and lsc. Quite often, for solving the convex optimization
problem (P) $\inf\{\Phi(x, 0) \mid x \in \mathbb{R}^n\}$, one can provide a suitable algorithm
for solving the dual problem (D) associated with (P) through the perturbation function Φ, namely

$$(D) \quad -V_D = \inf\{\Phi^*(0, y) \mid y \in \mathbb{R}^m\} = \inf\{\varphi^*(y) \mid y \in \mathbb{R}^m\},$$

where φ is the marginal function corresponding to Φ. We generate a stationary sequence $\{y_k\}$ associated with the subgradients $u_k \in \partial\varphi^*(y_k)$ such that
$u_k \to 0$. A key question that then emerges is, given the sequences $\{y_k, u_k\}$,
and assuming the basic assumption of weak coercivity of the problems
(P)–(D), (ensuring in particular the nonemptiness of the primal solution
set S_P), how can we choose $x_k \in \mathbb{R}^n$ such that the primal path $\{x_k\}$ converges to the primal set S_P of optimal solutions? This question will be the
main preoccupation of this section, in which we provide an explicit answer.

We begin by defining the following auxiliary functions, which will be used in the analysis developed below. Given the perturbation function Φ associated with (P) define

$$
\begin{aligned}
\forall\, v \in \mathbb{R}^n, \; \varphi^v(u) &= \inf\{\Phi(x,u) - \langle v,x \rangle \mid x \in \mathbb{R}^n\}, \quad E = \mathrm{aff}(\mathrm{dom}\,\varphi), \\
\forall\, u \in \mathbb{R}^m, \; \psi^u(v) &= \inf\{\Phi^*(v,y) - \langle u,y \rangle \mid y \in \mathbb{R}^m\}, \quad M = \mathrm{aff}(\mathrm{dom}\,\psi).
\end{aligned}
$$

Theorem 5.7.1 *Consider the pair of primal–dual problems (P)–(D) and let Φ be the associated perturbation function, which is assumed jointly convex, proper, and lsc.*
(a) Suppose that $\varphi(0) \in \mathbb{R}$ and $0 \in \mathrm{ri}(\mathrm{dom}\,\varphi)$. Then, for each $v \in \mathbb{R}^n$ one has $E = \mathrm{aff}(\mathrm{dom}\,\varphi^v)$ and $\psi(v) = \min_{y \in \mathbb{R}^m} \Phi^*(v,y)$, the minimum being attained with the nonempty dual optimal solution set given by

$$
S_D(v) = \begin{cases} \mathbb{R}^m & \textit{if } \psi(v) = +\infty, \\ \partial\varphi^v(0) = \partial\varphi_E^v(0) + E^\perp & \textit{otherwise.} \end{cases}
$$

Furthermore, one has $\Phi^*(v,y) = \Phi^*(v,y+z)$, $\forall y \in \mathbb{R}^m$, $\forall z \in E^\perp$.
(b) Suppose that $\psi(0) \in \mathbb{R}$ and $0 \in \mathrm{ri}(\mathrm{dom}\,\psi)$. Then for each $u \in \mathbb{R}^m$ one has $M = \mathrm{aff}(\mathrm{dom}\,\psi^u)$ and $\varphi(u) = \min_{x \in \mathbb{R}^n} \Phi(x,u)$, the minimum being attained with the nonempty primal optimal solution set given by

$$
S_P(v) = \begin{cases} \mathbb{R}^n & \textit{if } \varphi(u) = +\infty, \\ \partial\psi^u(0) = \partial\psi_M^u(0) + M^\perp & \textit{otherwise.} \end{cases}
$$

Furthermore, one has $\Phi(x,u) = \Phi(x+w,u)$, $\forall x \in \mathbb{R}^n$, $\forall w \in M^\perp$.

Proof. We prove only the statement in (a), since (b) is just the dual statement, which follows with a similar proof. Since $0 \in \mathrm{ri}(\mathrm{dom}\,\varphi)$ and $\varphi(0) \in \mathbb{R}$, by Theorem 5.1.4 the optimal dual set is given by $S_D(0) := S_D = \partial\varphi_E(0) + E^\perp$, with $\partial\varphi_E(0)$ nonempty and compact. Furthermore, these properties hold not only for the dual functional $\Phi^*(0,\cdot)$ but also for $\Phi^*(v,\cdot)$, that is, for all the perturbed problems $\psi(v) = \min_{y \in \mathbb{R}^m} \Phi^*(v,y)$ with $\varphi^v(0) \in \mathbb{R}$. To see this, it suffices to apply the same reasoning not to φ but to φ^v. Indeed, the conjugate of φ^v is precisely $\Phi^*(v,\cdot)$. On the other hand, since $\mathrm{dom}\,\varphi^v = \mathrm{dom}\,\varphi$, one has $0 \in \mathrm{ri}(\mathrm{dom}\,\varphi^v)$ as soon as $0 \in \mathrm{ri}(\mathrm{dom}\,\varphi)$, and this allows us to apply the previous results to φ^v. Moreover, it follows that the space $E_v = \mathrm{aff}(\mathrm{dom}\,\varphi^v)$ is a subspace, and does not depend on v, since $E_v = E$. Furthermore, since $0 \in \mathrm{ri}(\mathrm{dom}\,\varphi^v)$ and $\varphi^v(0) \in \mathbb{R}$, φ^v is convex and proper, and then by Corollary 2.5.5 it follows that $E^\perp$ is the constancy space of $\Phi^*(v,\cdot)$. Now, in order to end the proof we have to consider the case $\varphi^v(0) = -\infty$. Since

$$
\psi(v) = +\infty \iff \Phi^*(v,\cdot) = +\infty, \tag{5.9}
$$

we have only to prove that $\varphi^v(0) = -\infty \iff \psi(v) = +\infty$. Thus, suppose that $\varphi^v(0) \in \mathbb{R}$. Then since $0 \in \mathrm{ri}(\mathrm{dom}\,\varphi)$, we have $(\varphi^v)^{**}(0) = \varphi^v(0) \in$

$\mathbb{R}$, which implies in turn that $\Phi^*(v, \cdot) \not\equiv +\infty$. Conversely, suppose that $\psi(v) < +\infty$. Then $\varphi^v(0) \in \mathbb{R}$. Indeed, in the contrary case we would have $\varphi^v(0) = -\infty$, so that $(\varphi^v)^{**}(0) = -\infty$, which in turn implies that $(\varphi^v)^* \equiv +\infty$, i.e., since $(\varphi^v)^* = \Phi^*(v, \cdot)$, that $\psi(v) = +\infty$ by (5.9). $\qquad \square$

Lemma 5.7.1 *Suppose that $\varphi(0) \in \mathbb{R}$ and $0 \in \mathrm{ri}(\mathrm{dom}\,\varphi)$. Let $\{y_k\}$ be a stationary sequence for the dual problem (D); i.e., there exists $u_k \in \partial\varphi^*(y_k)$ with $u_k \to 0$. Then $u_k \in E$ and $\lim_{k\to\infty} \varphi(u_k) = \varphi(0)$.*

Proof. Since $\varphi(0) \in \mathbb{R}$ and $0 \in \mathrm{ri}(\mathrm{dom}\,\varphi)$, φ is a proper convex function (recall that Φ is assumed jointly convex throughout). Furthermore, one has $u_k \in \partial\varphi^*(y_k)$ if and only if $y_k \in \partial\varphi^{**}(u_k)$, so that $u_k \in \mathrm{dom}\,\varphi^{**}$ and $u^k \in E$. Since $\varphi^{**} = \mathrm{cl}\,\varphi$ by Proposition 1.2.5, it follows that aff $\mathrm{dom}\,\varphi^{**} =$ aff $\mathrm{dom}\,\varphi = E$. Then, since φ is relatively continuous at 0, it follows that $\varphi(u_k) \to \varphi(0)$. $\qquad \square$

Theorem 5.7.2 *Suppose that $\varphi(0) \in \mathbb{R}$, $0 \in \mathrm{ri}(\mathrm{dom}\,\varphi)$, $0 \in \mathrm{ri}(\mathrm{dom}\,\psi)$, and let S_P be the optimal set of problem (P). Then, S_P is nonempty and for each sequence $u_k \to 0$, with $\{u_k\} \subset E$ (in particular if $u_k \in \partial\varphi^*(y_k)$ with $u_k \to 0$), for k sufficiently large we have:*
(a) $S_P(u_k) := \mathrm{argmin}_x \Phi(x, u_k) \neq \emptyset$.
(b) $\mathrm{dist}(x_k, S_\mathrm{P}) \to 0$ with $x_k \in S_P(u_k)$.

Proof. (a) By Lemma 5.7.1 there exists k_0 such that for all $k \geq k_0$, $\varphi(u_k) \in \mathbb{R}$ with $\varphi(u_k) \to \varphi(0) = V_\mathrm{P}$. It follows from Theorem 5.7.1(b) that $S_P(u_k)$ is nonempty, and (a) holds. To prove (b) let $P_M(x_k)$ be the projection of x_k onto M and let us observe that $\mathrm{dist}(x_k, S_\mathrm{P}) = \mathrm{dist}(P_M(x_k), \partial\psi_M(0))$ and that $\psi(0) \in \mathbb{R}$, so that ψ is a proper, lsc, and convex function. Let us prove that the sequence $\{P_M(x_k)\}$ is bounded. Indeed, suppose the contrary. Then passing to a subsequence, we may assume without loss of generality that $\|P_M(x_k)\| \to +\infty$ and $\|P_M(x_k)\|^{-1} P_M(x_k) \to \bar{x} \neq 0$. Furthermore, since by Theorem 5.7.1 $\Phi(P_M(x_k), u_k) = \Phi(x_k, u_k) = \varphi(u_k) \to V_\mathrm{P}$, the sequence $\{\Phi(P_M(x_k), u_k)\}$ is bounded above by some $m \in \mathbb{R}$, and then we obtain

$$((\bar{x}, 0), 0) = \lim_{k\to\infty} \frac{(P_M(x_k), u_k), m)}{\|P_M(x_k)\|} \in \mathrm{epi}\,\Phi_\infty,$$

which is equivalent to $\Phi_\infty(\bar{x}, 0) \leq 0$, $0 \neq \bar{x} \in M$. Since $\mathrm{ri}\,\mathrm{dom}\,\psi = \mathrm{ri}\,\mathrm{dom}\,\psi^{**}$, by Theorem 3.2.1 one has that ψ^* is weakly coercive, $(\psi^*)_\infty(\bar{x}) = 0$, and $(\psi^*)_\infty(-\bar{x}) = 0$, so that $\bar{x} \in M^\perp$, a contradiction with the previously established fact $\Phi_\infty(\bar{x}, 0) \leq 0$, $0 \neq \bar{x} \in M$. Now, since we have proved that the sequence $\{P_M(x_k)\}$ is bounded, since $\Phi(P_M(x_k), u_k) \to V_\mathrm{P} \in \mathbb{R}$, the lower semicontinuity of Φ implies that every limit point x of the sequence $\{P_M(x_k)\}$ satisfies $\Phi(x, 0) \leq V_\mathrm{P}$, so that $x \in S_\mathrm{P} \cap M$ and therefore $\lim_{k\to\infty} \mathrm{dist}(P_M(x_k), S_\mathrm{P}) = 0$, which completes the proof. $\qquad \square$

We remark that finding an optimal solution x_k of $\min \Phi(\cdot, u_k)$ for the perturbed problem can be rewritten as a simpler problem. This is in particular equivalent to finding $(x_k, u_k) \in \partial \Phi^*(0, y_k)$ with $u_k \in \partial \varphi^*(y_k)$. Indeed, we have $y_k \in \partial \varphi(u_k)$, so that for each (x, u) we get

$$\Phi(x, u) - \Phi(x_k, u_k) \geq \varphi(u) - \varphi(u_k) \geq \langle y_k, u - u_k \rangle + \langle 0, x - x_k \rangle,$$

meaning that $(y_k, 0) \in \partial \Phi(x_k, u_k)$. As a consequence, $(x_k, u_k) \in \partial \Phi^*(0, y_k)$, which may be a simpler problem when Φ^* happens to be differentiable. In that case there are examples (see Notes and References) where x_k is explicitly given as a function of y_k.

5.8   Notes and References

Duality theory originated in the work of Von Neuman [132], and in Von Neuman and Morgenstern [133]. Later on, the proof of linear programming duality was given by Gale, Kuhn, and Tucker [76]. The extension of duality theory necessary to handle more general optimization problems leads to Lagrangian duality, Fenchel duality, and more generally to Rockafellar's perturbational–conjugate duality. Duality theory for optimization problems can be found in several books such as Auslender [6], Ekeland and Temam [67], Laurent [85], and Rockafellar [119]. The abstract scheme presented in Section 5.1 through general perturbations is based on [121], but with some refinements, in particular within Theorems 5.1.4 and 5.1.5, which play a central role in the derivations of all the results in this Chapter. Fenchel duality originated with the work of Fenchel [71], [72], and has been largely expanded by Rockafellar [119] and later on in [121]. Our proofs are somewhat different and based directly on the results of Section 5.1. The development of Lagrangian duality is well known and developed in the literature cited above. The material given is Section 5.4 concerning special classes of convex problems with zero duality gap given in Theorem 5.4.1 is due to Auslender [18], and Theorem 5.4.2, concerning weakly analytic functions is due to Kummer [84], with some new ideas in the proof. This last theorem is an extension of the result given by Rockafellar [120] for faithfully convex functions. The results in Section 5.5 on duality for optimization problems modeled through asymptotic functions complement those of Chapter 2, and are due to Auslender [17], extending some results of Ben-Tal and Teboulle [28]. The results of Section 5.6 on minimax problems are another application of the results presented in Section 5.1. While most of the results stated are well known, the presentation, analysis, and derivation of the results have been refined and simplified. The standard results for minimax theory can be found, for example, in Rockafellar [119],

[121], MacLinden [101], [102]. The last section, relating duality and stationary sequences, comes from work developed by Auslender, Cominetti, and Crouzeix [10], where more details and results can be found.

6

Maximal Monotone Maps and Variational Inequalities

Monotone maps are of fundamental importance in optimization and variational inequalities. A particularly important special class of monotone maps are those that are maximal monotone and that share several properties similar to those of subdifferentials of convex functions. In a certain way they naturally generalize the concept of subdifferentials and provide a unified framework to formulate convex optimization problems and more generally saddle points problem that arise in several important applications such as game theory and equilibrium problems in economy. This general framework permits us not only to recast these problems in the format of solving generalized equations, but also leads to the development of methods for their solutions. The purpose of this chapter is to give a self-contained introduction to the theory of maximal monotone maps, which are useful for analyzing and solving variational inequalities.

6.1   Maximal Monotone Maps

Monotone maps play a fundamental role in optimization and variational inequalities. A particularly important class of monotone maps is the class of maximal monotone maps, which represent a natural extension of subdifferentials of convex functions. We begin by defining the basic concept of monotonicity.

Definition 6.1.1 *A map $T : \mathbb{R}^n \rightrightarrows \mathbb{R}^n$ is called monotone if its graph is monotone, namely,*

$$\langle u - v, x - y \rangle \geq 0, \ \forall (x, u) \in \text{gph}\, T, \forall (y, v) \in \text{gph}\, T,$$

and strictly monotone if this inequality is strict when $x \neq y$.

When T is a single-valued map, the monotonicity property takes the simpler form

$$\langle T(x) - T(y), x - y \rangle \geq 0, \ \ \forall x, y \in \mathbb{R}^n.$$

An example for which the monotonicity relation holds is when $T = \partial f$, the subgradient of an lsc proper convex function on $\mathbb{R}^n$. Using the subgradient inequality for the convex function f (cf. Chapter 1) one has

$$f(x) - f(y) \geq \langle x - y, g(y) \rangle, \quad g(y) \ \in \partial f(y), \ \forall x, y \in \mathbb{R}^n,$$
$$f(y) - f(x) \geq \langle y - x, g(x) \rangle, \quad g(x) \ \in \partial f(x), \ \forall x, y \in \mathbb{R}^n.$$

Adding the above inequalities implies

$$\langle g(x) - g(y), x - y \rangle \geq 0, \ \forall x, y \in \mathbb{R}^n, \tag{6.1}$$

thus showing that ∂f is monotone. In fact, as we shall see later, the subgradient of lsc convex functions also enjoys the additional property of maximal monotonicity.

Some elementary operations preserving monotonicity are given in the next proposition. The proof follows immediately from Definition 6.1.1 and is thus left to the reader.

Proposition 6.1.1 *(a) $T : \mathbb{R}^n \rightrightarrows \mathbb{R}^n$ is monotone if and only if T^{-1} is monotone.*
(b)Let $T_i, i = 1, 2$, be (strictly) monotone. Then $\lambda T_1 + \mu T_2$ is also (strictly) monotone for all $\lambda, \mu > 0$. Moreover, if either T_1 or T_2 is strictly monotone, then $\lambda T_1 + \mu T_2$ is strictly monotone.
(c) For any matrix $A \in \mathbb{R}^{m \times n}$, and vector $b \in \mathbb{R}^m$, if $T : \mathbb{R}^n \rightrightarrows \mathbb{R}^m$ is monotone, then $S := A^T T(Ax + b)$ is monotone. In addition, if $\text{rank}\, A = n$ and T is strictly monotone, then S is strictly monotone.

We now introduce a fundamental notion related to monotone maps, called maximality. In essence, for monotone maps, maximality plays much the same role as continuity does for functions.

Definition 6.1.2 *A monotone map $T : \mathbb{R}^n \rightrightarrows \mathbb{R}^n$ is maximal if its graph is not properly contained in the graph of any other monotone operator, which can be equivalently stated as*

$$\langle u - v, x - y \rangle \geq 0, \ \forall v \in T(y), \ y \in \text{dom}\, T \implies u \in T(x).$$

From the definition, maximal monotonicity is preserved for the inverse map $T^{-1}(y) = \{x \in \mathbb{R}^n \mid y \in T(x)\}$; i.e., one has that T is maximal monotone if and only if T^{-1} is maximal monotone.

Maximal monotonicity is preserved for single-valued continuous maps.

Proposition 6.1.2 *A continuous monotone map $F : \mathbb{R}^n \to \mathbb{R}^n$ is maximal monotone.*

Proof. Let $x \in \mathbb{R}^n, u \in \mathbb{R}^n$ such that

$$\langle u - F(y), x - y \rangle \geq 0, \ \forall y \in \mathbb{R}^n.$$

By Definition 6.1.2, F is maximal monotone if one has $u = F(x)$. Take $y := x - t(z - x)$, with $t \in \mathbb{R}$ and $z \in \mathbb{R}^n$. Then the above inequality reduces to

$$\langle u - F(x - t(z - x)), z - x \rangle \geq 0, \ \forall z \in \mathbb{R}^n.$$

Letting $t \to 0$ in the latter inequality, together with the assumed continuity of F we then obtain $\langle u - F(x), z - x \rangle \geq 0, \ \forall z \in \mathbb{R}^n$ and therefore $u = F(x)$.
□

Proposition 6.1.3 *Let $T : \mathbb{R}^n \rightrightarrows \mathbb{R}^n$ be maximal monotone. Then:*
(a) $T(x)$ is closed convex for any $x \in \operatorname{dom} T$.
(b) The graph of T is closed.

Proof. (i) By Definition 6.1.2 one can write

$$T(x) = \bigcap_{(y,v) \in \operatorname{gph} T} \{u \in \mathbb{R}^n : \langle u - v, x - y \rangle \geq 0\},$$

i.e., $T(x)$ is the intersection of closed half-spaces, and hence is closed and convex.

(ii) Let $x_k \to x$, $u_k \in T(x_k)$, $u_k \to u$. Take $(y, v) \in \operatorname{gph} T$. Then passing to the limit in $\langle u_k - v, x_k - y \rangle \geq 0$ one obtains $\langle u - v, x - y \rangle \geq 0$, and hence by Definition 6.1.2, $u \in T(x)$.
□

Monotonicity and Nonexpansivity

There are some important relations between monotone and nonexpansive maps.

Definition 6.1.3 *A map $T : \mathbb{R}^n \rightrightarrows \mathbb{R}^n$ is nonexpansive if*

$$\|u - v\| \leq \|x - y\| \ \forall u \in T(x), v \in T(y),$$

and contractive if the inequality is strict when $x \neq y$.

Note that the above definition implies that $u = v$ whenever $x = y$. Thus, $T(x)$ consists of a single element for every $x \in \operatorname{dom} T$, which means that T is a single-valued map or a function.

Proposition 6.1.4 *If $T : \mathbb{R}^n \rightrightarrows \mathbb{R}^n$ is nonexpansive, then $I - T$ is monotone.*

Proof. Let $u \in T(x), v \in T(y)$. Then

$$
\begin{aligned}
\langle x - u - (y - v), x - y \rangle &= \|x - y\|^2 - \langle u - v, x - y \rangle, \\
&\geq \|x - y\|^2 - \|u - v\|\|x - y\|, \\
&\geq 0,
\end{aligned}
$$

by the assumed nonexpansivity of T. $\square$

Note that the converse statement is not necessarily true; namely, the monotonicity of $(I - T)$ does not imply the nonexpansivness of T.

Proposition 6.1.5 *The map $T : \mathbb{R}^n \rightrightarrows \mathbb{R}^n$ is monotone if and only if*

$$
\forall \lambda > 0, \ \forall u \in T(x), \ \forall v \in T(y) : \ \|x - y\| \leq \|x - y + \lambda(u - v)\|.
$$

Proof. For any $u \in T(x), v \in T(y)$, and $\lambda > 0$ one has

$$
\begin{aligned}
\|x - y + \lambda(u - v)\|^2 &= \|x - y\|^2 + \lambda^2 \|u - v\|^2 + 2\lambda \langle x - y, u - v \rangle \\
&\geq \|x - y\|^2,
\end{aligned}
$$

since T is monotone. Conversely, if the inequality of the proposition holds, then one obtains from the above inequality that

$$
\lambda^2 \|u - v\|^2 + 2\lambda \langle x - y, u - v \rangle \geq 0.
$$

Dividing by $\lambda > 0$ and letting $\lambda \to 0$, we get $\langle x - y, u - v \rangle \geq 0$, thus proving the monotonicity of T. $\square$

6.2   Minty Theorem

One of the most useful and fundamental characterizations of maximal monotone maps is given by the so-called theorem of Minty. It reduces the question of checking maximal monotonicity of a monotone map T to the checking of the surjectivity of the corresponding perturbed map $I + \lambda T$, $\lambda > 0$. Recall that if T is maximal monotone, then for any $\lambda > 0$ the map λT is also maximal monotone. Thus, without loss of generality we can replace T by λT; i.e., we can assume that $\lambda = 1$. To prove Minty's result we will rely on Definition 6.1.2. Thus, to prove that T maximal monotone implies $\mathrm{rge}(I + \lambda T) = \mathbb{R}^n$, one has to show that

$$\forall w \in \mathbb{R}^n, \; \exists x : \; T(x) + x \ni w.$$

Since maximal monotonicity is preserved under translation, i.e., since $T - w$ is also maximal monotone, it is enough the consider the case $w = 0$, namely, to show that $\exists x : T(x) + x \ni 0$. From Definition 6.1.2 one thus has to prove that

$$\exists x \in \mathbb{R}^n : \; \langle -x - v, x - y \rangle \geq 0 \; \forall (y, v) \in \mathrm{gph}\, T.$$

To prove this we first prove the following more general result, which is essentially the key to proving Minty's theorem.

Theorem 6.2.1 *Let $C \subset \mathbb{R}^n$ be a closed convex set and $T : \mathbb{R}^n \rightrightarrows \mathbb{R}^n$ a monotone map. Then*

$$\forall z \in \mathbb{R}^n \; \exists x \in C : \; \langle x + v, y - x \rangle \geq \langle z, y - x \rangle \; \forall (y, v) \in \mathrm{gph}\, T, \; y \in C.$$

Proof. Without loss of generality we can assume that $z = 0$. For any $(y, v) \in \mathrm{gph}\, T$, $y \in C$, we define

$$\begin{aligned} \phi(x; (y, v)) &:= \langle v + x, x - y \rangle = \|x\|^2 + \langle x, v - y \rangle - \langle v, y \rangle, \\ C(y, v) &:= \{ x \in C : \; \phi(x; (y, v)) \leq 0 \}. \end{aligned}$$

The statement of the theorem to be proved can thus be conveniently reformulated as

$$\bigcap_{y \in C, (y, v) \in \mathrm{gph}\, T} C(y, v) \neq \emptyset.$$

Clearly, the function $x \to \phi(x; (\cdot, \cdot))$ is a quadratic convex function on the closed convex set C. Moreover, one has $C_\infty(y, v) = \{0\}$, and therefore the set $C(y, v)$ is a convex closed and bounded set, and hence is compact. Thus, to show that the above intersection is nonempty, it suffices to show that for every finite family, say $(y_i, v_i) \in \mathrm{gph}\, T$, $y_i \in C$, $i = 1, \ldots, p$, one has $\cap_{i=1}^p C(y_i, v_i) \neq \emptyset$. Set $\Delta_p := \{\lambda \in \mathbb{R}^p : \sum_{i=1}^p \lambda_i = 1, \; \lambda_i \geq 0, \; i = 1, \ldots, p\}$ and define the function $f : \Delta_p \times \Delta_p \to \mathbb{R}$ by

$$f(\lambda, \mu) := \sum_{i=1}^p \mu_i \langle v_i + x(\lambda), x(\lambda) - y_i \rangle = \sum_{i=1}^p \mu_i \phi(x(\lambda); y_i, v_i))),$$

where $x(\lambda) := \sum_{j=1}^p \lambda_j y_j$. Since f is convex continuous in λ and linear in μ, invoking the classical minimax theorem (cf. Corollary 5.6.2), there exist λ^* and μ^* both in Δ_p, such that $f(\lambda^*, \mu) \leq f(\lambda, \mu^*)$, $\forall \lambda, \mu \in \Delta_p$, and thus we obtain $f(\lambda^*, \mu) \leq \max\{f(\lambda, \lambda) \; \lambda \in \Delta_p\}$, $\forall \mu \in \Delta_p$. Now, using

$x(\lambda) := \sum_{j=1}^{p} \lambda_j y_j$ we compute

$$
\begin{aligned}
f(\lambda, \lambda) &= \sum_{i=1}^{p} \lambda_i \phi(x(\lambda); y_i, v_i))) = \sum_{i=1}^{p} \lambda_i \langle x(\lambda) + v_i, x(\lambda) - y_i \rangle \\
&= \sum_{i,j=1}^{p} \lambda_i \lambda_j \langle x(\lambda) + v_i, y_j - y_i \rangle \\
&= \sum_{i,j=1}^{p} \lambda_i \lambda_j \langle v_i, y_j - y_i \rangle + \sum_{i,j=1}^{p} \lambda_i \lambda_j \langle x(\lambda), y_j - y_i \rangle = \sum_{i,j=1}^{p} \lambda_i \lambda_j \langle v_i, y_j - y_i \rangle \\
&= \frac{1}{2} \sum_{i,j=1}^{p} \lambda_i \lambda_j \langle v_i - v_j, y_j - y_i \rangle.
\end{aligned}
$$

Since T is monotone, we thus obtain for any $(y_i, v_i) \in \mathrm{gph}\, T$ that

$$
f(\lambda, \lambda) \leq 0,
$$

and hence for any $\mu \in \Delta_p$ it follows that $f(\lambda^*, \mu) \leq 0$, proving that $x(\lambda^*) \in \cap_{i=1}^{p} C(y_i, v_i)$. $\qquad\square$

We are now ready to prove Minty's theorem.

Theorem 6.2.2 *Let $T : \mathbb{R}^n \rightrightarrows \mathbb{R}^n$ be monotone and let $\lambda > 0$. Then T is maximal monotone if and only if $\mathrm{rge}(I + \lambda T) = \mathbb{R}^n$, or, equivalently, $\mathrm{dom}(I + \lambda T)^{-1} = \mathbb{R}^n$.*

Proof. As already mentioned above, it is enough to prove the result with $\lambda = 1$. We first show that if T is maximal monotone, then $\mathrm{rge}(I+T) = \mathbb{R}^n$. Invoking Theorem 6.2.1, with $C = \mathbb{R}^n$ for any $z \in \mathbb{R}^n$, there exists $x \in \mathbb{R}^n$ such that $\langle v - (z - x), y - x \rangle \geq 0$, $\forall (y, v) \in \mathrm{gph}\, T$, and therefore from the maximal monotonicity of T, one has $z - x \in T(x)$, i.e., $\mathrm{rge}(I + T) = \mathbb{R}^n$. To prove the converse statement, assume that $\mathrm{rge}(I + T) = \mathbb{R}^n$ and let $x \in \mathbb{R}^n, u \in \mathbb{R}^n$ such that

$$
\langle u - v, x - y \rangle \geq 0, \quad \forall (y, v)\, \mathrm{gph}\, T.
$$

From Definition 6.1.2, to prove that T is maximal monotone one has to show that $u \in T(x)$. Since $\mathrm{rge}(I+T) = \mathbb{R}^n$, then $\forall x, u \in \mathbb{R}^n \,\exists \xi : \xi + T(\xi) = u + x$. Take $y := \xi$, $v \in T(\xi)$ such that we get from the above inequality

$$
\|x - \xi\|^2 = \langle x - \xi, x - \xi \rangle = -\langle u - v, x - \xi \rangle \leq 0,
$$

and hence $x = \xi$ and $u = v \in T(\xi) = T(x)$.

$\qquad\square$

Resolvent and Yosida Approximation

Let $T : \mathbb{R}^n \rightrightarrows \mathbb{R}^n$ be monotone. The *resolvent* of T is the map defined by

$$J_\lambda := (I + \lambda T)^{-1}, \quad \text{for} \quad \lambda > 0.$$

The Yosida approximation (regularization) of any map T is defined by

$$T_\lambda := \lambda^{-1}(I - J_\lambda), \quad \lambda > 0.$$

These maps enjoy several interesting properties. We also remark that the Yosida regularization of T can be alternatively written as

$$T_\lambda = (\lambda I + T^{-1})^{-1}, \quad \lambda > 0. \tag{6.2}$$

The later representation is obtained by using the following useful identity, which is valid for any map $S : \mathbb{R}^n \rightrightarrows \mathbb{R}^n$ and can be verified by direct manipulation of the maps involved:

$$\lambda^{-1}[I - (I + \lambda S)^{-1}] = (\lambda I + S^{-1})^{-1} \ \forall \lambda > 0. \tag{6.3}$$

Proposition 6.2.1 *Let* $T : \mathbb{R}^n \rightrightarrows \mathbb{R}^n$ *be monotone and* $\lambda > 0$. *Then* $J_\lambda : \mathrm{rge}(I + \lambda T) \rightrightarrows \mathbb{R}^n$ *is monotone, nonexpansive, and single-valued. If in addition* T *is maximal, then* J_λ *is also maximal monotone and is a single-valued map from* $\mathbb{R}^n$ *into itself.*

Proof. Since by Proposition 6.1.1(ii) monotonicity of T is preserved under positive scalar multiplication, without loss of generality we assume that $\lambda = 1$ and we set $J_1 \equiv J$. By the same proposition, one also has that $(I + T)$ and hence $J = (I + T)^{-1}$ are also monotone. Now, let $x_i \in J(y_i)$, $i = 1, 2$. Then $y_i = x_i + z_i, z_i \in T(x_i)$, $i = 1, 2$, and we have, using the monotonicity of T,

$$
\begin{aligned}
\|y_1 - y_2\|^2 &= \|x_1 - x_2\|^2 + \|z_1 - z_2\|^2 + 2\langle x_1 - x_2, z_1 - z_2 \rangle, \\
&\geq \|x_1 - x_2\|^2 + \|z_1 - z_2\|^2.
\end{aligned}
$$

The last inequality implies

$$
\begin{aligned}
\|x_1 - x_2\| &\leq \|y_1 - y_2\|, \\
\|z_1 - z_2\| &\leq \|y_1 - y_2\|.
\end{aligned}
$$

Setting $y_1 = y_2$ in the first inequality proves that $x_1 = x_2$, and hence J is single-valued. Also, from the same inequality one has that J is nonexpansive, i.e., Lipschitz continuous with constant 1. If we assume that T is in addition maximal, by Minty's theorem one has $\mathrm{rge}\, J = \mathrm{rge}(I + T) = \mathrm{dom}(I + T)^{-1} = \mathbb{R}^n$. Therefore, since J is a map from $\mathrm{rge}\, J = \mathbb{R}^n$ into itself that is single-valued and Lipschitz continuous with constant 1, it follows that J is also maximal monotone. $\qquad\square$

Theorem 6.2.3 *Let $T : \mathbb{R}^n \rightrightarrows \mathbb{R}^n$ be maximal monotone and $\lambda > 0$. Then:*
(a) $T_\lambda(x) \in T(J_\lambda x)$, $\forall x \in \mathbb{R}^n$.
(b) T_λ is maximal monotone, in fact single valued and Lipschitz with constant λ^{-1}.
(c) Let $n(T(x)) := \min\{\|u\| \mid u \in T(x)\}$, i.e., the element of $T(x)$ with minimal norm. Then for any $x \in \operatorname{dom} T, \mu > 0$ the following hold:
(i) $\|T_\lambda(x) - n(T(x))\|^2 \le \|n(T(x))\|^2 - \|T_\lambda(x)\|^2$,
(ii) $T_{\mu+\lambda}(x) = (T_\mu)_\lambda(x)$,
(iii) $\lim_{\lambda \to 0^+} J_\lambda x = x$,
(iv) $\lim_{\lambda \to 0^+} T_\lambda x = n(T(x))$.

Proof. (a) By definition of the maps J_λ, T_λ, one has for any $x \in \mathbb{R}^n$ that $y \in T_\lambda(x)$ if and only if $y \in \lambda^{-1}(x - J_\lambda(x)) = T(J_\lambda(x))$. To prove (b), we compute

$$
\begin{aligned}
\langle T_\lambda(x_1) - T_\lambda(x_2), x_1 - x_2 \rangle &= \lambda \|T_\lambda(x_1) - T_\lambda(x_2)\|^2 \\
&+ \langle T_\lambda(x_1) - T_\lambda(x_2), J_\lambda(x_1) - J_\lambda(x_2) \rangle, \\
&= \lambda \|T_\lambda(x_1) - T_\lambda(x_2)\|^2 \\
&+ \langle T(J_\lambda(x_1)) - T(J_\lambda(x_2)), J_\lambda(x_1) - J_\lambda(x_2) \rangle, \\
&\ge \lambda \|T_\lambda(x_1) - T_\lambda(x_2)\|^2,
\end{aligned}
$$

where in the first equality we use the fact that for any x_i, $i = 1, 2$, one has $x_i \in J_\lambda(x_i) + \lambda T_\lambda(x_i)$; in the second equality we use (a); and in the last inequality we use the monotonicity of T. It follows from the latter inequality that T_λ is Lipschitz with constant λ^{-1} and maximal monotone. We prove now the assertions given in (c). We first establish the inequality (i). For any $x \in \operatorname{dom} T$ and $\lambda > 0$, we have

$$
\begin{aligned}
\|T_\lambda(x) - n(T(x))\|^2 &= \|T_\lambda(x)\|^2 + \|n(T(x))\|^2 - 2\langle T_\lambda(x), n(T(x)) \rangle, \\
&= \|n(T(x))\|^2 - \|T_\lambda(x)\|^2 \\
&- 2\langle T_\lambda(x), n(T(x)) - T_\lambda(x) \rangle, \\
&= \|n(T(x))\|^2 - \|T_\lambda(x)\|^2 \\
&- 2\lambda^{-1}\langle x - J_\lambda(x), n(T(x)) - T_\lambda(x) \rangle, \\
&\le \|n(T(x))\|^2 - \|T_\lambda(x)\|^2.
\end{aligned}
$$

In the second equality we use $x - J_\lambda(x) = \lambda T_\lambda(x)$, and the inequality follows from the fact that $n(T(x)) \in T(x), T_\lambda(x) \in T(J_\lambda(x))$, and from the monotonicity of T that implies, $\langle T_\lambda(x), n(T(x)) - T_\lambda(x) \rangle = \lambda^{-1}\langle x - J_\lambda(x), n(T(x)) - T_\lambda(x) \rangle \ge 0$. Therefore, (i) is proved, from which it follows that

$$\|T_\lambda(x)\|^2 \le \|n(T(x))\|^2; \quad \|T_\lambda(x)\|^2 \le \langle n(T(x)), T_\lambda(x) \rangle.$$

Since $\|x - J_\lambda(x)\| = \lambda\|T_\lambda(x)\| \le \lambda\|n(T(x))\|$, then $\lim_{\lambda \to 0^+} J_\lambda(x) = x$, proving the limit result (iii). To prove the identity (ii), for any $\lambda, \mu > 0$, using (6.2) one obtains

$$
\begin{aligned}
y \in T_{\mu+\lambda}(x) \iff{} & y \in ((\mu + \lambda)I + T^{-1})^{-1}(x) \\
\iff{} & x \in (\mu + \lambda)y + T^{-1}(y) = \lambda y + (\mu I + T^{-1})(y) \\
\iff{} & x \in (\lambda I + T_\mu^{-1})(y), \\
\iff{} & y \in (\lambda I + T_\mu^{-1})^{-1}(x) = (T_\mu)_\lambda(x).
\end{aligned}
$$

It remains to prove the limit result (iv). Using the inequality (i) applied to T_μ and using $n(T_\mu(x)) \le T_\mu(x)$ we obtain

$$
\|T_{\mu+\lambda}(x) - T_\mu(x)\|^2 \le \|T_\mu(x)\|^2 - \|T_{\lambda+\mu}(x)\|^2,
$$

from which it follows that the sequence $\{\|T_\mu(x)\|\}_{\mu>0}$ is decreasing and bounded below and hence converges, and therefore $\lim_{\lambda,\mu\to 0^+} \|T_{\mu+\lambda}(x) - T_\mu(x)\|^2 = 0$, implying that $\{T_\mu(x))\}_{\mu\in\mathbb{N}}$ is a Cauchy sequence, converging to some $u \in \mathbb{R}^n$. But since the graph of T is closed and by (a) $T_\lambda(x) \in T(J_\lambda(x))$, we thus have by (c) $u \in T(x)$ and $\|u\| = \lim_{\lambda\to 0^+} \|T_\lambda(x)\| \le \|n(T(x))\|$, from which it follows that $u = n(T(x))$ and $\lim_{\lambda\to 0^+} T_\lambda(x) = n(T(x))$.

$\square$

6.3   Convex Functionals and Maximal Monotonicity

Let $f : \mathbb{R}^n \to \mathbb{R} \cup \{+\infty\}$ be a proper lsc convex function. The set-valued subdifferential map $\partial f : \mathbb{R}^n \rightrightarrows \mathbb{R}^n$ (cf. Chapter 1) is one of the most fundamental objects, behind the more general concept of maximal monotone maps.

Theorem 6.3.1 *Let $f : \mathbb{R}^n \to \mathbb{R} \cup \{+\infty\}$ be a proper lsc convex function. Then $\partial f : \mathbb{R}^n \rightrightarrows \mathbb{R}^n$ is maximal monotone.*

Proof. The monotonicity of ∂f was already established in (6.1). To prove the maximality of ∂f we use Theorem 6.2.2, that is we will show that $\mathrm{rge}(I + \partial f) = \mathbb{R}^n$. Fix any $d \in \mathbb{R}^n$ and define the function $x \longrightarrow h_d(x) := f(x) + \frac{1}{2}\|x\|^2 - \langle x, d \rangle$. Clearly, since f is a proper lsc convex function, then h_d shares the same properties. Moreover, since for any $y \in \mathrm{ri\,dom}\, f$ and $c \in \partial f(y)$ we have $h_d(x) \ge f(y) + \langle c, x - d \rangle - \langle c, y \rangle + \frac{1}{2}\|x\|^2$, one obtains $\lim_{\|x\|\to\infty} h_d(x) = +\infty$; i.e., h_d is coercive, and hence the minimum of h_d over $x \in \mathbb{R}^n$ is attained at some x^*, which means that $0 \in \partial h_d(x^*)$. Since here $\partial h_d(x^*) = \partial f(x^*) + x^* - d$, one thus has obtained that $\forall d \in \mathbb{R}^n$, $\exists x^*$ such that $d \in (I + \partial f)(x^*)$, and hence $\mathrm{rge}(I + \partial f) = \mathbb{R}^n$.

$\square$

An important and useful application of Theorem 6.3.1 is to the indicator function δ_C of a closed convex set $C \subset \mathbb{R}^n$. Recall that $\delta_C(x) = 0$ if $x \in C$ and $+\infty$ otherwise is then a proper lsc convex function and using the definition of the subdifferential one has

$$\partial \delta_C(x) = \{z \in \mathbb{R}^n \mid \langle z, y - x \rangle \leq 0, \ \forall y \in C\}.$$

It is easy to verify that $\partial \delta_C(x)$ is a closed convex cone, which is the normal cone of C at x and is denoted by $N_C(x)$; cf. Chapter 1. Accordingly, one can view the normal cone as a set-valued map $N_C : \mathbb{R}^n \rightrightarrows \mathbb{R}^n$ by setting

$$N_C(x) := \begin{cases} \{z \mid \langle z, y - x \rangle \leq 0 \ \forall y \in C\} & \text{if } x \in C, \\ \emptyset & \text{otherwise.} \end{cases}$$

Clearly, one has dom $N_C = C$ and $N_C(x) = \{0\}$ when $C \equiv \mathbb{R}^n$ or $x \in \operatorname{int} C$. From the above development, invoking Theorem 6.3.1 with $f \equiv \delta_C$, we thus obtain the following result.

Corollary 6.3.1 *Let $C \subset \mathbb{R}^n$ be a nonempty closed convex set. Then the normal cone map $N_C : \mathbb{R}^n \rightrightarrows \mathbb{R}^n$ is maximal monotone.*

Example 6.3.1 Whenever $T = \partial f$ with f proper, lsc, and convex on $\mathbb{R}^n$, the resolvent map $J_\lambda = (I + \lambda \partial f)^{-1}$ coincides with the *proximal map* uniquely defined by

$$\operatorname{Prox}(\lambda, f)(x) = \operatorname*{argmin}_u \{f(u) + (2\lambda)^{-1} \|u - x\|^2\}.$$

This can be seen directly by applying the necessary and sufficient optimality condition given in Theorem 4.1.1 for the strictly convex and coercive function $u \to f(u) + (2\lambda)^{-1}\|u - x\|^2$, which reduces to $0 \in \lambda \partial f(u) + (u - x)$ and can be written as $x \in (I + \lambda \partial f)(u)$ or equivalently $u \in (I + \lambda \partial f)^{-1}(x) = J_\lambda(x)$. In the special case $f = \delta_C$ with C closed convex, the proximal map coincides with the projection operator P_C onto C, which is thus single-valued and hence maximal monotone.

There is an interesting connection between the resolvent map associated with the maximal monotone map N_C and the projection maps onto convex sets.

Proposition 6.3.1 *Let $C \subset \mathbb{R}^n$ be a nonempty closed convex set. Then*

$$\forall \lambda > 0, \ \forall x \in \mathbb{R}^n \ \ (I + \lambda N_C)^{-1}(x) = P_C(x),$$

and therefore the projection P_C onto C is a single-valued maximal monotone map.

Proof. Since $N_C = \partial \delta_C$, the resolvent associated with the proper lsc convex function δ_C is by definition

$$J_\lambda := (I + \lambda \partial \delta_C)^{-1} = (I + \lambda N_C)^{-1} = P_C.$$

The latter equality follows by Example 6.3.1.   □

Specializing the above to linear subspaces we get the following result.

Corollary 6.3.2 *Let $M \subset \mathbb{R}^n$ be a subspace with orthogonal complement $M^\perp$. Then:*
(a)

$$N_M(x) = \begin{cases} M^\perp & \text{if } x \in M, \\ \emptyset & \text{if } x \notin M; \end{cases} \quad N_M^{-1}(y) = \begin{cases} M & \text{if } y \in M^\perp, \\ \emptyset & \text{if } y \notin M^\perp; \end{cases}$$

with both maps being maximal monotone.
(b) The projection maps onto a subspace and its orthogonal complement are maximal monotone, nonexpansive and given respectively by

$$P_M = (I + N_M)^{-1}, \quad P_{M^\perp} = (I + N_M^{-1})^{-1},$$

and $P_M + P_M^\perp = I$.

Proof. Statements (a) and (b) are immediate consequences of Theorem 6.3.1 and Proposition 6.3.1, while the last identity is a special case of the general identity (6.3) with $S \equiv N_M$.   □

One can verify that this result can be extended over closed convex cones. Let $K \subset \mathbb{R}^n$ be a closed convex cone with polar cone K^*. Then, any $z \in \mathbb{R}^n$ can be uniquely represented in the form

$$z = x + v, \quad x \in K, \quad v \in K^*, \quad \langle x, v \rangle = 0,$$

namely with $x = P_K(z), v = P_{K^*}(z)$ the projections of z onto the cones K, K^*, respectively. The resulting projection maps are maximal monotone and nonexpansive.
We now turn to the more general case of convex–concave functions that appears in saddle point problems and minimax theory (cf. Chapter 5). Let $C \subset \mathbb{R}^n$ and $D \subset \mathbb{R}^m$ be nonempty convex sets, and let $K : \mathbb{R}^n \times \mathbb{R}^m \to \overline{\mathbb{R}}$ be a convex–concave closed function as defined in Definition 5.6.3, namely $K : C \times D \to \mathbb{R}$ and $K(x, y) = -\infty$, for $x \in C$, $y \notin D$, $K(x, y) = +\infty$ for $x \notin C$, and such that for each $(x, y) \in C \times D$, the functions $K(\cdot, y)$ and $-K(x, \cdot)$ are lsc, proper, and convex. Let $Z := \{(x, y) : \partial_x K(x, y) \neq \emptyset, \partial_y[-K](x, y) \neq \emptyset\} \subset C \times D$ and consider the map $T : \mathbb{R}^n \times \mathbb{R}^m \rightrightarrows \mathbb{R}^n \times \mathbb{R}^m$ defined by

$$T(x, y) = \begin{cases} \partial_x K(x, y) \times \partial_y[-K](x, y) & \text{if } (x, y) \in Z, \\ \emptyset & \text{otherwise.} \end{cases} \tag{6.4}$$

Note that we thus have $\operatorname{dom} T = Z \neq \emptyset$, since $\operatorname{ri} C \times \operatorname{ri} D \subset Z$. Using the subgradient inequalities for each convex (concave) function K it is easy to verify the monotonicity of the saddle map T given in (6.4). In fact, the saddle map is also maximal monotone.

Theorem 6.3.2 *Let $K : \mathbb{R}^n \times \mathbb{R}^m \to \overline{\mathbb{R}}$ be a closed convex-concave function. Then the set-valued map T defined in (6.4) is maximal monotone.*

Proof. The proof is established in a similar fashion to the one given in Theorem 6.3.1. For $d := (d_1, d_2) \in \mathbb{R}^n \times \mathbb{R}^m$, define

$$H_d(x, y) := K(x, y) + \frac{1}{2}\|x\|^2 - \langle x, d_1 \rangle - \frac{1}{2}\|y\|^2 + \langle d_2, y \rangle.$$

Then H_d is a closed convex–concave function, and with the same arguments as those used in the proof of Theorem 6.3.1, we deduce that for each $(x, y) \in C \times D$, the functions $H_d(\cdot, y)$ and $-H_d(x, \cdot)$ are coercive. Then, using Corollary 5.6.3, it follows that the set of saddle points of H_d is nonempty and compact and the minimax is attained at some (x^*, y^*), which by the optimality conditions means that $(0, 0) \in \partial H_d(x^*, y^*)$, or, equivalently,

$$0 \in \partial_x K(x^*, y^*) + x^* - d_1 \quad \text{and} \quad 0 \in \partial_y[-K(x^*, y^*)] - y^* + d_2.$$

Therefore, for any $d = (d_1, d_2) \in \mathbb{R} \times \mathbb{R}^m$, $\exists (x^*, y^*)$ such that $d \in (I + T)(x^*, y^*)$, and hence $\operatorname{rge}(I + T) = \mathbb{R}^n \times \mathbb{R}^m$, proving the maximal monotonicity of T. $\square$

Example 6.3.2 Consider the convex program defined in Section 5.3 by

$$\text{(P)} \quad \inf\{f_0(x) \mid f_i(x) \leq 0, \, x \in \mathbb{R}^n\},$$

where the functions $f_i : \mathbb{R}^n \to \mathbb{R} \cup \{+\infty\}$, $i = 0, \ldots, m$, are supposed proper, convex, and lsc and such that $\operatorname{dom} f_0 \subset \operatorname{dom} f_i$, $\operatorname{ri} \operatorname{dom} f_0 \subset \operatorname{ri} \operatorname{dom} f_i$, $i = 1, \ldots, m$. Let $L : \mathbb{R}^n \times \mathbb{R}_+^m \to \overline{\mathbb{R}}$ be the Lagrangian associated with (P) defined by

$$L(x, y) = \begin{cases} f_0(x) + \sum_{i=1}^m y_i f_i(x) & \text{if } x \in \operatorname{dom} f_0, y \in \mathbb{R}_+^m, \\ -\infty & \text{if } x \in \operatorname{dom} f_0, y \notin \mathbb{R}_+^m, \\ +\infty & \text{if } x \notin \operatorname{dom} f_0. \end{cases}$$

As already noted in Chapter 5, L is a closed convex–concave function. Let $F(x) := (f_1(x), \ldots, f_m(x))^T$. Then, the map T defined in (6.4) associated with L reduces to a map defined on $\mathbb{R}^n \times \mathbb{R}_+^m$, and given by

$$T(x, y) = \left\{ (y, w) \mid y \in \partial f_0(x) + \sum_{i=1}^m y_i \partial f_i(x), w \in -F(x) + N_{\mathbb{R}_+^m}(y) \right\}$$

if $(x, y) \in \operatorname{dom} T = \operatorname{dom} f_0 \times \mathbb{R}_+^m \neq \emptyset$, and $\emptyset$ otherwise, and is maximal monotone by Theorem 6.3.2.

6.4   Domains and Ranges of Maximal Monotone Maps

Domains and Ranges: Basic Properties

We begin with the following general result, which is fundamental in the course of the analysis of this Section.

Theorem 6.4.1 *Let* $A : \mathbb{R}^n \rightrightarrows \mathbb{R}^n$ *be maximal monotone,* $C \subset \mathbb{R}^n$ *such that*

$$[\mathrm{BH}] \quad \forall v \in C, \; \exists y \in \mathbb{R}^n \; : \quad \sup_{(x,u)\in \mathrm{gph}\, A} \langle u - v, y - x \rangle < +\infty.$$

Then:
(a) $\mathrm{conv}\, C \subset \mathrm{cl}\, \mathrm{rge}\, A$,
(b) $\mathrm{ri}\, \mathrm{conv}\, C \subset \mathrm{rge}\, A$ *whenever* $\mathrm{rge}\, A \subset \mathrm{aff}\, C$.

Proof. We first show that we can replace C by its convex hull in the hypothesis [BH] of the theorem. For that, let $v := \sum_{i=1}^m \lambda_i v_i$ be a convex combination of elements $v_i \in C$. By the hypothesis of the theorem, there exist $y_i \in \mathbb{R}^n$, $c_i \in \mathbb{R}$ such that for all $i = 1, \ldots, m$,

$$\langle u - v_i, y_i - x \rangle \leq c_i, \; \forall\, (x, u) \in \mathrm{gph}\, A.$$

This can be rewritten as

$$\langle u, y_i \rangle - \langle u, x \rangle + \langle v_i, x \rangle \leq c_i + \langle v_i, y_i \rangle, \; \forall i = 1, \ldots, m.$$

Let $y := \sum_{i=1}^m \lambda_i y_i$. Then multiplying by $\lambda_i \geq 0$ and summing the latter inequalities over $i = 1, \ldots, m$ it follows that

$$\langle u, y \rangle - \langle u, x \rangle + \langle v, x \rangle \leq \sum_{i=1}^m \lambda_i c_i + \sum_{i=1}^m \lambda_i \langle v_i, y_i \rangle,$$

and hence $\langle u - v, y - x \rangle \leq \sum_{i=1}^m \lambda_i c_i + \sum_{i=1}^m \lambda_i \langle v_i, y_i \rangle - \langle v, y \rangle \equiv c < \infty$. This shows that [BH] remains valid with C replaced by $\mathrm{conv}\, C$, and therefore without loss of generality we can now assume that C is convex, so that we have to prove

$$(i) \; C \subset \mathrm{cl}\, \mathrm{rge}\, A \quad \text{and} \quad (ii) \; \mathrm{ri}\, C \subset \mathrm{rge}\, A.$$

For any $v \in C$ and any $\varepsilon > 0$, let x_ε be the solution of $v \in \varepsilon x_\varepsilon + A(x_\varepsilon)$. Such a solution always exists by Minty's Theorem 6.2.2, since we assume here that A is maximal monotone. Let $y \in \mathbb{R}^n$, $c \in \mathbb{R}$ satisfy the hypothesis of the theorem; i.e.,

$$\langle u - v, y - x \rangle \leq c \; \forall (x, u) \in \mathrm{gph}\, A.$$

Setting $(x := x_\varepsilon, u := v - \varepsilon x_\varepsilon) \in \operatorname{gph} A$, we obtain $\langle \varepsilon x_\varepsilon, x_\varepsilon - y \rangle \leq c$ and thus

$$\varepsilon \|x_\varepsilon\|^2 \leq \varepsilon \langle x_\varepsilon, y \rangle + c \leq \frac{\varepsilon}{2} \|x_\varepsilon\|^2 + \frac{\varepsilon}{2} \|y\|^2 + c,$$

which implies $\|\sqrt{\varepsilon} x_\varepsilon\|^2 \leq \varepsilon \|y\|^2 + 2c := r^2$, showing that $\sqrt{\varepsilon} x_\varepsilon$ remains in the ball $r\mathbb{B}$. Therefore, for any $\varepsilon > 0$, $v \in \sqrt{\varepsilon} \sqrt{\varepsilon} x_\varepsilon + A(x_\varepsilon) \subset \sqrt{\varepsilon} r\mathbb{B} + \operatorname{rge} A$, thus showing that $v \in \operatorname{cl} \operatorname{rge} A$ and completing the proof of (a). To prove (b), namely $\operatorname{ri} C \subset \operatorname{rge} A$, let $v \in \operatorname{ri} C$. Then there exists $\rho > 0$ such that for any $z \in \operatorname{aff}(C) - v$ with $\|z\| \leq \rho$, one has $v + z \in C$. By the hypothesis [BH] of the theorem, we thus obtain that there exist $y \in \mathbb{R}^n$, $c \in \mathbb{R}$ such that

$$\langle u - v - z, y - x \rangle \leq c, \quad \forall (x, u) \in \operatorname{gph} A.$$

Now let x_ε be the solution of $v \in \varepsilon x_\varepsilon + A(x_\varepsilon)$. Set $(x = x_\varepsilon, u = v - \varepsilon x_\varepsilon) \in \operatorname{gph} A$, we then obtain for any $z \in \operatorname{aff}(C) - v$,

$$\langle -\varepsilon x_\varepsilon - z, y - x_\varepsilon \rangle \leq c,$$

from which it follows that $\forall z \in \operatorname{aff}(C) - v$ with $\|z\| \leq \rho$,

$$\begin{aligned}
\langle z, x_\varepsilon \rangle &\leq c + \langle z, y \rangle + \varepsilon \langle x_\varepsilon, y \rangle - \varepsilon \|x_\varepsilon\|^2 \\
&\leq c + \langle z, y \rangle + \frac{\varepsilon}{2} (\|y\|^2 - \|x_\varepsilon\|^2) \\
&\leq c + \langle z, y \rangle + \frac{\varepsilon}{2} \|y\|^2,
\end{aligned}$$

showing that $\langle z, x_\varepsilon \rangle$ remains bounded as $\varepsilon \to 0$. Moreover, since under (b) we assume $\operatorname{rge} A \subset \operatorname{aff} C$, then one has

$$x_\varepsilon \in \varepsilon^{-1}(v - A(x_\varepsilon)) \subset \varepsilon^{-1}(C - \operatorname{rge} A) \subset \operatorname{aff}(C - C),$$

and it follows that $\{x_\varepsilon\}$ remains bounded as $\varepsilon \to 0$. After reindexing if necessary let the subsequence x_ε converges to x^*. Since $v - \varepsilon x_\varepsilon \in A(x_\varepsilon)$ and $v - \varepsilon x_\varepsilon$ converges to v, using the fact that A is maximal monotone, Proposition 6.1.3(b) says that $\operatorname{gph} A$ is closed, and hence one obtains $v \in A(x^*)$, that is $v \in \operatorname{rge} A$.

$$\square$$

Note that without the additional hypothesis made in (b) $\operatorname{rge} A \subset \operatorname{cl} C$ one still has $\operatorname{int} \operatorname{conv} C \subset \operatorname{int} \operatorname{rge} A$.

An important consequence of the above theorem are the following convexity properties.

Proposition 6.4.1 *Let $A : \mathbb{R}^n \rightrightarrows \mathbb{R}^n$ be maximal monotone. Then:*
(a) $\operatorname{cl} \operatorname{dom} A$ and $\operatorname{ri} \operatorname{dom} A$ are convex.
(b) $\operatorname{cl} \operatorname{rge} A$ and $\operatorname{ri} \operatorname{rge} A$ are convex.

Proof. Take $C := \operatorname{rge} A$ in Theorem 6.4.1. Then since A is maximal monotone, the hypothesis [BH] is satisfied and one thus obtains $\operatorname{conv} \operatorname{rge} A \subset \operatorname{cl} \operatorname{rge} A$ and $\operatorname{ri} \operatorname{conv} \operatorname{rge} A \subset \operatorname{rge} A$, which implies that $\operatorname{cl} \operatorname{rge} A$ and $\operatorname{ri} \operatorname{rge} A$ are nonempty convex sets, proving (b). Since A^{-1} is also maximal monotone and $\operatorname{dom} A = \operatorname{rge} A^{-1}$, the proof of (a) follows immediately.

$\square$

Remark 6.4.1 A set $C \subset \mathbb{R}^n$ satisfying $\operatorname{ri} \operatorname{conv} C \subset C$ is called almost convex. Since for the affine hull of C, $\operatorname{aff} C$, it always holds that $\operatorname{aff} C = \operatorname{aff} \operatorname{conv} C$, it follows that for any almost convex set C one has $\operatorname{ri} \operatorname{conv} C = \operatorname{ri} C$. We also note that from Proposition 6.4.1, for any maximal monotone map A we have that $\operatorname{dom} A$ and $\operatorname{rge} A$ are almost convex sets, and thus the following useful relations holds

$$\operatorname{ri} \operatorname{conv} \operatorname{rge} A = \operatorname{ri} \operatorname{rge} A, \quad \operatorname{ri} \operatorname{conv} \operatorname{dom} A = \operatorname{ri} \operatorname{dom} A. \tag{6.5}$$

6.5 Asymptotic Functionals of Maximal Monotone Maps

The Asymptotic Function of a Maximal Monotone Map
For any maximal monotone map $A : \mathbb{R}^n \rightrightarrows \mathbb{R}^n$ we can associate an asymptotic function.

Definition 6.5.1 *Let $A : \mathbb{R}^n \rightrightarrows \mathbb{R}^n$ be a set-valued map. The asymptotic function associated with A is defined by*

$$f_\infty^A(d) := \sup\{\langle c, d \rangle \,|\, c \in \operatorname{rge} A\} = \sigma_{\operatorname{rge} A}(d). \tag{6.6}$$

We recall that for any set $C \subset \mathbb{R}^n$ the support function satisfies (cf. 1.3.2(c))

$$\sigma_C = \sigma_{\operatorname{cl} C} = \sigma_{\operatorname{conv} C} = \sigma_{\operatorname{cl} \operatorname{conv} C},$$

and therefore the definition (6.6) remains valid when the range of A is replaced by its closure or its convex hull. This fact is often useful and will be used when needed without further mention.

The motivation behind the above formulation is partly supported by the fact that in the special case where $A := \partial f$ is the subdifferential map of an lsc convex function, we recover the corresponding asymptotic function characterized in Proposition 2.5.2 and given in (2.20) by

$$f_\infty(d) = \sup_{t>0} \frac{f(x + td) - f(x)}{t}, \quad \forall d \in \mathbb{R}^n, \ \forall x \in \operatorname{dom} f.$$

Proposition 6.5.1 *Let $f : \mathbb{R}^n \to \mathbb{R} \cup \{+\infty\}$ be a proper, convex, and lsc function and let $T := \partial f$ its subdifferential map. Then $f^T_\infty = f_\infty$.*

Proof. For any proper lsc convex function f, by (1.1) the range of its subdifferential satisfies $\operatorname{ri} \operatorname{dom} f^* \subset \operatorname{rge} \partial f \subset \operatorname{dom} f^*$, where f^* denotes the conjugate of f. Hence $\operatorname{cl} \operatorname{ri} \operatorname{dom} f^* \subset \operatorname{cl} \operatorname{rge} \partial f \subset \operatorname{cl} \operatorname{dom} f^*$, and since $\operatorname{cl} \operatorname{ri} \operatorname{dom} f^* = \operatorname{cl} \operatorname{dom} f^*$, it follows that $\sigma_{\operatorname{rge} \partial f}(d) = \sigma_{\operatorname{dom} f^*}(d) = f_\infty(d)$, where the latter equality is from Theorem 2.5.4(b). $\qquad\square$

A natural and fundamental question involving asymptotic functionals associated with maximal monotone operators is to know when $f^A_\infty + f^B_\infty = f^{A+B}_\infty$, where A, B are given maximal monotone maps on $\mathbb{R}^n$. In general, this relation does not hold.

Example 6.5.1 Let A be the map defined as the rotation through the angle $\pi/2$ in $\mathbb{R}^2$ and let $B := N_C$ be the normal cone of the set $C = \mathbb{R} \times \{0\}$. Then A and $A + B$ are maximal monotone, yet one has for $d \neq 0$, $f^A_\infty(d) + f^B_\infty(d) = +\infty$, while

$$f^A_\infty(d) + f^B_\infty(d) \neq f^{A+B}_\infty(d) = \begin{cases} 0 & \text{if } 0 \neq d = (d_1, 0), \\ +\infty & \text{otherwise.} \end{cases}$$

The above question is directly related to questions about the range of the sum of maximal monotone maps. In fact, as we shall see, useful and important formulas for the asymptotic function of a sum of two maximal monotone maps can be obtained through the use of corresponding results on the range of a sum of maximal monotone maps.

Range of the Sum of Two Maximal Monotone Maps

Let A, B be two maximal monotone maps on $\mathbb{R}^n$. We study the relations between the range of their sum; namely, we are interested in relating the following two operations:

$$\operatorname{rge}(A + B) \quad := \bigcup_{x \in \operatorname{dom} A \cap \operatorname{dom} B} Ax + Bx,$$

$$\operatorname{rge} A + \operatorname{rge} B \quad := \bigcup_{x \in \operatorname{dom} A, y \in \operatorname{dom} B} Ax + By.$$

For that purpose we now introduce a fundamental class of monotone maps.

Definition 6.5.2 *A monotone map $A : \mathbb{R}^n \rightrightarrows \mathbb{R}^n$ is called star-monotone if*

$$\forall w \in \operatorname{rge} A, \ \forall y \in \operatorname{dom} A, \quad \sup_{(x,u) \in \operatorname{gph} A} \langle u - w, y - x \rangle < +\infty.$$

A useful way to determine when a monotone map is star-monotone is via the following result.

Proposition 6.5.2 *Any monotone map* $A : \mathbb{R}^n \rightrightarrows \mathbb{R}^n$ *satisfying the inequality*

$$\langle u, y - x \rangle + \langle v, z - y \rangle + \langle w, x - z \rangle \leq 0, \ \forall u \in A(x), v \in A(y), w \in A(z),$$

is star-monotone.

Proof. The inequality given in the proposition can be written equivalently as

$$\langle u - w, y - x \rangle \leq \langle v - w, y - z \rangle \ \forall u \in A(x), v \in A(y), w \in A(z).$$

Since $w \in A(z) \subset \text{rge } A$, it follows from the last inequality that

$$\forall w \in \text{rge } A, \ \forall y \in \text{dom } A, \quad \sup_{(x,u) \in \text{gph } A} \langle u - w, y - x \rangle \leq c < +\infty,$$

with $c := \langle v - w, y - z \rangle$ a finite constant that depends only on (y, w), thus proving from Definition 6.5.2 that A is star-monotone. $\qquad \square$

In the following we list a number of important maps that are star-monotone.

Example 6.5.2 *(Star-monotone maps)* (i) The subdifferential of a proper lsc convex function f is star-monotone. This follows from the subgradient inequality and is just a particular case of Proposition 6.5.2. In fact, for $A = \partial f$, one has a stronger property than star-monotonicity, which reads

$$\forall w \in \text{dom } f^*, \ \forall y \in \text{dom } f, \quad \sup_{u \in \partial f(x)} \langle u - w, y - x \rangle < \infty.$$

This can be verified using the subgradient inequality for the convex function f together with Fenchel inequality for the pair (f, f^*).
(ii) Let $A : \mathbb{R} \rightrightarrows \mathbb{R}^n$ be a monotone map. The map A is said to be strongly coercive if it satisfies

$$\forall y \in \text{dom } A : \quad \lim_{z \in \text{dom } A, \|z\| \to \infty} \frac{\langle Az, z - y \rangle}{\|z\|} = +\infty.$$

Then A is star-monotone if dom A is bounded or if A is strongly coercive. Indeed, given $y \in \text{dom } A$, $w \in \text{rge } A$, there exists $\rho > 0$ such that for any $u \in A(x)$ with $\|x\| \geq \rho$ one has $\langle u - w, y - x \rangle \leq 0$, whereas for $\|x\| < \rho$ one has with $v \in A(y)$,

$$\langle u - w, y - x \rangle \leq v - w, y - x \rangle \leq (\|v\| + \|w\|)(\|y\| + \rho) \equiv c.$$

As an interesting particular case, one thus has that for any monotone map A, the map $A + \varepsilon I$ is star-monotone for any $\varepsilon > 0$.

(iii) If a monotone map A is star-monotone, then A^{-1} is star-monotone. Therefore, we also have in particular that for any $\lambda > 0$, its resolvent $J_\lambda = (I + \lambda A)^{-1}$ and Yosida approximation A_λ are star-monotone. If, in addition, A is also assumed maximal, then A is star-monotone whenever rge A is bounded.

We are now ready to answer the questions posed at the beginning of this subsection on the image of the sum of monotone maps.

Theorem 6.5.1 *Let A, B be two monotone maps on $\mathbb{R}^n$, which are assumed star-monotone. If $A + B$ is maximal monotone, then:*
(a) $\operatorname{int} \operatorname{rge}(A + B) = \operatorname{int}(\operatorname{rge} A + \operatorname{rge} B)$.
(b) $\operatorname{ri} \operatorname{conv}(\operatorname{rge} A + \operatorname{rge} B) \subset \operatorname{rge}(A + B)$.
(c) $\operatorname{cl} \operatorname{rge}(A + B) = \operatorname{cl}(\operatorname{rge} A + \operatorname{rge} B)$.
Moreover, the relation (c) is equivalent to

$$f_\infty^{A+B} = f_\infty^A + f_\infty^B.$$

Proof. The relation $\operatorname{rge}(A + B) \subseteq \operatorname{rge} A + \operatorname{rge} B$ holds trivially. Therefore, to prove (a) and (c), it is sufficient to show that

$$\operatorname{rge} A + \operatorname{rge} B \subseteq \operatorname{cl} \operatorname{rge}(A + B), \quad \text{and} \quad \operatorname{int}(\operatorname{rge} A + \operatorname{rge} B) \subseteq \operatorname{rge}(A + B).$$

To prove this we apply Theorem 6.4.1 to the maximal monotone map $A+B$ and the set $C := \operatorname{rge} A + \operatorname{rge} B$. We thus need to check that the condition [BH] holds in that case, which then also immediately implies the proof of (b), since here one always has $\operatorname{rge}(A + B) \subset \operatorname{cl}(\operatorname{rge} A + \operatorname{rge} B)$. For this, let $v \in C$, i.e., $v = v_A + v_B$ with $v_A \in \operatorname{rge} A, v_B \in \operatorname{rge} B$. Since $A + B$ is maximal monotone, $\operatorname{dom} A \cap \operatorname{dom} B \neq \emptyset$. Take $y \in \operatorname{dom} A \cap \operatorname{dom} B$. Since A and B are star-monotone, there exist $c, d \in \mathbb{R}$ such that

$$\langle u_A - v_A, y - x \rangle \; \leq \; c, \; \forall x \in \operatorname{dom} A \cap \operatorname{dom} B, \; \forall u_A \in A(x), \quad (6.7)$$
$$\langle u_B - v_B, y - x \rangle \; \leq \; d, \; \forall x \in \operatorname{dom} A \cap \operatorname{dom} B, \; \forall u_B \in B(x). \quad (6.8)$$

Adding these two inequalities, it follows that

$$\sup_{(x,u) \in \operatorname{gph}(A+B)} \langle u - v, y - x \rangle < \infty.$$

The proof of the last statement of the theorem follows at once from the validity of relation (c) in the theorem and the definition of the asymptotic function (6.6). □

Theorem 6.5.2 *Let A, B be two monotone maps on $\mathbb{R}^n$ satisfying:*
(a) $\operatorname{dom} A \subset \operatorname{dom} B$.

(b) $A + B$ is maximal monotone.
(c) B is star-monotone.
Then,

$$
\begin{aligned}
\operatorname{int} \operatorname{rge}(A + B) &= \operatorname{int}(\operatorname{rge} A + \operatorname{rge} B), \\
\operatorname{ri} \operatorname{conv}(\operatorname{rge} A + \operatorname{rge} B) &\subset \operatorname{rge}(A + B), \\
\operatorname{cl} \operatorname{rge}(A + B) &= \operatorname{cl}(\operatorname{rge} A + \operatorname{rge} B).
\end{aligned}
$$

Moreover, the last relation is equivalent to

$$
f_\infty^{A+B} = f_\infty^A + f_\infty^B.
$$

Proof. The proof is similar to the proof of Theorem 6.5.1. We need to check that condition [BH] holds for the maximal monotone map $A + B$ and the set $C := \operatorname{rge} A + \operatorname{rge} B$. For this, let $v \in C$, i.e., $v = v_A + v_B$ with $v_A \in \operatorname{rge} A$, $v_B \in \operatorname{rge} B$. Take $y \in \operatorname{dom} A \subset \operatorname{dom} B$ such that $v_A \in A(y)$. Using the monotonicity of A one obtains $\langle u_A - v_A, y - x \rangle \leq 0, \quad \forall x \in \operatorname{dom} A, \forall u_A \in A(x)$. Since B is assumed star-monotone, there exists $c \in \mathbb{R}$ such that

$$
\langle u_B - v_B, y - x \rangle \leq c, \quad \forall x \in \operatorname{dom} A \subset \operatorname{dom} B \; \forall u_B \in B(x).
$$

Adding the two inequalities, we obtain $\forall u_A \in A(x)$, $u_B \in B(x)$, i.e., for all $u := u_A + u_B \in (A + B)(x)$ that $\langle u - v, y - x \rangle \leq c$, and hence

$$
\sup_{(x,u) \in \operatorname{gph}(A+B)} \langle u - v, y - x \rangle < \infty.
$$

$\square$

Two important and useful variants of Theorems 6.5.1–6.5.2 are now given.

Theorem 6.5.3 *Let A, B be two monotone maps on $\mathbb{R}^n$ such that $A + B$ is maximal monotone. Consider the following hypothesis:*
(a) A is star-monotone.
(b) B is star-monotone.
(c) $\operatorname{dom} A \subset \operatorname{conv} \operatorname{dom} B$.
Then under either (a) and (b) or (b) and (c) one has

$$
\begin{aligned}
\operatorname{ri}(\operatorname{conv} \operatorname{rge} A + \operatorname{conv} \operatorname{rge} B) &\subset rge(A + B), \\
\operatorname{cl}(\operatorname{conv} \operatorname{rge} A + \operatorname{conv} \operatorname{rge} B) &= \operatorname{cl} \operatorname{rge}(A + B),
\end{aligned}
$$

and equality holds in the first inclusion whenever the relative interior is replaced by the interior.

Proof. The proof is essentially based on the proofs of Theorems 6.5.1–6.5.2. Under the hypothesis (a) and (b), from the proof of Theorem 6.5.1

one obtains $\operatorname{conv}(\operatorname{rge} A + \operatorname{rge} B) \subset \operatorname{cl} \operatorname{rge}(A+B)$ and $\operatorname{ri} \operatorname{conv}(\operatorname{rge} A + \operatorname{rge} B) \subset \operatorname{rge}(A+B)$. Since for any sets $C, D \subset \mathbb{R}^n$ one has the identity $\operatorname{conv}(C+D) = \operatorname{conv} C + \operatorname{conv} D$, the required result follows. Consider now the hypothesis (b) and (c). First recall that star-monotonicity holds when we replace $\operatorname{rge} A$ and $\operatorname{dom} A$ by their convex hulls. Thus, as in the proof of Theorem 6.5.2, since by (c) $y \in \operatorname{dom} A \implies y \in \operatorname{conv} \operatorname{dom} B$, one arrives at the desired results. $\qquad\square$

Theorem 6.5.4 *Let $h : \mathbb{R}^n \to \mathbb{R} \cup \{+\infty\}$ be a proper, lsc, convex function and let A be a monotone map with $\operatorname{dom} A \subset \operatorname{dom} h$ such that $A + \partial h$ is maximal monotone. Then*

$$
\begin{aligned}
\operatorname{cl}(\operatorname{rge} A + \operatorname{dom} h^*) &= \operatorname{cl} \operatorname{rge}(A + \partial h), \\
\operatorname{ri} \operatorname{conv}(\operatorname{rge} A + \operatorname{dom} h^*) &\subset \operatorname{rge}(A + \partial h),
\end{aligned}
$$

and equality holds in the second inclusion whenever the relative interior is replaced by the interior.

Proof. The relation $\operatorname{rge}(A + \partial h) \subset \operatorname{rge} A + \operatorname{rge} \partial h$ trivially holds, and since h is a proper lsc convex, function one has $\operatorname{rge} \partial h \subset \operatorname{dom} h^*$ and therefore

$$
\operatorname{rge}(A + \partial h) \subset \operatorname{rge} A + \operatorname{dom} h^*. \tag{6.9}
$$

Once again the proof will be completed by using Theorem 6.4.1, applied to $C := \operatorname{rge} A + \operatorname{dom} h^*$ and the maximal monotone map $A + \partial h$. Thus we need to verify condition [BH] in that case. As shown in Example 6.5.2, the subdifferential map ∂h is star-monotone, and one has

$$
\forall w \in \operatorname{dom} h^*, \quad \forall y \in \operatorname{dom} h, \quad \sup_{(x,u)\in \operatorname{gph} \partial h} \langle u - w, y - x \rangle < \infty,
$$

and hence, since A is monotone, it follows that [BH] is satisfied, and thus we have

$$
\operatorname{conv}(\operatorname{rge} A + \operatorname{dom} h^*) \subset \operatorname{cl} \operatorname{rge}(A + \partial h),
$$

which together with (6.9) implies the desired results. $\qquad\square$

Another variant of these types of results is given in the next proposition with proof once again based on the use of Theorem 6.4.1 and left to the reader.

Proposition 6.5.3 *Let A, B be two monotone maps on $\mathbb{R}^n$ satisfying:*
(a) $\operatorname{rge} B = \mathbb{R}^n$.
(b) $A + B$ is maximal monotone.
(c) B is star monotone.
Then $\operatorname{rge}(A + B) = \mathbb{R}^n$.

We consider now the important special case when one of the maps involved is the subdifferential of a given lower semicontinuous proper convex function and derive the corresponding asymptotic functional formula. As we shall see, this type of result is important for establishing existence of solutions of generalized equations with such structure; see Section 6.8.

Theorem 6.5.5 *Let $h : \mathbb{R}^n \to \mathbb{R} \cup \{+\infty\}$ be a proper lsc convex function and let A be monotone with $\operatorname{dom} h \subset \operatorname{dom} A$ and such that $A + \partial h$ is maximal monotone. Then*

$$f_\infty^{A+\partial h}(d) = \sup\{\langle c, d \rangle | c \in A(x), \ x \in \operatorname{dom} h\} + h_\infty(d). \tag{6.10}$$

Proof. Define $\hat{A}(x) := A(x)$ if $x \in \operatorname{dom} h$ and $\emptyset$ otherwise. Then, since $\operatorname{dom} \hat{A} = \operatorname{dom} h$ and $\operatorname{dom} \partial h \subset \operatorname{dom} h$, we obtain

$$\operatorname{dom}(A + \partial h) = \operatorname{dom}(\hat{A} + \partial h) = \operatorname{dom} \partial h, \ \operatorname{rge}(A + \partial h) = \operatorname{rge}(\hat{A} + \partial h),$$

and hence

$$f_\infty^{A+\partial h}(d) = f_\infty^{\hat{A}+\partial h}(d), \ \forall d. \tag{6.11}$$

Now, since $\hat{A} + \partial h$ is also maximal monotone, with $\operatorname{dom} \hat{A} = \operatorname{dom} h$, then invoking Theorem 6.5.4 one has

$$\operatorname{cl} \operatorname{rge}(\hat{A} + \partial h) = \operatorname{cl}(\operatorname{rge} \hat{A} + \operatorname{dom} h^*).$$

Using the latter relation in the definition (6.6) and (6.11) we then have

$$\begin{aligned} f_\infty^{A+\partial h}(d) &= \sup\{\langle c, d \rangle | c \in \operatorname{cl} \operatorname{rge}(\hat{A} + \partial h)\} \\ &= \sup\{\langle c, d \rangle | c \in \operatorname{cl}(\operatorname{rge} \hat{A} + \operatorname{dom} h^*)\} \\ &= \sup\{\langle c, d \rangle | c \in \operatorname{rge} \hat{A} + \operatorname{dom} h^*\} \\ &= \sup\{\langle u, d \rangle | u \in \operatorname{rge} \hat{A}\} + \sup\{\langle v, d \rangle | v \in \operatorname{dom} h^*\} \\ &= \sup\{\langle u, d \rangle | u \in A(x), \ x \in \operatorname{dom} h\} + h_\infty(d), \end{aligned}$$

where in the last equality we use the formula $h_\infty = \sigma_{\operatorname{dom} h^*}$ given in Theorem 2.5.4(b). □

Preserving Maximal Monotonicity
The sum of monotone mappings remains monotone. However, maximal monotonicity is not in general preserved without assuming some kind of constraint qualification on the domain of the given maps, much in the spirit of the conditions used in convex analysis in dealing with the sum of subdifferentials of convex functions (cf. Chapter 3). We begin with the sum theorem and then give a more general result involving composition of maps.

Theorem 6.5.6 *Let $A_i : \mathbb{R}^n \rightrightarrows \mathbb{R}^n$, $i = 1, 2$, be maximal monotone. If $\mathrm{ri}(\mathrm{dom}\, A_1) \cap \mathrm{ri}(\mathrm{dom}\, A_2) \neq \emptyset$, then $A_1 + A_2$ is maximal monotone.*

Proof. By Minty's theorem, showing that $A_1 + A_2$ is maximal monotone reduces to the question of showing that $\mathrm{rge}(I + A_1 + A_2) = \mathbb{R}^n$; i.e., one has to prove that

$$\forall y \in \mathbb{R}^n, \ \exists x : \ y \in (I + A_1 + A_2)(x).$$

Since translation preserves maximal monotonicity, i.e., $A_1 - \{y\}$ is still maximal monotone, it is enough to consider the case $y = 0$ in the above inclusion. Equivalently, one has

$$\exists x : \quad 0 \in (I + A_1 + A_2)(x)$$

$$\Longleftrightarrow \quad \exists x : \quad 0 \in \left(\frac{I}{2} + A_1\right)(x) + \left(\frac{I}{2} + A_2\right)(x),$$

$$\Longleftrightarrow \quad \exists z : \quad z \in \left(\frac{I}{2} + A_1\right)(x); \ -z \in \left(\frac{I}{2} + A_2\right)(x),$$

$$\Longleftrightarrow \quad \exists z : \quad 0 \in \left(\frac{I}{2} + A_1\right)^{-1}(z) - \left(\frac{I}{2} + A_2\right)^{-1}(-z).$$

Let $S_1(z) := (\frac{I}{2} + A_1)^{-1}(z)$ and $S_2(z) := -(\frac{I}{2} + A_2)^{-1}(-z)$. Then both maps S_1, S_2 are maximal monotone, in fact single-valued and continuous (cf. Proposition 6.2.1), with domains equal to $\mathbb{R}^n$. Therefore, $S := S_1 + S_2$ is maximal monotone, and the latter inclusion is equivalent to $\exists z : \ 0 \in S(z)$, and thus we have reduced the question to showing that $S^{-1}(0) \neq \emptyset$; i.e., one has to show that $0 \in \mathrm{rge}\, S = \mathrm{rge}(S_1 + S_2)$. By Example 6.5.2(iii) one has that S_1 and S_2 are also star-monotone. Therefore invoking Theorem 6.5.3 one has

$$\mathrm{ri}(\mathrm{conv}\,\mathrm{rge}\, S_1 + \mathrm{conv}\,\mathrm{rge}\, S_2) \subset \mathrm{rge}(S_1 + S_2).$$

But since $\mathrm{conv}\,\mathrm{rge}\, S_i$ are convex sets and $\mathrm{rge}\, S_i$ are almost convex, it follows from the latter inclusion that $\mathrm{ri}\,\mathrm{rge}\, S_1 + \mathrm{ri}\,\mathrm{rge}\, S_2 \subset \mathrm{rge}(S_1 + S_2)$. Now, since $\mathrm{rge}\, S_1 = \mathrm{dom}\, A_1$ and $\mathrm{rge}\, S_2 = -\mathrm{dom}\, A_2$, then together with the hypothesis in the theorem $\mathrm{ri}(\mathrm{dom} A_1) \cap \mathrm{ri}(\mathrm{dom} A_2) \neq \emptyset$, it follows that

$$0 \in \mathrm{ri}\,\mathrm{dom}\, A_1 - \mathrm{ri}\,\mathrm{dom}\, A_2 = \mathrm{ri}\,\mathrm{rge}\, S_1 + \mathrm{ri}\,\mathrm{rge}\, S_2 \subset \mathrm{rge}(S_1 + S_2),$$

i.e., $0 \in \mathrm{rge}(S_1 + S_2)$, and the proof is completed.

$\square$

Theorem 6.5.7 *Let $S : \mathbb{R}^m \rightrightarrows \mathbb{R}^m$ be maximal monotone and define $T(x) = A^T S(Ax + b)$, where $A \in \mathbb{R}^{m \times n}$ is a given matrix and $b \in \mathbb{R}^m$ a given vector. If $(\mathrm{rge}\, A + b) \cap \mathrm{ri}\,\mathrm{dom}\, S \neq \emptyset$, then T is maximal monotone from $\mathbb{R}^n$ to $\mathbb{R}^n$.*

Proof. Without loss of generality we can take $b = 0$. Let $R := S + N_{\mathrm{rge}\,A}$. Since $\mathrm{dom}\,N_{\mathrm{rge}\,A} = \mathrm{rge}\,A$ and $\mathrm{ri}\,\mathrm{rge}\,A = \mathrm{rge}\,A$, the hypothesis in the theorem is the same as $\mathrm{ri}\,\mathrm{rge}\,A \cap \mathrm{ri}\,\mathrm{dom}\,S \neq \emptyset$, which by Theorem 6.5.6 implies the maximal monotonicity of R. To show that T is maximal monotone, by Definition 6.1.2 one has to verify that for each $x \in \mathbb{R}^n, t \in \mathbb{R}^n$ satisfying

$$\langle t - v, x - y \rangle \geq 0, \quad \forall (y, v)\,\mathrm{gph}\,T, \tag{6.12}$$

one obtains $t \in T(x)$. Consider such an $(x, t) \in \mathbb{R}^n \times \mathbb{R}^n$, and let us prove first that $t \in \mathrm{rge}\,A^T$. Since $\ker A$ is the complementary subspace of $\mathrm{rge}\,A^T$, there exist c and $s \in \ker A$ with $t = A^T c + s$. Now, since $\mathrm{rge}\,A \cap \mathrm{dom}\,S \neq \emptyset$, there exists $p_0 \in \mathrm{dom}\,T$. Let $z_0 \in S(Ap_0)$, $\lambda \in \mathbb{R}$ and define $p_\lambda = p_0 + \lambda s$. Since $s \in \ker A$, we have $A^T z_0 \in T(p_\lambda)$, so that using (6.12) we get

$$\begin{aligned}
0 \leq \langle x - p_\lambda, t - A^T z_0 \rangle &= \langle x - p_0, t - A^T z_0 \rangle - \lambda \langle s, A^T(c - z_0) + s \rangle \\
&= \langle x - p_0, t - A^T z_0 \rangle - \lambda \|s\|^2.
\end{aligned}$$

Taking $\lambda \to +\infty$, this implies that $s = 0$ and therefore $t = A^T c$. By definition of R one has

$$R(z) = S(z) + N_{\mathrm{rge}\,A}(z) = \begin{cases} S(z) + \ker A^T & \text{if } z \in \mathrm{rge}\,A, \\ \emptyset & \text{otherwise.} \end{cases}$$

Let (z, u) be any pair in the graph of R. Then $u = d + e$ with $d \in S(Ah), e \in \ker A^T$ and $z = Ah$ for some h. As a consequence, $A^T d \in T(h)$, and using (6.12) we get

$$0 \leq \langle x - h, A^T c - A^T d \rangle = \langle x - h, A^T c - A^T u \rangle = \langle Ax - z, c - u \rangle,$$

and since R is maximal monotone, it follows that $c = R(Ax) = SAx + g$ with $g \in \ker A^T$. As a consequence, $t = A^T c = A^T SAx$, showing that $t \in T(x)$ and completing the proof. $\qquad\square$

We end this section with the following useful consequence.

Proposition 6.5.4 *Let $C : \mathbb{R}^n \times \mathbb{R}^d \rightrightarrows \mathbb{R}^n \times \mathbb{R}^d$ be maximal monotone. Fix $x \in \mathbb{R}^n$ and define $B : \mathbb{R}^d \to \mathbb{R}^d$ by*

$$B(z) := \{w | \exists v : (v, w) \in C(x, z)\}.$$

If x is such that there exists $y \in \mathbb{R}^d$ with $(x, y) \in \mathrm{ri}(\mathrm{dom}\,C)$, then B is also maximal monotone.

Proof. Applying Theorem 6.5.7 with $S := C$ and the affine map A defined by $z \to (x, z)$, the result follows.

$\qquad\square$

6.6   Further Properties of Maximal Monotone Maps

Structure of Maximal Monotone Maps

We begin with some general results on the structure of maximal monotone maps. In particular, we will often use the dimension reducing argument given below, which is useful whenever the domain of a maximal monotone map is not the whole space, i.e., whenever one might have $\operatorname{int} \operatorname{dom} T = \emptyset$. First, in the next result we give a useful formula for the asymptotic cone of the closed convex set $T(x)$.

Proposition 6.6.1 Let $T : \mathbb{R}^n \rightrightarrows \mathbb{R}^n$ be maximal monotone. Then for each $x \in \operatorname{dom} T$ one has $(T(x))_\infty(x) = N_{\operatorname{cl} \operatorname{dom} T}(x)$.

Proof. Recall that since T is maximal monotone, by Proposition 6.1.3 one has for each $x \in \operatorname{dom} T$ that $T(x)$ is a closed convex set given by

$$T(x) = \{u \in \mathbb{R}^n \mid \langle u - v, x - y \rangle \geq 0 \;\; \forall (y, v) \in \operatorname{gph} T\}.$$

Therefore, using Corollary 2.5.4 we have

$$
\begin{aligned}
T(x)_\infty &= \{d \in \mathbb{R}^n \mid \langle d, x - y \rangle \geq 0 \;\; \forall (y, v) \in \operatorname{gph} T\} \\
&= \{d \in \mathbb{R}^n \mid \langle d, x - y \rangle \geq 0 \;\; \forall y \in \operatorname{cl} \operatorname{dom} T\} = N_{\operatorname{cl} \operatorname{dom} T}(x),
\end{aligned}
$$

where the last equality follows from the definition of the normal cone for the nonempty convex set $\operatorname{cl} \operatorname{dom} T$.

$\square$

Proposition 6.6.2 Let $T : \mathbb{R}^n \rightrightarrows \mathbb{R}^n$ be maximal monotone and let $x \in \operatorname{dom} T$. Then $T(x)$ is a convex compact set if and only if $x \in \operatorname{int} \operatorname{dom} T$.

Proof. The convexity and closedness of $T(x)$ was proven in Proposition 6.1.3. By Proposition 2.1.2, $T(x)$ is compact if and only if $(T(x))_\infty = \{0\}$. But by Proposition 6.6.1 one has for any $x \in \operatorname{dom} T$, $(T(x))_\infty = N_{\operatorname{cl} \operatorname{dom} T}(x)$, and therefore one has to show that $N_{\operatorname{cl} \operatorname{dom} T}(x) = \{0\}$, but the latter means precisely that $x \in \operatorname{int} \operatorname{cl} \operatorname{dom} T = \operatorname{int} \operatorname{dom} T$. $\square$

The next result gives the structure of a maximal monotone map A whenever $\operatorname{int} \operatorname{dom} A = \emptyset$. We denote by M the affine hull of the domain of a given maximal monotone map A and by L its parallel subspace such that $M = x + L$ for any $x \in \operatorname{dom} A$. The projection map on L is denoted by P_L. Since $M + L^\perp = \mathbb{R}^n$, every $x \in \mathbb{R}^n$ can be uniquely decomposed as follows:

$$x = P_M(x) + x_{L^\perp}, \quad \text{with} \;\; x_{L^\perp} \in L^\perp.$$

Proposition 6.6.3 *Let $A : \mathbb{R}^n \rightrightarrows \mathbb{R}^n$ be maximal monotone, and let $M = \text{aff dom } A$ with L the subspace parallel to M. Then for any $x \in \text{dom } A$:*
(a) $A(x) = A(x) + L^\perp$.
(b) $A(x) \cap L \neq \emptyset$. More precisely, $P_L(u) \in A(x) \cap L$ for each $u \in A(x)$.

Proof. (a) For any $x \in \text{cl dom } A$ we have $L^\perp \subset N_{\text{cl dom } A}(x)$, so using Proposition 6.6.1 one obtains $A(x) + L^\perp \subset A(x) + N_{\text{cl dom } A}(x) = A(x) + (A(x))_\infty = A(x)$, since $A(x)$ is convex, and the result follows. To prove (b), let $u \in A(x), v \in A(z)$ and set $w = P_L(u)$. Since A is monotone and $z - x \in L$, we have

$$\langle v - w, z - x \rangle = \langle v - u, z - x \rangle \geq 0, \quad \forall z \in \text{dom } A.$$

Since A is a maximal map, we obtain $w \in A(x)$, and together with the fact that $w \in L$ it follows that $A(x) \cap L \neq \emptyset$ for each $x \in \text{dom } A$. $\qquad \square$

The last proposition, part (b), leads us to define an extension map A_L of A as follows:

$$A_L(x) = A(P_M(x)) \cap L, \quad \forall x \in \mathbb{R}^n.$$

Therefore, from Proposition 6.6.3(b) it follows that

$$A_L(x) = A(x) \cap L, \ \forall x \in \text{dom } A \quad \text{and} \quad \text{dom } A_L = \text{dom } A + L^\perp. \quad (6.13)$$

The next results gives the properties satisfied by the extension map A_L.

Proposition 6.6.4 *Let $A : \mathbb{R}^n \rightrightarrows \mathbb{R}^n$ be maximal monotone, and let $M = \text{aff dom } A$ with L the subspace parallel to M. Then:*
(a) A_L is maximal monotone.
(b)
$$A(x) = \begin{cases} A_L(x) + L^\perp & \text{if } x \in M, \\ \emptyset & \text{otherwise.} \end{cases}$$
(c) If $0 \in \text{dom } A$, one has $\text{dist}(u, A(0)) = \text{dist}(P_L(u), A_L(0)), \quad \forall u$.
(d) $\text{int dom } A_L = \text{ri dom } A + L^\perp$.
(e) If $0 \in \text{ri dom } A$, then $A(0) = A_L(0) + L^\perp$, with $A_L(0)$ a nonempty compact convex set.

Proof. (a) We first show that A_L is monotone. Let $u \in A_L(x), v \in A_L(y)$. By definition of the extension map A_L we have $u \in A(P_M(x)) \cap L, v \in A(P_M(y)) \cap L$, and thus using the monotonicity of A we obtain

$$\langle u - v, x - y \rangle = \langle u - v, P_M(x) - P_M(y) \rangle \geq 0,$$

thus showing the monotonicity of A_L. To prove the maximality of A_L, by Minty's theorem 6.2.2, it is enough to show that $\text{rge}(A_L + I) = \mathbb{R}^n$; that is, one has to prove that

$$\forall y \in \mathbb{R}^n, \ \exists x : y \in A_L(x) + x. \quad (6.14)$$

Let $y = P_M(y) + y_{L^\perp}$. But since A is assumed maximal monotone, by the same theorem one has that there exists $z \in M$ such that $P_M(y) \in A(z) + z$, and hence $x = z + y_{L^\perp}$ solves (6.14).

(b) Define the map

$$B(x) = \begin{cases} A_L(x) + L^\perp & \text{if } x \in M, \\ \emptyset & \text{otherwise.} \end{cases}$$

For $x \in M \setminus \text{dom } A$ we have $A(x) = B(x) = \emptyset$. Now let $x \in \text{dom } A$, and take $u \in B(x)$. Then $u = u_1 + u_2$ with $u_1 \in A_L(x)$, $u_2 \in L^\perp$. Since by definition $A_L(x) = A(x) \cap L$, it follows from Proposition 6.6.3(a) that $B(x) \subset A(x)$. To prove the converse, let $u \in A(x)$. Take $u := u_1 + u_2$, $u_1 \in L, u_2 \in L^\perp$. Using again Proposition 6.6.3(a) it follows that $u_1 \in A(x)$ and hence $A(x) \subset B(x)$.

(c) Under the assumption $0 \in \text{dom } A$, one has $L = M$. Let $v = v_1 + v_2$ with $v_1 \in A_L(0), v_2 \in L^\perp$. Then

$$\|u - v\|^2 = \|P_L(u) - v_1\|^2 + \|P_{L^\perp}(u) - v_2\|^2.$$

Therefore, using (b) the relation (c) follows.

(d) An immediate consequence of (6.13) leads to the relations (d). Moreover, since $\text{ri dom } A$ is a nonempty convex set (cf. Proposition 6.4.1), it follows that $\text{int dom } A_L$ is a nonempty convex set as well.

(e) The required formula is a consequence of (b). Moreover, since we assume that $0 \in \text{ri dom } A$, from (d) we obtain $0 \in \text{int dom } A_L$, which implies, using Proposition 6.6.2, that $A_L(0)$ is a compact convex set.

$\square$

Upper Semicontinuity and Directional Local Boundedness

The subdifferential of a proper lsc convex function f is a very special maximal monotone map that possesses several useful and important properties. For example, if the interior of the effective domain $\text{int dom } f$ is nonempty and if C is a given compact subset of $\text{inf dom } f$, then we know that the set $\cup\{\partial f(x) : x \in C\}$ is compact. Furthermore, the map ∂f is upper semicontinuous at every $x \in \text{int dom } f$. Other directional boundedness properties are also parts of the useful properties of ∂f. It is thus natural to know under which conditions similar properties can be established for general maximal monotone maps.

The set of points $y \in T(x)$ such that $d \in \mathbb{R}^n$ is normal to $T(x)$ is denoted by $T(x)_d$; i.e.,

$$T(x)_d := \{y \in T(x) \mid \sigma_{T(x)}(d) = \langle y, d \rangle\}.$$

Proposition 6.6.5 Let $T : \mathbb{R}^n \rightrightarrows \mathbb{R}^n$ be maximal monotone and let $x \in \text{dom } T$. Assume that $\text{int dom } T \neq \emptyset$ and let $\varepsilon_0 > 0$ and $d \in \mathbb{R}^n$ be such that $x + \varepsilon d \in \text{int dom } T$ for every $\varepsilon \in (0, \varepsilon_0]$. Then $T(x)_d$ is nonempty and compact.

Proof. We first show that $T(x)_d$ is nonempty. For that, suppose that $T(x)_d$ is empty. Then there exists a sequence $\{u_k\}$ with

$$u_k \in T(x), \quad \lim_{k\to\infty} \|u_k\| = +\infty, \quad \lim_{k\to\infty} u_k \|u_k\|^{-1} = u \neq 0,$$

and such that $\lim_{k\to\infty} \langle u_k, d\rangle = \sup\{\langle c, d\rangle \mid c \in T(x)\} := \alpha$. From this, we deduce immediately that $0 \neq u \in (T(x)_\infty$ and $\langle u, d\rangle \geq 0$. Now take a nonzero element $u \in (T(x))_\infty$, which by Proposition 6.6.1 means $u \in N_{\mathrm{cl\,dom\,}T}(x)$. Then using the definition of the normal cone one has

$$\sigma_{\mathrm{cl\,dom\,}T}(u) - \langle u, x\rangle = \sigma_{\mathrm{dom\,}T}(u) - \langle u, x\rangle \leq 0.$$

One the other hand, since we assumed that $x + \varepsilon d \in \mathrm{int\,dom\,}T$, then using Theorem 1.3.2(iii) it follows that

$$\langle u, d\rangle < \varepsilon^{-1}(\sigma_{\mathrm{dom\,}T}(u) - \langle u, x\rangle),$$

and hence combining the above two inequalities one has

$$\langle u, d\rangle < 0, \quad \forall\, 0 \neq u \in T(x)_\infty, \tag{6.15}$$

which yields a contradiction. To prove compactness of $T(x)_d$, we note that this set can be written as $T(x)_d = S \cap T(x)$, where $S := \{z \mid \langle z, d\rangle \geq \alpha\}$. Since $T(x)_d \neq \emptyset$, then $\alpha \in \mathbb{R}$, and hence $S_\infty = \{u \mid \langle u, d\rangle \geq 0\}$. Invoking Proposition 2.1.9, one obtains $(T(x)_d)_\infty = S_\infty \cap (T(x))_\infty$, and since as just shown above, for any $0 \neq u \in (T(x))_\infty$ one also has $\langle u, d\rangle < 0$, it follows that $(T(x)_d)_\infty = \{0\}$, which by Proposition 2.1.2 proves that $T(x)_d$ is compact. □

The next result is an extension of properties shared by subdifferential maps of convex functions and is useful in the analysis of approximate solutions of generalized equations; cf. Section 6.8.

Theorem 6.6.1 *Let $T : \mathbb{R}^n \rightrightarrows \mathbb{R}^n$ be maximal monotone and let $x \in \mathrm{dom\,}T$. Suppose that $\mathrm{int\,dom\,}T \neq \emptyset$ and let $d \in \mathbb{R}^n, \varepsilon_0 > 0$ be such that $x + \varepsilon d \in \mathrm{int\,dom\,}T$ for all $\varepsilon \in (0, \varepsilon]$. Suppose that $\{x_k\}$ is a sequence in $\mathrm{dom\,}T$ converging to x such that*

$$\lim_{k\to\infty} \frac{x - x_k}{\|x - x_k\|} = d.$$

Then for any $\varepsilon > 0$, there exists an index k_0 such that

$$T(x_k) \subset T(x)_d + \varepsilon\mathbb{B}, \quad \forall k \geq k_0.$$

Proof. Let $\{x_k\}$ be a sequence converging to x and satisfying the hypothesis of the theorem and let $u_k \in T(x_k)$. Then the sequence $\{u_k\}$ is bounded. Indeed, suppose the contrary. Then there exist $u^* \neq 0$ and a subsequence $\{u_{k(i)}\}$ such that

$$\lim_{i \to \infty} \|u_{k(i)}\| = +\infty, \quad \lim_{i \to \infty} u_{k(i)} \|u_{k(i)}\|^{-1} = u^*.$$

Since T is monotone, we have

$$\left\langle \frac{u_{k(i)} - v}{\|u_{k(i)}\|}, x_{k(i)} - w \right\rangle \geq 0 \quad \forall v \in T(w),$$

and passing to the limit, we obtain $\langle u^*, x - w \rangle \geq 0$, $\forall v \in \operatorname{dom} T$, implying that $u^* \in N_{\operatorname{cl dom} T}(x)$, namely, by Proposition 6.6.1 that $u^* \in (T(x))_\infty$. Furthermore, using again the monotonicity of T we also have

$$\left\langle \frac{u_{k(i)} - y}{\|u_{k(i)}\|}, \frac{x_{k(i)} - x}{\|x_{k(i)} - x\|} \right\rangle \geq 0 \quad \forall y \in T(x),$$

and passing to the limit we get $\langle u^*, d \rangle \geq 0$, which is impossible, as already proved in Proposition 6.6.5 (cf. (6.15)). Now suppose that the conclusion of the theorem is not satisfied. Then there would exist $\varepsilon > 0$ and a subsequence $u_{k(i)} \in T(x_{k(i)})$ such that $\operatorname{dist}(u_{k(i)}, T(x)_d) \geq \varepsilon$. Since we have shown that the sequence $\{u_{k(i)}\}$ is bounded, without loss of generality we can assume that $u_{k(i)}$ converges to the point u^*, and then passing to the limit in the latter inequality we obtain $\operatorname{dist}(u^*, T(x)_d) \geq \varepsilon$. However, by the monotonicity of T we have

$$\left\langle u_{k(i)} - y, \frac{x_{k(i)} - x}{\|x_{k(i)} - x\|} \right\rangle \geq 0 \quad \forall y \in T(x)_d,$$

and passing to the limit we obtain $\langle u^* - y, d \rangle \geq 0$, $\forall y \in T(x)_d$, which in turn is equivalent to $\langle u^*, d \rangle \geq \sigma_{T(x)}(d)$. But since T is closed on $\operatorname{dom} T$, then $u^* \in T(x)$, and it follows that $u^* \in T(x)_d$, which yields a contradiction to $\operatorname{dist}(u^*, T(x)_d) \geq \varepsilon > 0$, and the proof is completed.

$\square$

Using the reduction arguments of Proposition 6.6.4, this result can be improved by working on the extension map T_L, and we obtain a similar property even when $\operatorname{int} \operatorname{dom} T = \emptyset$.

Corollary 6.6.1 *Let $T : \mathbb{R}^n \rightrightarrows \mathbb{R}^n$ be maximal monotone, $M = \operatorname{aff}(\operatorname{dom} T)$, and L its parallel subspace. Let $x \in \operatorname{dom} T$ and let $d \in \mathbb{R}^n, \varepsilon_0 > 0$ such that $x + \varepsilon d \in \operatorname{ri} \operatorname{dom} T$ for all $\varepsilon \in (0, \varepsilon]$. Then:*
(a) $T(x)_d = T_L(x)_d + L^\perp$, and $T_L(x)_d$ is a nonempty convex compact set.
(b) For a sequence satisfying the hypothesis of Theorem 6.6.1, the conclusion of that theorem holds.

Proof. Using Proposition 6.6.4(a) and (d), it follows that the map T_L is maximal monotone and $x + \varepsilon d \in \operatorname{int} \operatorname{dom} T_L$. We can then apply Theorem 6.6.1 to conclude that $T_L(x)_d$ is a nonempty convex compact set. Moreover, for any $\varepsilon > 0$, there exists k_0 such that

$$T_L(x_k) \subset T_L(x)_d + \varepsilon \mathbb{B}, \quad \forall k \geq k_0.$$

Now using Proposition 6.6.4(b) we obtain for each $y \in \operatorname{dom} T$ that $\sigma_{T(y)}(d) = \sigma_{T_L(y)}(d)$ and by definition of $T(y)_d$ that $T(y)_d = T_L(y)_d + L^\perp$. Then adding $L^\perp$ to the inclusion above and the last equation we obtain the desired result.
$\square$

Proposition 6.6.6 *Let* $T : \mathbb{R}^n \rightrightarrows \mathbb{R}^n$ *be maximal monotone and let* $x \in \operatorname{dom} T$. *Then* T *is locally bounded at* x *if and only if* $x \in \operatorname{int} \operatorname{dom} T$.

Proof. Suppose that T is not locally bounded at x, then there would exist d and sequences $\{x_k\}, \{u_k\}$ such that

$$x_k \to x, \quad \frac{x_k - x}{\|x_k - x\|} \to d, \quad u_k \in T(x_k), \quad \|u_k\| \to +\infty.$$

If $x \in \operatorname{int} \operatorname{dom} T$, then $x + \varepsilon d \in \operatorname{dom} T$ for $\varepsilon \in (0, \varepsilon_0], \varepsilon_0 > 0$. Applying Theorem 6.6.1 it follows that $\{u_k\}$ is bounded, which is a contradiction. To prove the reverse assertion, let T be locally bounded at x and suppose that $x \notin \operatorname{int} \operatorname{dom} T$. Then by Proposition 6.6.2, one has $T(x)$ unbounded, a contradiction.
$\square$

Proposition 6.6.7 *Let* $T : \mathbb{R}^n \rightrightarrows \mathbb{R}^n$ *be maximal monotone and* $C \subset \operatorname{int} \operatorname{dom} T$ *be compact. Then the set* $T(C) = \cup_{x \in C} T(x)$ *is compact.*

Proof. Let $u_k \in T(x_k)$ with $x_k \in C$. Since C is compact, there exists a subsequence $\{x_{k(l)}\} \to x \in C$. Then $x \in \operatorname{int} \operatorname{dom} T$, and from Proposition 6.6.6, T is locally bounded at x, so that $\{u_{k(l)}\}$ is bounded. Since T is closed at x, it follows that there exists a subsequence $u_{k(j)} \to u \in T(x)$, and hence $T(C)$ is compact.
$\square$

Proposition 6.6.8 *Any maximal monotone map* $T : \mathbb{R}^n \rightrightarrows \mathbb{R}^n$ *is upper semicontinuous at every* $x \in \operatorname{int} \operatorname{dom} T$.

Proof. This follows from Proposition 6.6.6, Proposition 6.1.3, and Theorem 1.4.1.
$\square$

6.7 Variational Inequalities Problems

Let $f : \mathbb{R}^n \to \mathbb{R} \cup \{+\infty\}$ be a proper lsc convex function. Then $x^* \in \mathbb{R}^n$ is a solution of the convex problem $\min\{f(x) \mid x \in \mathbb{R}^n\}$ if and only if the inclusion $0 \in \partial f(x^*)$ holds; i.e., solving a convex problem is equivalent to finding the *zero* of the maximal monotone map ∂f. More generally, given a set-valued map $T : \mathbb{R}^n \rightrightarrows \mathbb{R}^n$ we can look at solving the inclusion (often called *a generalized equation*) $0 \in T(x^*)$. In general, it is not always possible to associate an optimization problem with a generalized equation. However, a general and important class of problems that can be associated with monotone maps is that of *variational inequalities*.

Given a set-valued map T, and a closed convex subset C of $\mathbb{R}^n$, the variational inequality problem $\mathrm{VI}(T, C)$ consists in finding a pair $x^* \in C$ and $g^* \in T(x^*)$ such that

$$\langle x - x^*, g^* \rangle \geq 0, \ \forall x \in C, \tag{6.16}$$

Solving $\mathrm{VI}(T, C)$ can be reduced to finding the zero of an appropriate generalized equation. Indeed, from (6.16), x^* solves $\mathrm{VI}(T, C)$ if and only if one has

$$-g^* \in N_C(x^*), \ \text{and} \ g^* \in T(x^*),$$

which in turn is equivalent to finding x^* such that $0 \in N_C(x^*) + T(x^*)$. Therefore, defining the set-valued map

$$A(x) := \begin{cases} N_C(x) + T(x) & \text{if } x \in C, \\ \emptyset & \text{if } x \notin C, \end{cases}$$

the variational inequality problem reduces to finding the zero of the generalized equation $0 \in A(x^*)$.

Note that if T is assumed monotone, then since N_C is also monotone, the set-valued map is monotone as well. If T is also maximal, then since N_C is maximal monotone by Corollary 6.3.1, it follows from Theorem 6.5.6 that the maximal monotonicity of A is preserved under the condition $\mathrm{ri}\,\mathrm{dom}\,T \cap \mathrm{ri}\,C \neq \emptyset$. The interplay between variational inequalities and the more general framework of maximal monotone inclusions will be useful in the analysis to follow. Meanwhile, we now give some interesting examples.

Example 6.7.1 (i) Let $C = \mathbb{R}^n$, so that $N_C(x) = \{0\}$ and $\mathrm{VI}(T, \mathbb{R}^n)$ reduces to the equation $0 \in T(x)$. If T also happens to be single-valued on $\mathbb{R}^n$, then the variational inequality problem is nothing else but solving a system of nonlinear equations $T(x) = 0$.

(ii) Let $T := F$ with $F : \mathbb{R}^n \to \mathbb{R}^n$ single-valued. Then $\mathrm{VI}(F, C)$ reduces to finding $x^* \in C$ such that $\langle F(x^*), x - x^* \rangle \geq 0, \ \forall x \in C$. If $C := \mathbb{R}^n_+$, then $\mathrm{VI}(F, \mathbb{R}^n_+)$ reduces to the *complementarity* problem: Find x^* satisfying

$$x \in \mathbb{R}^n_+, \ F(x^*) \in \mathbb{R}^n_+, \ \langle x^*, F(x^*) \rangle = 0.$$

(iii) Let $X \subset \mathbb{R}^n$, $Y \subset \mathbb{R}^m$ be nonempty closed convex sets and let $L : X \times Y \to \mathbb{R}$ be a convex–concave continuously differentiable function. Then the saddle point condition

$$x^* \in X, y^* \in Y : \ L(x^*, y) \leq L(x^*, y^*) \leq L(x, y^*), \quad \forall (x, y) \in X \times Y,$$

is equivalent to the variational inequality problem $\text{VI}(F, C)$ with

$$C := X \times Y \quad \text{and} \quad F : (x, y) \to (\nabla_x L((x, y), -\nabla_y L(x, y)),$$

which in turn is also equivalent to the generalized equation $0 \in T(x^*, y^*)$ with

$$T(x, y) := (\nabla_x L((x, y) + N_X(x), -\nabla_y L(x, y)) + N_Y(y)).$$

(iv) *Normal Maps.* Let $C \subset \mathbb{R}^n$ be a nonempty closed convex set and let $\phi : \mathbb{R}^n \to \mathbb{R}^n$ be a continuous function. A problem that arises very often in optimization and equilibrium analysis is that of finding a point x satisfying the equation

$$\phi(P_C(x)) + (x - P_C(x)) = 0,$$

where P_C is the projection map onto C. The latter equation is in fact equivalent to the generalized equation $0 \in \phi(y) + N_C(y)$. Indeed, recall from Proposition 6.3.1 that $P_C(x) = (I + N_C)^{-1}(x)$ and that P_C is single-valued. Therefore, by direct manipulation one has

$$0 \in \phi(y) + N_C(y) \iff 0 \in \phi(y) - y + y + N_C(y)$$
$$\iff 0 \in \phi(y) - y + (I + N_C)(y),$$

and thus letting $x := (I + N_C)(y)$ in the latter inclusion and recalling that $y = (I + N_C)^{-1}(x) = P_C(x)$ is single-valued, the latter inclusion is nothing else but $\phi(P_C(x)) - P_C(x) + x = 0$.

(v) *Gap functions.* Let $F : \mathbb{R}^n \to \mathbb{R}^n$ be a continuous monotone map, and let $C \subset \mathbb{R}^n$ be closed and convex. Define for any $x \in C$ the function $g(x) = \sup\{\langle x - y, F(y) \rangle \mid y \in C\}$. Then if the solution set S of the optimization problem $\inf\{g(x) \mid x \in C\}$ is nonempty, one has $x^* \in S$ with $g(x^*) = 0$ if and only if x^* solves the variational inequality problem $\text{VI}(F, C)$. Another possible gap function sharing the latter properties and valid for any map F can be defined via $\gamma(x) = \sup\{\langle y - x, F(x) \rangle \mid y \in C\}$.

An interesting class of variational inequality problems that covers a wide range of applications is the case where T is a maximal monotone mapping from $\mathbb{R}^n$ into itself and the constraint set C is explicitly defined by

$$C := \{x \in \mathbb{R}^n \mid F(x) \leq 0\},$$

where $F(x) := (f_1(x), \ldots, f_m(x))^T$, with $f_i : \mathbb{R}^n \to \mathbb{R} \cup \{+\infty\}$, $i = 1, \ldots, m$, proper closed convex functions. This type of problem will be analyzed in greater detail in Section 6.9.

6.8   Existence Results for Variational Inequalities

The asymptotic function of a maximal monotone map introduced in Section 6.5 is particularly important in establishing existence of solutions for variational problems described by generalized equations involving maximal monotone maps.

Proposition 6.8.1 *Let* $A : \mathbb{R}^n \rightrightarrows \mathbb{R}^n$ *be any maximal monotone map. Then the solution set* $A^{-1}(0)$ *of the generalized equation* $0 \in A(x)$ *is nonempty and compact if and only if* $0 \in \text{int rge } A$, *which in turn can be equivalently written as*

$$\forall d \neq 0, \quad f_\infty^A(d) > 0. \tag{6.17}$$

Proof. Since A is maximal monotone, so is A^{-1}, and thus applying Proposition 6.6.1 to the maximal monotone map A^{-1} one has, using $\text{dom } A^{-1} = \text{rge } A$,

$$(A^{-1}(0))_\infty = N_{\text{cl dom } A^{-1}}(0) = N_{\text{cl rge } A}(0).$$

By Proposition 6.4.1(ii), the closure of the range of the maximal monotone map A is convex, and thus from Proposition 2.1.2, it follows that the solution set $A^{-1}(0)$ is nonempty and bounded (hence compact) if and only if $N_{\text{cl rge } A}(0) = \{0\}$, but the latter is equivalent to $0 \in \text{int cl rge } A$; i.e., by Theorem 1.3.2 to $\sigma_{\text{cl rge } A}(d) > 0$, $\forall d \neq 0$. By definition of the asymptotic function associated with A given in (6.6) the last inequality means exactly (6.17).

$\square$

A maximal monotone map A satisfying (6.17) is often called a *coercive* map.

Let $h : \mathbb{R}^n \to \mathbb{R} \cup \{+\infty\}$ be a proper lsc convex function and let A be a monotone map A with $\text{dom } h \subset \text{dom } A$. Applying Theorem 6.5.5 to the special case $\partial h := \partial \delta(\cdot|C) = N_C$, with $\text{dom } h = C \subset \text{dom } A$, and recalling that $\delta_\infty(\cdot|C) = \delta(\cdot|C_\infty)$, using (6.6) together with Proposition 6.5.1 we obtain the following result.

Proposition 6.8.2 *Let* $A : \mathbb{R}^n \rightrightarrows \mathbb{R}^n$ *be maximal monotone, and* C *a closed convex set such that* $C \subset \text{dom } A$. *Then*

$$f^{A,C}(d) := f_\infty^{A+N_C}(d) := \begin{cases} \sup\{\langle c, d\rangle | c \in A(x), x \in C\} & \text{if } d \in C_\infty, \\ +\infty & \text{otherwise.} \end{cases}$$

As an application to the VI problem, which corresponds to finding the zero of $T + N_C$, we obtain immediately, using Proposition 6.8.2, the following corollary.

Corollary 6.8.1 *Let* $T : \mathbb{R}^n \rightrightarrows \mathbb{R}^n$ *be maximal monotone and* $C \subset \mathbb{R}^n$ *closed convex such that* $T + N_C$ *is maximal monotone and* $C \subset \operatorname{dom} T$. *Then the solution set of* $VI(T, C)$, *i.e.,* $(T + N_C)^{-1}(0)$, *is nonempty and compact if and only if*

$$f_\infty^{T+N_C}(d) := f_\infty^{T,C}(d) > 0, \ \forall d \neq 0,$$

where $f_\infty^{T,C} = \sup\{\langle c, d \rangle \mid c \in T(x), \ x \in C\}$.

In similarity to optimization problems, we would like to be able to consider variational inequalities with noncompact solutions sets, namely to relax the assumption $0 \in \operatorname{int} \operatorname{rge} A$. A natural extension (cf. Chapter 3) would thus be to ask that $0 \in \operatorname{ri} \operatorname{rge} A$, in which case we shall say that A is *weakly coercive*. In terms of asymptotic functions, using Theorem 1.3.2(ii), this is equivalent to requiring that

$$f_\infty^A(d) \geq 0, \ \forall d \ \text{and that} \ L_A := \{d \mid f_\infty^A(d) = 0\} \ \text{be a linear space},$$

or more concisely, this is equivalent to

$$f_\infty^A(d) > 0 \ \forall 0 \neq d \in L_A^\perp, \tag{6.18}$$

where $L_A^\perp$ is the orthogonal of L_A, assumed to be a linear space.

Example 6.8.1 Suppose T is maximal monotone and coercive and let F be an $n \times m$ matrix with maximal rank. Then by Theorem 6.5.7, the map $S := F^T T F$ is maximal monotone. Furthermore, since T is coercive, one can verify that S is weakly coercive and that $L_S = \ker F$.

We now characterize weakly coercive maximal monotone maps via the explicit structure of the set of solutions of the generalized equation $0 \in T(x)$. We let E be the parallel subspace to the affine hull of $\operatorname{dom} T^{-1}$.

Theorem 6.8.1 *Let* $T : \mathbb{R}^n \rightrightarrows \mathbb{R}^n$ *be a maximal monotone map. Then* $0 \in \operatorname{ri} \operatorname{rge} T$ *if and only if the solution set of* $0 \in T(x)$, *i.e.,* $T^{-1}(0)$, *is given by*

$$T^{-1}(0) = T_E^{-1}(0) + E^\perp, \tag{6.19}$$

where $T_E^{-1}(0)$ *is a nonempty and compact set, or equivalently if and only if*

$$f_\infty^T(d) > 0 \ \forall d \in E^\perp = L_T = \{d \mid \langle c, d \rangle = 0, \ \forall c \in \operatorname{rge} T\}.$$

Proof. If $0 \in \operatorname{ri} \operatorname{rge} T = \operatorname{ri} \operatorname{dom} T^{-1}$, then formula (6.19) follows from Proposition 6.6.4(v), with $T_E^{-1}(0)$ nonempty and compact. Conversely, assume that (6.19) holds and that $T_E^{-1}(0)$ is nonempty and compact. The latter means that $(T_E^{-1}(0))_\infty = \{0\}$ and by Proposition 6.6.1 is thus equivalent to $N_{\operatorname{cl} \operatorname{dom} T_E^{-1}}(0) = \{0\}$, i.e., to $0 \in \operatorname{int} \operatorname{cl} \operatorname{dom} T_E^{-1} = \operatorname{int} \operatorname{dom} T_E^{-1}$ (recall that $\operatorname{int} \operatorname{dom} T_E^{-1}$ is nonempty and convex). Since by Proposition

6.6.4(iv) one has $\operatorname{int} \operatorname{dom} T_E^{-1} = \operatorname{ri} \operatorname{dom} T^{-1} + E^\perp$, we thus deduce that $0 \in \operatorname{ri} \operatorname{dom} T^{-1} = \operatorname{ri} \operatorname{rge} T$. Finally, using the definition of f_∞^T, and since L_T is a linear space, one obtains

$$L_T = (\operatorname{rge} T)^\perp = (\operatorname{dom} T^{-1})^\perp = E^\perp$$

and the desired equivalent statement in terms of the asymptotic functional associated with the map T.

$\square$

As an application, we can then obtain a relaxed condition that will guarantee the existence of solutions for the variational inequality problem, and thus extending Corollary 6.8.1.

Corollary 6.8.2 *Let* $T : \mathbb{R}^n \rightrightarrows \mathbb{R}^n$ *be a maximal monotone map and* $C \subset \mathbb{R}^n$ *a nonempty closed convex set. Set* $S := T + N_C$ *and let* $X := (T + N_C)^{-1}(0)$ *denote the solution set of the variational inequality problem* $\mathrm{VI}(T, C)$. *Suppose that* $C \subset \operatorname{dom} T$ *and that* S *is maximal monotone. Then* $0 \in \operatorname{ri} \operatorname{rge} S$ *if and only if* $X = S_E^{-1}(0) + E^\perp$ *with* $E = L_S$ *and* $S_E^{-1}(0)$ *nonempty and compact, or equivalently, if and only if*

$$f_\infty^{T,C}(d) > 0 \quad 0 \neq d \in L_S^\perp,$$

with $L_S = \{d \in C_\infty \cap -C_\infty \mid \langle c, d \rangle = 0, \ \forall c \in T(x), \ \forall x \in C\}$.

Proof. The statement is a result of formula (6.19) of Theorem 6.8.1 when applied to the maximal monotone map S. The equivalence in terms of asymptotic functions associated with S is derived by explicit computations. By definition, $L_S = \{d : \ f_\infty^S(d) = f_\infty^{T+N_C}(d) = 0\}$, and this is a subspace. Using Corollary 6.8.2 it follows that

$$L_S = \{d \mid f_\infty^{T,C}(d) = f_\infty^{T,C}(-d) = 0\}.$$

Using the formula for $f_\infty^{T,C}$ as given in Corollary 6.8.2, the desired explicit formulation for L_S follows.

$\square$

Two Existence Results for Special Generalized Equations

We now give two important existence results involving composition of maps and related to generalized equations with special structures often encountered in the context of numerical methods for solving VI.

Consider a class of functions $f : \mathbb{R}^p \to \mathbb{R} \cup \{+\infty\}$ satisfying the following properties:

(i) f is a closed proper convex function with $\operatorname{dom} f$ open.

(ii) f is differentiable on $\operatorname{dom} f$.

(iii) $f_\infty(d) = +\infty, \ \forall d \neq 0$.

We denote by $\mathcal{F}$ the class of functions satisfying (i), (ii), and (iii).

Example 6.8.2 Consider the convex function $\varphi(t) := t^2 - t - \log t$, $t > 0$, and $+\infty$ otherwise. Then one can verify that the function $f(x) = \sum_{i=1}^{n} \varphi(x_i)$ is in the class $\mathcal{F}$.

Proposition 6.8.3 *Let $f \in \Im$. Then:*
(a) The gradient mapping ∇f is onto.
(b) Let A be a (p, n) matrix with $p \geq n$ of rank n, $b \in \mathbb{R}^p$ with $(b - A(\mathbb{R}^p)) \cap \operatorname{dom} f \neq \emptyset$, and set $h(x) := f(b - Ax)$. Then:
(i) $h \in \Im$.
(ii) Let T be a maximal monotone map such that $\operatorname{dom} T \cap \operatorname{dom} h \neq \emptyset$ and set

$$U(x) := \begin{cases} T(x) + \nabla h(x) & \text{if } x \in \operatorname{dom} T \cap \operatorname{dom} \nabla h, \\ \emptyset & \text{otherwise.} \end{cases} \tag{6.20}$$

Then there exists at least a solution x to the generalized equation

$$0 \in U(x), \tag{6.21}$$

which is unique if in addition f is supposed to be strictly convex on its domain.

Proof: (a) Let $y \in \mathbb{R}^p$ and set $v(x) = f(x) - \langle y, x \rangle$. Since $v_\infty(d) = f_\infty(d) - \langle y, d \rangle$, we have

$$v_\infty(d) = +\infty \quad \forall d \neq 0. \tag{6.22}$$

As a consequence of (6.22), if we minimize v on $\mathbb{R}^p$, the optimal set is nonempty, and since $\operatorname{dom} v$ is open, each optimal solution x satisfies $\nabla f(x) = y$, so that $\operatorname{rge} \nabla f = \mathbb{R}^p$.
(b) (i) h is obviously a closed proper convex function. Since $\operatorname{dom} h = A^{-1}(b - \operatorname{dom} f)$ (where A^{-1} denotes the inverse mapping), and since $\operatorname{dom} f$ is open, then $\operatorname{dom} h$ is open and h is differentiable on $\operatorname{dom} h$ by the chain-rule differentiation theorem. Finally, since $h_\infty(d) = f_\infty(-Ad)$ and since A is of maximal rank, we have

$$h_\infty(d) = +\infty \quad \forall d \neq 0,$$

and therefore $h \in \Im$.
(b) Let ∂g be the subdifferential of a closed proper convex function defined on $\mathbb{R}^p$ and such that $\operatorname{dom} T \cap \operatorname{int} \operatorname{dom} g \neq \emptyset$. Since ∂g is star-monotone (cf. Definition 6.5.2) and we assume $\operatorname{rge} \partial g = \mathbb{R}^n$, we can apply Proposition 6.5.3, which implies that $\operatorname{rge}(T + \partial g) = \mathbb{R}^n$. Since ∇f is onto and A is of maximal rank, we thus have $\operatorname{rge} \nabla h = \mathbb{R}^n$ and ∇h is onto. Therefore, it follows that the generalized equation (6.20) admits at least a solution. If in addition f is strictly convex on its effective domain, since A is of maximal rank, this implies that h is strictly convex on its domain, and then $T + \nabla h$ is strictly monotone, which obviously implies uniqueness. $\square$

An interesting class of variational inequality problems that covers a wide range of applications is the case where T is a maximal monotone mapping from $\mathbb{R}^n$ into itself and the constraint set C is explicitly defined by

$$C := \{x \in \mathbb{R}^n \mid F(x) \leq 0\}, \tag{6.23}$$

where $F(x) := (f_1(x), \ldots, f_m(x))^T$, with $f_i : \mathbb{R}^n \to \mathbb{R} \cup \{+\infty\}$, $i = 1, \ldots, m$, given proper closed convex functions.
We make the following standing assumptions on the problem's data:

Assumption A. (a) T is a maximal monotone operator with $\cap_{i=1}^m \mathrm{dom} f_i$ an open subset of $\mathrm{int}\,\mathrm{dom}\,T$.

Assumption B. (a) The solution set of (VI) is nonempty and compact.
(b) Slater's condition holds for $z \in \mathrm{dom}\,T$.

Let $\psi : (-\infty, +\infty) \to \mathbb{R}$ be a convex nondecreasing function with $\psi_\infty(1) = +\infty$ and $\psi_\infty(-1) = 0$, let $\lambda > 0$, and define for $u > 0$ the set-valued map

$$H(x, u, \lambda) := \begin{cases} T(x) + \sum_{i=1}^m u_i(\psi)'(\lambda f_i(x)/u_i)\partial f_i(x) & \text{if } x \in \cap_{i=1}^m \mathrm{dom} f_i, \\ \emptyset & \text{otherwise.} \end{cases}$$
$$\tag{6.24}$$

The maximal monotone map H can be viewed as a kind of generalized Lagrangian for the variational inequality problem with convex constraints, with an appropriate choice of the function ψ.

Example 6.8.3 An interesting example of such a function ψ can be derived as follows. Let $\nu > \mu > 0$ be given fixed parameters, and define

$$\varphi(t) = \begin{cases} \frac{\nu}{2}(t - 1)^2 + \mu(t - \log t - 1) & \text{if } t > 0, \\ +\infty & \text{otherwise.} \end{cases}$$

The conjugate φ^* of φ denoted by ψ can be explicitly computed and is given by

$$\psi(s) = \frac{\nu}{2}t^2(s) + \mu \log t(s) - \frac{\nu}{2},$$
$$t(s) := (2\nu)^{-1}\{(\nu - \mu) + s + \sqrt{((\nu - \mu) + s)^2 + 4\mu\nu}\} = \psi'(s).$$

One can verify that the function ψ shares the following properties: $\mathrm{dom}\psi = \mathbb{R}$, and $\psi \in C^\infty(\mathbb{R})$. Moreover, $\psi'(s)$ is Lipschitz for all $s \in \mathbb{R}$, with constant ν^{-1} and $(\psi''(s) \leq \nu^{-1}$, $\forall s \in \mathbb{R}$.

The result given below is then fundamental for designing a corresponding numerical method based on solving the generalized equation $0 \in H(x)$.

Proposition 6.8.4 *Suppose that assumptions A and B hold for problem (VI) and ψ is as defined above. Then $\forall \lambda > 0$, $\forall u \in \mathbb{R}_{++}^m$:*
(a) The operator $H(\cdot, u, \lambda)$ is maximal monotone on $\mathbb{R}^n$.
(b) The solution set of $0 \in H(x, u, \lambda)$, namely $H^{-1}(0, u, \lambda)$, is nonempty and compact.

Proof: Fix $\lambda > 0, u > 0$, and define $g(x) := \lambda^{-1} \sum_{i=1}^{m} u_i^2 \psi(\lambda f_i(x)/u_i)$ if $x \cap_{i=1}^{m} \text{dom } f_i$, $+\infty$ otherwise. Since ψ satisfies $(\psi)_\infty(-1) = 0, (\psi)_\infty(1) = +\infty$, we can thus apply Proposition 2.6.4 to conclude that g is a closed proper convex function with $\text{dom} g = \cap_{i=1}^{m} \text{dom} f_i \neq \emptyset$. Furthermore,

$$g_\infty(d) = \begin{cases} 0 & \text{if } (f_i)_\infty(d) \leq 0, \ \forall i, \\ +\infty & \text{otherwise.} \end{cases} \qquad (6.25)$$

Now, since $\cap_{i=1}^{m} \text{dom} f_i$ is open, using subdifferential calculus one can verify that $H = T + \partial g$, and then by Assumption A, from Theorem 6.5.6 it follows that H is maximal monotone. Furthermore, since $\text{dom} g \subset \text{dom} T$ we can apply Theorem 6.5.5 to obtain

$$f_\infty^H(d) = \sup\{\langle c, d\rangle | c \in T(x), \ x \in \text{dom} g\} + g_\infty(d).$$

To show that the solution set $H^{-1}(0, u, \lambda)$ is nonempty and compact, it suffices to show that $f_\infty^H(d) > 0$, for $d \neq 0$; i.e., using the above representation of f_∞^H it suffices to show that

$$\sup\{\langle c, d\rangle | c \in T(x), \ x \in \text{dom} g\} > 0 \ \text{ when } \ (f_i)_\infty(d) \leq 0, \ \forall i.$$

But since $T + N_C$ is maximal monotone and we assumed that the solution set of (VI) is nonempty and compact, and $C \subset \cap_{i=1}^{m} \text{dom} f_i \subset \text{dom} T$, we also have, using Corollary 6.8.2,

$$\beta := \sup\{\langle c, d\rangle | c \in T(x), x \in C\} > 0,$$

and hence since $C \subset \text{dom} g$, then $\sup\{\langle c, d\rangle | c \in T(x), \ x \in \text{dom} g\} \geq \beta$, and the proof is completed. $\qquad \square$

Stationary Sequences

Let T be a maximal monotone map on $\mathbb{R}^n$ and consider the problem of finding a zero of T. A question that arises very often in the study of algorithms for solving the inclusion $0 \in T(x)$ is as follows. Suppose one has an algorithm producing a sequence $\{x_k\} \in \mathbb{R}^n$ such that $\text{dist}(0, T(x_k)) \to 0$ as k approaches ∞. Let $S := T^{-1}(0)$ denote the solution set of our problem. Then naturally, such an algorithm can be considered as satisfactory if one has $\text{dist}(x_k, S) \to 0$. Incidentaly such a natural property does not hold in general, without requiring further assumptions on the problem's data. Even in the simplest case, when T is the subdifferential of a proper lsc convex function, the difficulty is present when the algorithm produces unbounded sequences, as already seen in Chapter 4.

Definition 6.8.1 *Let $T : \mathbb{R}^n \rightrightarrows \mathbb{R}^n$ be maximal monotone. A sequence $\{x_k\}$ satisfying $\text{dist}(0, T(x_k)) \to 0$ as $k \to \infty$ will be called a stationary sequence. The sequence $\{x_k\}$ is said to have good asymptotic behavior if $\text{dist}(x_k, T^{-1}(0)) \to 0$ as $k \to \infty$. We then say that T is asymptotically well behaved if each stationary sequence has good asymptotic behavior.*

Our aim is to derive sufficient conditions that enssure that T is asymptotically well behaved. We use the same notation as that introduced in Section 6.6 and set $A := T^{-1}$. In particular, recall that E stands for the parallel subspace to the affine hull of $\operatorname{dom} T^{-1}$.

Theorem 6.8.2 *Let $T : \mathbb{R}^n \rightrightarrows \mathbb{R}^n$ be maximal monotone and such that $0 \in \operatorname{dom} T^{-1}$. Let $\{x_k\}$ be a stationary sequence for T and $\{u_k\}$ a sequence with*

$$u_k \in T(x_k), \quad u_k \to 0, \quad u_k \|u_k\|^{-1} \to d. \tag{6.26}$$

Suppose there exists $\varepsilon_0 > 0$ such that $\varepsilon d \in \operatorname{ri} \operatorname{dom} T^{-1}$, $\forall \varepsilon \in (0, \varepsilon_0]$. Then the sequence $\{x_k\}$ has good asymptotic behavior. Furthermore, the sequence $\{P_E(x_k)\}$ is bounded, and all its limit points belong to $T^{-1}(0) \cap E$.

Proof. By Proposition 6.6.4(b) any element $x_k \in A(u_k)$ can be written as

$$x_k = y_k + z_k, \quad \text{with } y_k \in A_E(u_k), \ z_k \in E^\perp.$$

Since $y_k \in E$, then $y_k = P_E(x_k)$, and from Proposition 6.6.4(c) we obtain $\operatorname{dist}(x_k, A(0)) = \operatorname{dist}(y_k, A_E(0))$, and thus it remains to show that $\operatorname{dist}(y_k, A_E(0)) \to 0$. Since $\varepsilon d \in \operatorname{int} \operatorname{dom} A_E$, $\forall \varepsilon \in (0, \varepsilon_0]$, it follows from Theorem 6.6.1 and Proposition 6.6.5 that the sequence $\{y_k\}$ is bounded, and since A_E is closed, every limit point $\bar{y}$ of this sequence satisfies $\bar{y} \in A_E(0)_d + \varepsilon \mathbb{B}(0, 1)$, $\forall \varepsilon \in (0, \varepsilon_0]$, thus implying that $\bar{y} \in A_E(0)_d$. $\qquad \square$

Corollary 6.8.3 *Let $T : \mathbb{R}^n \rightrightarrows \mathbb{R}^n$ be maximal monotone and such that $0 \in \operatorname{ri} \operatorname{dom} T^{-1}$. Then T is asymptotically well behaved, and if $\{x_k\}$ is a stationary sequence*
(a) $\operatorname{dist}(x_k, T^{-1}(0)) \to 0$.
(b) The sequence $\{P_E(x_k)\}$ is bounded with all its limit points in $T^{-1}(0) \cap E$.

Proof. Under the assumption $0 \in \operatorname{ri} \operatorname{dom} T^{-1}$, each stationary sequence satisfying condition (6.26) implies that there is an $\varepsilon_0 > 0$ such that d satisfies $\varepsilon d \in \operatorname{ri} \operatorname{dom} T^{-1}$, $\forall \varepsilon \in (0, \varepsilon_0]$, and hence the desired conclusions follow immediately from Theorem 6.8.2. $\qquad \square$

As a consequence of Theorem 6.8.2 when $0 \in \operatorname{dom} T^{-1}$ but $0 \notin \operatorname{ri} \operatorname{dom} T^{-1}$, we can approximate the problem $0 \in T(x)$ by a *stable* problem $0 \in T_\varepsilon(x)$, where we define $T_\varepsilon(x) := T(x) - \varepsilon d$ with $\varepsilon d \in \operatorname{ri} \operatorname{dom} T^{-1}$, $\forall \varepsilon \in (0, \varepsilon_0]$. From Theorem 6.8.2 we then have that $\forall \varepsilon \in (0, \varepsilon_0]$, T_ε is maximal monotone with $\operatorname{dom} T_\varepsilon^{-1} = \operatorname{dom} T^{-1} - \varepsilon d$ and $0 \in \operatorname{ri} \operatorname{dom} T_\varepsilon^{-1}$, and therefore Corollary 6.8.3 applies to T_ε.

6.9   Duality for Variational Inequalities

As shown in the previous sections, variational inequality problems can be recast as solving an appropriate generalized equation. We will first develop a duality framework for a general set-valued map formulation of variational inequalities. This abstract framework shares similarity with the Fenchel–Rockafellar conjugate duality scheme used in convex optimization. We then develop a Lagrangian duality scheme for variational inequality problems that is particularly appropriate for the design of numerical methods for their solutions.

An Abstract Duality Scheme

Consider the following general primal problem involving composition of set-valued maps and that consists in finding $x \in \mathbb{R}^n$ satisfying

$$(\mathrm{PI}) \quad 0 \in A(x) + QBP(x),$$

where $A : \mathbb{R}^n \rightrightarrows \mathbb{R}^n$, $P : \mathbb{R}^n \rightrightarrows \mathbb{R}^m$, $B : \mathbb{R}^m \rightrightarrows \mathbb{R}^m$, $Q : \mathbb{R}^m \rightrightarrows \mathbb{R}^n$ are arbitrary set-valued maps. A simple duality principle follows by observing that the primal inclusion (PI) can be rewritten in an equivalent form involving inverse set-valued maps.

Proposition 6.9.1 *The inclusion (PI) is solvable if and only if the following inclusion is solvable:*

$$(\mathrm{DI}) \quad 0 \in (-P)A^{-1}(-Q)(u) + B^{-1}(u).$$

Proof. By direct manipulation one has

$$
\begin{aligned}
0 \in A(x) + QBP(x) \quad &\Longleftrightarrow \quad 0 \in A(x) + Qu, \ u \in BP(x), \\
&\Longleftrightarrow \quad x \in A^{-1}(-Q)(u), \ 0 \in -P(x) + B^{-1}(u), \\
&\Longleftrightarrow \quad 0 \in (-P)A^{-1}(-Q)(u) + B^{-1}(u).
\end{aligned}
$$

$\square$

The inclusion (DI) is called the dual inclusion associated with (PI). Note that (DI) is now formulated with a set-valued map defined on $\mathbb{R}^m$. We will point out below some of the aspects and usefulness of this dual formulation, but first we begin with a simple example, which also leads us to define a dual problem associated with variational inequalities.

Example 6.9.1 (i) Let $m = n$ and let $P = Q = I$ be the identity map. Then we obtain the pair of problems

$$0 \in A(x) + B(x); \quad 0 \in -A^{-1}(-u) + B^{-1}(u).$$

Suppose now that B is the subdifferential map of an lsc proper convex function $b : \mathbb{R}^n \to \mathbb{R} \cup \{+\infty\}$. Then, recalling the fact that $(\partial b)^{-1} = \partial b^*$, one obtains the pair of problems

$$0 \in A(x) + \partial b(x), \;\; 0 \in -A^{-1}(-u) + \partial b^*(u).$$

This is known as the duality scheme of Mosco. Note that the variational inequality problem $\mathrm{VI}(T, C)$ is a special case of the primal inclusion with the choice $A = T, b = \delta(\cdot | C)$, and thus the second problem can be viewed as a dual for $\mathrm{VI}(T, C)$.

In the rest of the section we concentrate on one of the most prominent and useful special cases of the above abstract framework. We let $P := M, Q := M^T$, where M is a linear map from $\mathbb{R}^n$ to $\mathbb{R}^m$. The pair of problems (PI) and (DI) then reduces in that case to

(PI) $0 \in A(x) + M^T B M(x)$,   (DI) $0 \in (-M)A^{-1}(-M^T)(u) + B^{-1}(u)$.

The dual terminology is justified from the fact that the above scheme is akin to the conjugate duality scheme used in convex optimization problems, see Chapter 5. Indeed, suppose that both A and B are the subdifferential of some lsc proper convex functions f, g defined on $\mathbb{R}^n$ and $\mathbb{R}^m$, respectively. Then, under appropriate regularity assumptions (cf. Chapter 5), one can express the pair of problems (PI)–(DI) as equivalent to solving the following pair of convex optimization problems, which are nothing else but the primal–dual pair derived from the Fenchel conjugate duality scheme:

$$\min\{f(x) + g(Mx) \mid x \in \mathbb{R}^n\}, \qquad \min\{f^*(-M^T u) + g^*(u) \mid u \in \mathbb{R}^m\}.$$

Another natural formulation is the *primal–dual* formulation that is obtained by reformulating the pair (PI)–(DI) as, to find $(x, u) \in \mathbb{R}^n \times \mathbb{R}^m$ satisfying

$$0 \in A(x) + M^T u, \;\; 0 \in -Mx + B^{-1}(u).$$

Define the following set-valued map L on $\mathbb{R}^n \times \mathbb{R}^m$:

$$L(x, u) := (A(x) \times B^{-1}(u)) + \mathcal{M}(x, u), \tag{6.27}$$

where $\mathcal{M}$ is the matrix

$$\mathcal{M} = \begin{pmatrix} 0 & M^T \\ -M & 0 \end{pmatrix}.$$

Then, the above pair is equivalent to solving $(0, 0) \in L(x, y)$. In the special case of convex optimization as discussed above, this primal–dual formulation corresponds to finding a saddle point of the Lagrangian l (see Definition 5.6.1) given by $l(x, y) = f(x) + \langle Mx, y \rangle - g^*(y)$. The following elementary proposition summarizes the relationships among the three formulations.

Proposition 6.9.2 *For a pair $(x, u) \in \mathbb{R}^n \times \mathbb{R}^m$ the following statements are equivalent:*
(a) $(0,0) \in L(x, u)$ where L is defined in (6.27).
(b) $(x, -M^T u) \in \operatorname{gph} A$, $(Mx, u) \in \operatorname{gph} B$.
Furthermore the following assertions hold:
$x \in \mathbb{R}^n$ *solves* $0 \in A(x) + M^T BM(x)$ *if and only if there exists* $u \in \mathbb{R}^m$ *such that (a)–(b) hold. Dually,* $u \in \mathbb{R}^m$ *solves* $0 \in -MA^{-1}(-M^T u) + B^{-1}(u)$ *if and only if there exists* $x \in \mathbb{R}^n$ *such that (a)–(b) hold.*

Proof. The equivalence between the statements (a) and (b) is immediate from the definition of L and the manipulations developed just above the proposition. We prove only the next primal assertion, since the dual one follows in a similar way. One has

$$
\begin{aligned}
0 \in A(x) + M^T BM(x) \quad &\Longleftrightarrow \quad z \in A(x), \ -z \in M^T BM(x), \\
&\Longleftrightarrow \quad z \in A(x), \ -z = M^T u, \ u \in BM(x) \\
&\Longleftrightarrow \quad -M^T u \in A(x), \ u \in BM(x),
\end{aligned}
$$

and the last relations are exactly (b). $\square$

A key question that arises with the above inclusions is under what conditions each of the three formulations preserves maximality when the data are assumed maximal monotone. Indeed, most well-known algorithms designed for solving generalized equations rely on this minimal assumption to guarantee their convergence. The sum of maximal monotone maps remains maximal monotone whenever each map is maximal monotone and satisfies some kind of regularity assumption (cf. Section 6.5). Both the primal and dual formulations involve composition of set-valued maps. If A, B are maximal monotone, then one can use Theorem 6.5.7 to obtain the required conditions that guarantee maximal monotonicity of $M^T BM$ and $-MA^{-1}(-M)^T$. On the other hand, the primal–dual formulation has the advantage that the resulting operator L remains maximal monotone with no further assumptions.

Proposition 6.9.3 *Let A and B be maximal monotone on $\mathbb{R}^n$ and $\mathbb{R}^m$, respectively, and let $M : \mathbb{R}^n \to \mathbb{R}^m$ be any linear map. Then the map L is maximal monotone on $\mathbb{R}^n \times \mathbb{R}^m$.*

Proof. Let $S(x, u) := (A(x) \times B^{-1}(u)$ and $T(x, u) := \mathcal{M}(x, u)$. Then one has $L = S + T$. Since A, B are given maximal monotone, then S is maximal monotone, and T being a single-valued linear map is also maximal monotone. It follows from Theorem 6.5.6 that the sum $S + T$ is maximal monotone. $\square$

We now return to the variational inequality problem; that is, in problem (PI) we set $M = I$ and $A := T$, $B = \partial\delta_C = N_C$. Then the dual variational problem is

$$0 \in -T^{-1}(-u) + N_C^{-1}(u),$$

and the primal–dual formulation consists in finding the zero of

$$T_{\mathrm{PD}}(x, u) = (T(x) \times N_C^{-1}(u)) + (u, -x)^T.$$

A main difficulty with the above framework is that it requires the construction of the inverse operators T^{-1} and N_C^{-1} to formulate a dual problem, a task that can be very difficult unless some specific structure exists such as, for example, the interesting special case of box constraints; e.g., when $C = \{x \in \mathbb{R}^n \mid l \le x \le u\}$, an explicit computation of N_C^{-1} is available.

Example 6.9.2 Let $C := \{x \in \mathbb{R}^n \mid l \le x \le u\}$, with $l \in [-\infty, \infty)^n$ $u \in (-\infty, \infty)^n$, and $l \le u$. In that case the variational inequality problem is a complementarity problem. It is sufficient to compute for each i the ith component of the corresponding normal cones, N_i, N_i^{-1}, since given the structure of the set C one clearly has that $N_C = \Pi_{i=1}^n N_i$ and $N_C^{-1} = \Pi_{i=1}^m N_i^{-1}$. One can verify by direct computation that

$$
\begin{aligned}
N_i &= [(\{l_i\} \times (-\infty, 0) \cup ([l_i, u_i] \times \{0\}) \cup (\{u_i\} \times (0, \infty))] \cap \mathbb{R}^2, \\
N_i^{-1} &= [((-\infty, 0) \times \{l_i\}) \cup (\{0\} \times [l_i, u_i]) \cup ((0, \infty) \times \{u_i\})] \cap \mathbb{R}^2.
\end{aligned}
$$

Lagrangian Duality for VI

We are now interested in formulating a dual problem for the variational inequality problem with convex constraints cf. (6.23). As indicated above, for that class of problems, an explicit computation of the inverse normal cone map is not at hand to generate a dual problem via the previous abstract framework. We thus consider now an alternative duality scheme that is in the spirit of the classical Lagrangian duality framework for constrained optimization problems. It will permit us to take advantage of the particular structure of the set C described by convex inequalities.

The starting point is the simple and well-known observation that x^* is a solution of (VI) if and only if

$$x^* \in \operatorname{argmin}\{\langle g^*, x - x^* \rangle \mid x \in C\}, \tag{6.28}$$

where $g^* \in T(x^*)$, $T : \mathbb{R}^n \rightrightarrows \mathbb{R}^n$, and $C = \{x \mid f_i(x) \le 0, \ i = 1, \ldots, m\}$. In what follows we assume that each $f_i : \mathbb{R}^n \to \mathbb{R}$ is a convex function. Formally, we can thus associate with the *convex* optimization problem

$$\min\{\langle g^*, x - x^* \rangle \mid f_i(x) \le 0, \ i = 1, \ldots, m\} \tag{6.29}$$

a Lagrangian defined by $L : \mathbb{R}^n \times \mathbb{R}^m_+ \to [-\infty, +\infty)$,

$$L(x, u; x^*) := \begin{cases} \langle g^*, x - x^* \rangle + \sum_{i=1}^m u_i f_i(x) & \text{if } x \in \mathbb{R}^n, u \in \mathbb{R}^m_+, \\ -\infty & \text{if } u \notin \mathbb{R}^m_+, \end{cases}$$

(6.30)

where $u \in \mathbb{R}^m_+$ is the dual multiplier attached to the constraints, and a dual problem defined by

$$\sup_{u \geq 0} \inf \{ L(x, u, x^*) \mid x \in D \}.$$

(6.31)

By the standard saddle point optimality theorem (see, Chapter 5) we know that under the usual Slater's condition one has that $(x^*, u^*) \in \mathbb{R}^n \times \mathbb{R}^m_+$ is a saddle point of L if and only if $x^* \in C$ and $u^* \geq 0$ are respectively optimal for the primal and dual problems (6.29)–(6.31), that is,

$$0 \in g^* + \sum_{i=1}^m u_i^* \partial f_i(x^*),$$

(6.32)

$$0 \in -F(x^*) + N_{\mathbb{R}^m_+}(u^*),$$

(6.33)

with $g^* \in T(x^*)$. The relations (6.32)-(6.33) are just the Karush–Khun–Tucker optimality conditions for (6.29), which are necessary and sufficient for optimality, and thus u^* can be interpreted as the solution of a *Lagrangian dual variational inequality* which can be defined as follows. For each $u \in \mathbb{R}^m_+$, set

$$M(u) := \left\{ x \in \mathbb{R}^n \mid 0 \in T(x) + \sum_{i=1}^m u_i \partial f_i(x) \right\},$$

(6.34)

$$G(u) := \{ -F(x) \mid x \in M(u) \},$$

(6.35)

$$T_D(u) := G(u) + N_{\mathbb{R}^m_+}(u).$$

(6.36)

The dual variational inequality problem associated with (VI) is then

$$\text{find } u^* \in \mathbb{R}^m_+, \ d^* \in G(u^*) : \ \langle d^*, u - u^* \rangle \geq 0, \ \forall u \in \mathbb{R}^m_+,$$

which can also be written using (6.36) as

$$0 \in T_D(u^*).$$

(6.37)

Likewise, we can then associate a primal–dual formulation of (VI) via equations (6.33)–(6.35):

$$(0, 0) \in S(x^*, u^*),$$

(6.38)

where the operator S is defined on $\mathbb{R}^n \times \mathbb{R}^m_+$ by

$$S(x, u) := \left\{ (y, w) \mid y \in T(x) + \sum_{i=1}^m u_i \partial f_i(x), w \in -F(x) + N_{\mathbb{R}^m_+}(u) \right\},$$

(6.39)

if $(x, u) \in \text{dom} S = \text{dom} T \times \mathbb{R}^m_+ \neq \emptyset$, and $\emptyset$ otherwise.

From the above discussion we have thus shown that we have essentially three equivalent representations for the variational inequality problem (VI), which consist of the primal, dual, and primal–dual problems defined above. More precisely we have proved the following:

Theorem 6.9.1 *Suppose that Slater's condition holds for the constraint set C. Then $x^* \in \mathbb{R}^n$ solves (VI) if and only if there exists $u^* \in \mathbb{R}^m_+$ such that (x^*, u^*) solves the primal–dual equation $0 \in S(x^*, u^*)$.*

We end this part with a result that will be useful later on.

Theorem 6.9.2 *Suppose $0 \in T_{\mathrm{D}}(u^*)$. Then there exists x^* solving (VI) such that $u^* + F(x^*) = 0$ and $\langle u^*, F(x^*) \rangle = 0$.*

Proof. By definition, there exists x^* such that for each $u \geq 0$, $\langle u - u^*, -F(x^*) \rangle \geq 0$. Taking $u = u^* + \Delta u$, with $\Delta u \geq 0$, it follows that $F(x^*) \leq 0$, and then with $u = 0$, we get $\langle u^*, F(x^*) \rangle = 0$, which clearly implies that x^* solves (VI). $\qquad\square$

Maximal Monotonicity of the Three Formulations

For convenience we often use the following notation

$$
\begin{aligned}
T_{\mathrm{P}} &:= T + N_C, \\
T_{\mathrm{D}} &:= G + N_{\mathbb{R}^m_+}, \\
T_{\mathrm{S}} &:= S.
\end{aligned}
$$

The primal operator poses no problems. From Theorem 6.5.7(a) the maximal monotonicity of T_{P} is preserved under the condition $\text{dom}\, T \cap \text{int}\, C \neq \emptyset$. We now turn to the dual operator T_{D}. First, we show that T_{D} in monotone on $\mathbb{R}^m_+$.

Proposition 6.9.4 *Let $T : \mathbb{R}^n \rightrightarrows \mathbb{R}^n$ be monotone. Then the dual operator $T_{\mathrm{D}} = G + N_{\mathbb{R}^m_+}$ is monotone on $\mathbb{R}^m_+$.*

Proof: Since $N_{\mathbb{R}^m_+}$ is monotone, the monotonicity of T_{D} will follow by proving that G monotone. Let $(u, u'), (v, v')$ be arbitrary points in G. By definition of G given in (6.35), $\exists (x, u), (y, v)$ such that

$$u' = -F(x), x \in M(u); \quad v' = -F(y), y \in M(v), u, v \geq 0.$$

Since f_i are convex, using the subgradient inequality for each f_i, the fact that $u, v \geq 0$, and $F(x) = (f_1(x), \ldots, f_m(x))^T$, one easily obtains

$$\langle u - v, u' - v' \rangle = \langle u - v, F(y) - F(x) \rangle$$

$$\geq \left\langle y - x, \sum_{i=1}^{m} u_i \partial f_i(x) - \sum_{i=1}^{m} v_i \partial f_i(y) \right\rangle$$
$$= \langle y - x, y' - x' \rangle, \ x' \in T(x), y' \in T(y)$$
$$\geq 0,$$

where the third equality follows by using $x \in M(u), y \in M(v)$ with $M(\cdot)$ defined in (6.34), and the last inequality is from the monotonicity of T. $\square$

To establish the maximal monotonicity of T_{D} we first establish that T_{S} is maximal monotone.

Proposition 6.9.5 *Let* $T : \mathbb{R}^n \rightrightarrows \mathbb{R}^n$ *be maximal monotone. Then the primal–dual operator* $T_S = S$ *defined in (6.39) is maximal monotone.*

Proof: The operator T_S defined in (6.39) can be decomposed as follows. Let

$$A(x, u) = \begin{cases} T(x) \times \{0\} & \text{if } x \in \text{dom}\, T, \\ \emptyset & \text{otherwise,} \end{cases}$$

and

$$B(x, u) = \begin{cases} \{\sum_{i=1}^{m} u_i \partial f_i(x)\} \times \{-F(x) + N_{\mathbb{R}_+^m}(u)\} & \text{if } x \in \mathbb{R}^n, u \in \mathbb{R}_+^m, \\ \emptyset & \text{otherwise.} \end{cases}$$

Then we have $S = A + B$. Since T is maximal monotone, it is easy to see that A is also maximal monotone. On the other hand, defining $l : \mathbb{R}^n \times \mathbb{R}_+^m \to \mathbb{R}$ by

$$l(x, u) = \begin{cases} \sum_{i=1}^{m} u_i f_i(x) & \text{if } u \in \mathbb{R}_+^m, \\ -\infty & \text{if } u \notin \mathbb{R}_+^m, \end{cases}$$

we have

$$B(x, u) = \begin{cases} \partial_x l(x, u) \times \partial_u[-l(x, u)] & \text{if } u \in \mathbb{R}_+^m, \\ \emptyset & \text{if } u \notin \mathbb{R}_+^m, \end{cases}$$

and by Theorem 6.3.2, B is maximal monotone. Invoking Theorem 6.5.6 on the maps A, B then gives the desired result. $\square$

We also need the following result, establishing the boundedness of the dual solution set.

Proposition 6.9.6 *Let* $T : \mathbb{R}^n \rightrightarrows \mathbb{R}^n$ *be maximal monotone. Suppose that the solution set of (VI) is nonempty, and there exists* $z \in \text{dom}\, T$ *satisfying Slater's condition. Then the solution set of the dual problem* $0 \in T_{\mathrm{D}}(u)$ *is nonempty. In addition, if the solution set of (VI) is bounded, then the solution set* $T_{\mathrm{D}}^{-1}(0)$ *of the dual is also bounded.*

Proof: Under the given assumptions, from Theorem 6.9.1 we have $x^* \in T_{\mathrm{P}}^{-1}(0)$ and $\exists u^* \geq 0$ such that (6.32)–(6.33) hold, which by (6.34) means that $x^* \in M(u^*)$. As a consequence, we have using (6.33),

$$\langle G(u^*), u - u^* \rangle = \langle -F(x^*), u - u^* \rangle \geq 0, \ \forall u \geq 0,$$

and hence $u^* \in T_{\mathrm{D}}^{-1}(0)$. Now, if $T_{\mathrm{D}}^{-1}(0)$ is not bounded, thanks to Theorem 6.9.2, there would exist a sequence $\{u^k, x^k, g^k, g_i^k, i = 1, \ldots, m\}$ with $u^k \geq 0, x^k \in T_P^{-1}(0), g^k \in T(x^k), g_i^k \in \partial f_i(x^k), i = 1, \ldots, m$, such that

$$\|u^k\| \to \infty, \quad u^k \|u^k\|^{-1} \to \bar{u} \neq 0, \quad x^k \to \bar{x},$$

and (6.32)–(6.33) hold at (x^k, u^k); i.e.,

$$0 \in g^k + \sum_{i=1}^{m} u_i^k g_i^k, \quad u_i^k \geq 0, \quad u_i^k f_i(x^k) = 0, \quad f_i(x^k) \leq 0, \ i = 1, \ldots, m.$$

Using the subgradient inequality for the convex function f_i, multiplying by $u_i^k \geq 0$, and summing we obtain

$$\sum_{i=1}^{m} u_i^k f_i(z) \geq \left\langle z - x^k, \sum_{i=1}^{m} u_i^k g_i^k \right\rangle, \quad \forall z \in \mathrm{dom} T.$$

But since $x^k \in M(u^k)$, the above inequality reduces to

$$\sum_{i=1}^{m} u_i^k f_i(z) \geq \langle z - x^k, -g^k \rangle, \forall z \in \mathrm{dom} T$$

$$\geq \langle x^k - z, g \rangle, \ g \in T(z), z \in \mathrm{dom} T,$$

since T is monotone. Since we assumed that $T_P^{-1}(0)$ is bounded, dividing the latter inequality by $\|u^k\|$ and passing to the limit we obtain $\sum_{i=1}^{m} \bar{u}_i f_i(z) \geq 0$, for all z, and hence with z satisfying Slater's condition for C, and recalling that $\bar{u} \neq 0, \bar{u} \geq 0$, which is impossible. □

We can now establish the desired result on the maximal monotonicity of the dual map T_{D}.

Proposition 6.9.7 *Suppose that T is maximal monotone, that the solution set of (VI) is nonempty and bounded and that there exists $z \in \mathrm{dom}\, T$ satisfying Slater's condition. Then the dual operator T_{D} is maximal monotone.*

Proof: Let $X := (T + N_C)^{-1}(0)$ and $U := T_D^{-1}(0)$ denote the solution set of the primal (VI) and dual problems, respectively. We have T_{D} maximal monotone if and only if T_D^{-1} is maximal monotone. Using the definition of T_D, (6.33), (6.34), and (6.39) we have

$$
\begin{aligned}
T_D^{-1}(w) &= \{u | w \in G(u) + N_{\mathbb{R}_+^m}(u)\} \\
&= \{u | \exists x : w \in -F(x) + N_{\mathbb{R}_+^m}(u), x \in M(u)\} \\
&= \{u | \exists x : (x, u) \in S^{-1}(0, w)\}.
\end{aligned}
$$

Since from Proposition 6.9.5 $T_S := S$ is maximal monotone, invoking Proposition 6.5.4 we thus have that T_D^{-1} is maximal monotone if $(0, 0) \in \operatorname{ri} \operatorname{dom} S^{-1}$. By Proposition 6.8.1(i), the latter condition will be satisfied if the solution set $S^{-1}(0, 0)$ is nonempty and bounded. But by Theorem 6.9.1 we have $S^{-1}(0, 0) \subset T_P^{-1}(0) \times T_D^{-1}(0)$ and since we assumed that $T_P^{-1}(0)$ is bounded, then by Proposition 6.9.6 $T_D^{-1}(0)$ is also bounded, and the result is proved. $\square$

We end with an asymptotic result that provides conditions under which a sequence is minimizing for problem (VI). This is, in fact, the general version of Theorem 4.4.3 stated in Chapter 4 for convex optimization problems.

Proposition 6.9.8 *Let $T : \mathbb{R}^n \rightrightarrows \mathbb{R}^n$ be maximal monotone such that $T_P = T + N_C$ is maximal monotone. Suppose that $T_P^{-1}(0)$ is nonempty and compact and that $C \subset \operatorname{dom} T$. Let u^k be a bounded sequence in $\mathbb{R}_+^m$ and consider a sequence $\{x^k, g^k, g_i^k, i = 1, \dots, m\}$ with $g^k \in T(x^k)$, $g_i^k \in \partial f_i(x^k), i = 1, \dots, m$, such that*

$$
\varepsilon^k := g^k + \sum_i^m u_i^k g_i^k \quad \to \quad 0, \tag{6.40}
$$

$$
\limsup_{k \to \infty} f_i(x^k) \quad \leq \quad 0, \ \forall i, \tag{6.41}
$$

$$
u_i^k f_i(x^k) \quad \to \quad 0, \ \forall i. \tag{6.42}
$$

Then the sequence $\{x^k\}$ is bounded, and each limit point of the sequence $\{x^k\}$ solves (VI).

Proof: The proof is by contradiction. Suppose that the sequence $\{x^k\}$ is not bounded. Then without loss of generality one can assume that

$$
\|x^k\| \to +\infty, \quad \frac{x^k}{\|x^k\|} \to \bar{x} \neq 0.
$$

Let $\varepsilon > 0$. Then from (6.41) for k sufficiently large we have

$$
f_i\left(\frac{x^k}{\|x^k\|}\|x^k\|\right)\|x^k\|^{-1} \leq \varepsilon\|x^k\|^{-1}. \tag{6.43}
$$

Since for any function f we have

$$f_\infty(d) = \inf \left\{ \lim_{n \to +\infty} \inf f(t_n x_n)/t_n \mid t_n \to +\infty, x_n \to d \right\},$$

passing to the limit in (6.43), we then obtain that $(f_i)_\infty(\bar{x}) \leq 0, \forall i = 1, \ldots, m$, which means that $\bar{x} \in C_\infty$. Now, $\forall g \in T(x)$, $x \in C$, using arguments similar to the one used in Proposition 6.9.6 we obtain using the definition of ε_k given in (6.40) that

$$\langle g, x^k \rangle \leq \langle g, x \rangle + \langle g^k, x^k - x \rangle$$
$$\leq \langle g, x \rangle + \langle \varepsilon^k, x^k - x \rangle + \sum_{i=1}^{m} u_i^k (f_i(x^k) - f_i(x))$$

Dividing the latter inequality by $\|x^k\|$, passing to the limit, and using (6.40)–(6.42) we thus obtain $\langle g, \bar{x} \rangle \leq 0$, i.e., from Corollary 6.8.2 that $f_\infty^{T,C}(\bar{x}) \leq 0$, which contradicts the assumption that $T_P^{-1}(0)$ is compact. Now let x^∞ be a limit point of $\{x^k\}$. Then using $\varepsilon_k = g^k + \sum_{i=1}^{m} u_i^k g_i^k$, the convexity of f_i and the monotonicity of T we obtain

$$\langle g_1, x - x^k \rangle \geq \sum_{i=1}^{m} u_i^k f_i(x^k) + \langle \varepsilon_k, x - x^k \rangle \ \forall x \in C, \ \forall g_1 \in T(x).$$

Then passing to the limit in the above inequality, together with (6.40)–(6.42) we thus get

$$\langle x - x^\infty, g_1 \rangle \geq 0, \ \forall g_1 \in T(x).$$

But since $x^\infty \in C$, we also have, $\langle x - x^\infty, g_2 \rangle \geq 0, \ \forall g_2 \in N_C(x), x \in C$. Therefore, it follows that

$$\langle g, x - x^\infty \rangle \geq 0, \ \forall g \in T_P(x) = T(x) + N_C(x), \ \forall x \in C.$$

Since T_P is maximal monotone, this implies that $0 \in T_P(x^\infty)$; i.e., $x^\infty \in T_P^{-1}(0)$.

□

6.10   Notes and References

A classical text including most of the definitions and operations involving set-valued maps is the book of Berge [31]. More results and details can be found in the recent book by Rockafellar and Wets [123]. The concept of maximal monotone map is due to Minty [104]. Most of the results on maximal monotone maps have been developed for infinite-dimensional spaces, in particular for Hilbert spaces. The classical book of Brezis [41] covers the principal results given in Section 6.3, which can also be found in numerous papers of many other authors. Other key contributions in the infinite-dimensional setting can be found in the work of Browder [43]. One of the most fundamental results, is the characterization theorem of maximal monotonicity due to Minty [104]. Another fundamental work in the theory of maximal monotone maps is the one due to Haraux and Brezis [40], where they introduce the concept of star monotonicity, which leads to several key results on the domain and range of maximal monotone maps and in particular on the image sum of two maximal monotone maps as developed in Section 6.5. The first attempt to define a notion of asymptotic function for monotone maps to characterize existence of solutions for variational inequality problems can be found in the works of Brezis and Nirenberg [42], Lions [90], and Browder [44]. More recently, various definitions of asymptotic functions associated with maximal monotone maps have been compared in Attouch, Chbani, and Moudafi [2], where they showed that the most appropriate definition is the one given in terms of the support function of the range of the operator, and which we have used in this book. The formula for the asymptotic functional of the sum of a monotone map with the subdifferential of a convex function was recently established by Auslender and Teboulle [20] and extends a result established earlier by Auslender [16]. Preservation of maximal monotonicity involving the sum of maximal monotone maps was proven by Rockafellar [118]. The proof given here in Theorem 6.5.6 makes use of the Haraux–Brezis results. The extension of this result involving composition of maximal monotone maps with affine mapping was recently given in the book of Rockafellar and Wets [123], where further results and properties on maximal monotone maps such as Proposition 6.5.4 can also be found. The proof of Theorem 6.5.7 is from the recent work of Robinson [115], where further results on composition of maximal monotone maps can also be found. The results given in Section 6.6 on the structure of maximal monotone maps and directional local boundedness as well as the results on stationary sequences are from Auslender [9], except for Propositions 6.6.1, 6.6.2, 6.6.6, 6.6.7, 6.6.8, which are classical. Variational inequalities and generalized equations have a long history and can be found in several works; see, e.g., Lions and Stampacchia [91] and Robinson [114]. The existence results for variational inequalities given in Section 6.8 using asymptotic functionals are relatively new and have been established in the works of Auslender [16] and Auslender–Teboulle [20]. Corollary 6.8.1 was given earlier by Rockafellar and Wets [123] with another proof. Proposition 6.8.3 is from Auslender, Teboulle, and Ben-Tiba [19]. The first work on duality for variational inequalities seems to be the paper by Mosco [105]. The abstract duality principle of Section 6.9 started with the work of Attouch and Thera [3], and was extended by Robinson [115] and Eckstein–Ferris [68], which also gives the formula for the inverse normal cone of box constraints as given in Example 6.9.2. Partial results on Lagrangian duality for variational inequality can be found in Bensoussan, Lions, and Temam [30], Rockafellar [122], and Auslender [16]. The Lagrangian duality framework and associated results developed in Section 6.9 are from the recent work of Auslender–Teboulle [20], which also includes more details and results on multiplier methods for solving variational inequality problems.

References

[1] P. Angleraud. Caractérisation duale du bon comportement de fonctions convexes. *C.R. Académie des Sciences*, **t. 314**, Serie I, 19 92, 583–586.

[2] H. Attouch, Z. Chbani, and A. Moudafi. Recession operators and solvability of variational problems. *Series on Advances in Mathematics for Applied Sciences*, Vol.18, World Scientific, 1994, 51–67.

[3] H. Attouch and M. Thera. A general duality principle for the sum of two operators. *J. of Convex Analysis*, **3**, 1996, 1–24.

[4] J.P. Aubin and I. Ekeland. *Applied Nonlinear Analysis*. Wiley, New-York, 1984.

[5] J.P. Aubin and H. Frankowska. *Set-Valued Analysis*. Birkhäuser, Boston, 1990.

[6] A. Auslender. *Optimisation: Méthodes Numériques*, Masson, Paris, 1976.

[7] A. Auslender and J.P. Crouzeix. Global regularity theorems. *Mathematics of Operations Research*, **13**, 1988, 243–253.

[8] A. Auslender and J.P. Crouzeix. Well behaved asymptotical convex functions. *Ann. Inst. Poincar. Anal. Non Linéaire*, **6**, 1989, 101–122.

[9] A. Auslender. Convergence of stationary sequences for variational inequalities with maximal monotone operators. *Applied Math and Optimization*, **28**, 1993, 161–172.

[10] A. Auslender, R. Cominetti, and J.P. Crouzeix. Convex fuctions with unbounded level sets and applications to duality theory. *SIAM J. Optimization*, **3**, 1993, 669–687.

[11] A. Auslender and P. Coutat. On closed convex sets without boundary rays and asymptotes. *Set Valued Analysis*, **2**, 1994, 19–33.

[12] A. Auslender. Noncoercive Optimization. *Mathematics of Operations Research*, **21**, 1996, 769–782.

[13] A. Auslender. Closedness criteria for the image of a closed set by a linear operator. *Numer. Funct. Anal. Optim.*, **17**, 1996, 503–515.

[14] A. Auslender, R. Cominetti,and M. Haddou. Asymptotic analysis of penalty and barrier methods in convex and linear programming. *Mathematics of Operations Research*, **22**, 1997, 43–62.

[15] A. Auslender. How to deal with the Unbounded in optimization: theory and algorithms. *Mathematical Programming*, **79**, Serie B, 1997, 3–18.

[16] A. Auslender. Asymptotic analysis for penalty and barrier methods in noncoercive variational inequalities. *SIAM J. Control and Optimization*, **37**, 1999, 653–671.

[17] A. Auslender. Penalty and barrier methods: A unified framework. *SIAM J. Optimization*, **10**, 1999, 211–230.

[18] A. Auslender. Existence of optimal solutions and duality results under weak conditions. *Mathematical Programming*, **88**, Serie A, 2000, 45–59.

[19] A. Auslender, M. Teboulle, and S. Ben-Tiba. A logarithmic–quadratic proximal method for variational Inequalities. *Computational Optimization and Applications*, **12**, (1998), 31–40.

[20] A. Auslender and M. Teboulle. Lagrangian duality and related multiplier methods for variational inequality problems. *SIAM J. Optimization*, **10**, 2000, 1097–1115.

[21] D. Aze and L. Michel. Computable dual characterizations of asymptotically well-behaved convex functions. *Comm. Appl. Nonlinear Analysis*, **6**, 1999, 39–49.

[22] C. Baiocchi, G. Buttazo, F. Gastaldi, and F. Tomarelli. General existence theorems for unilateral problems in continuum mechanics. *Arch. Rational Mech. Anal.*, **100**, 1998, 149–189.

[23] B. Bank, J. Guddat, D. Klatte, B. Kummer, and K. Tammer. *Non-Linear Parametric Optimization*. Akademie Verlag, Berlin, 1982.

[24] R. Bellman and K. Fan. On systems of linear inequalities in Hermitian matrix variables. In *Convexity*, edited by V. Klee, Proceedings of Symposia in Pure Mathematics, American Mathematical Society, Vol. 7, 1963, 1–11.

[25] G.G. Belousov. *Introduction to Convex Analysis and Integer Optimisation*. University Publisher, Moscow 1977 (in Russian).

[26] J. Benoist and J.B. Hirriart-Urruty. What is the subdifferential of the closed convex hull of a funtion. *SIAM Journal of Mathematical Analysis*, **27**, 1996, 1661–1679.

[27] A. Ben-Tal, A. Ben-Israel, and M. Teboulle. Certainty equivalent and information measures: duality and extremal principles. *Journal of Mathematical Analysis and Applications*, **157** (1991), 211–236.

[28] A. Ben-Tal and M. Teboulle. A smoothing technique for nondifferentiable optimization problems. In *Optimization*, Fifth French–German Conference, Lecture Notes in Mathematics 1405, Springer-Verlag (1989), 1–11.

[29] A. Ben-Tal, M. Teboulle, and Wei H. Hang. A least squares-based method for a class of nonsmooth minimization problems with applications in plasticity. *Applied Mathematics and Optimization*, **24**, 1991, 273–288.

[30] A. Bensoussan, J.-L. Lions, and R. Temam. Sur les méthodes de décomposition, de décentralisation, de coordinations et applications. In *Méthodes Numériques en Sciences Physiques et Economiques*. Eds. J.-L. Lions and G.I. Marchouk, 133–257, Dunod-Bordas, Paris, 1974.

[31] C. Berge. *Topological Spaces*. Macmillan, New York, 1963.

[32] C. Bergthaller and I. Singer. The distance to a polyhedron. *Linear Algebra and Applications*, **169**, 1992, 111–129.

[33] D. Bertsekas. Approximation procedures based on method of multipliers. *Journal of Optimization Theory and Applications*, **23**, 1977, 487–510.

[34] D. Bertsekas. *Constrained Optimization and Lagrange Multiplier Methods*. Academic Press, NY, 1982.

[35] D. Bertsekas. A note on error bounds for convex and nonconvex programs. *Computational Optimization and Applications*, **12**, 1999, 41–51.

[36] D. Bertsekas. *Nonlinear Programming*. Athena Scientific, Belmont, MA, 1999.

[37] J.F. Bonnans and A. Shapiro. *Perturbation Analysis of Optimization Problems*. Springer-Verlag, New York, 2000.

[38] J.M. Borwein and A.S. Lewis. *Convex Analysis and Nonlinear Optimization: Theory and Examples*. Springer-Verlag, New York, 2000.

[39] N. Bourbaki. *Espace Vectoriels Topologiques*. Herman, Paris, 1966.

[40] H. Brezis and A. Haraux. Images d'une somme d'opérateurs monotones et applications. *Israel Journal of Mathematics*, **23**, 1976, 165–186.

[41] H. Brezis. *Opérateurs Maximaux Monotones et Semigroupes de Contractions dans les espaces de Hilbert*. North-Holland, Amsterdam, 1973.

[42] H. Brezis and S. Niremberg. Characterization of ranges of some nonlinear operators and applications to boundary value problems. *Ann. Scuola. Norm. Sup. Pisa. Cl. Sci.*, **5**, 1978, 225–236.

[43] F. Browder. Nonlinear maximal monotone operators. *Math. Annalen*, **175**, 1968, 89–113.

[44] F. Browder. On the range of the sum of nonlinear operators and Landesman–Lazer principle. *Bulletin Un. Math. Ital.*, **B5**, 1979, 305–314.

[45] J.V. Burke and P. Tseng. A unified analysis of Hoffman's bound via Fenchel duality. *SIAM J. Optimization*, **6**, 1996, 265–282.

[46] G. Buttazo and F. Tomarelli. Nonlinear Neumann problems. *Advances in Mathematics*, **89**, 126–142, 1991.

[47] G. Choquet. Ensembles et cones faiblement complets. *C.R. Académie des Sciences*, Serie A, **254**, 1962, 1908–1910.

[48] C.C. Chou, K.F. Ng, and J.-S. Pang. Minimizing and stationary sequences for optimization problems. *SIAM J. Control and Optimization*, **36**, 1998, 1908–1936.

[49] R. Cominetti. Some remarks in convex duality in normed spaces without compactness. *Control and Cybernetics*, **23**, 1994, 123–137.

[50] J.N. Corvellec. A note on coercivity of lower semi-continuous functions and nonsmooth critical point theory. *Serdica Math. Journal*, **22** 1996, 57–68.

[51] D.G. Costa and E.A. Silva. The Palais Smale condition versus coercivity. *Nonlinear Analysis, Theory, Methods and Applications*, **16**, 1991, 371–381.

[52] R.W. Cottle, F. Gianessi, and J.-L. Lions. *Variational Inequalities and Complementary Problems*. Wiley, New-York, 1980.

[53] R.W. Cottle, J.-S. Pang, and R.E. Stone. *The Linear Complementary Problem*. Academic Press, Boston, 1992.

[54] P. Coutat. More on continuous closed convex sets. *Optimization*, **37**, 1996, 99–112.

[55] C. Davis. All convex invariant functions of Hermitian matrices. *Arkiv der Mathematik*, **8**, 1957, 276–278.

[56] G. Debreu. *Theory of Value*. Yale University Press, New Haven, 1975.

[57] J.P. Dedieu. Critères de fermeture pour l'image d'un fermé non convexe par une multiapplication. *Compte Rendus de L'Académie des Sciences*, Serie A, **287**, 1978, 941–943.

[58] J.P. Dedieu. Cone asymptotique d'un ensemble non convexe. Application à l'optimisation. *C.R. Académie des Sciences*, **287**, 1977, 91–103.

[59] J.P. Dedieu. Cones asymptotiques d' ensembles non convexes. *Bulletin Société Mathématiques de France*, Analyse non convexe, Mémoire 60, 1979, 31–44.

[60] J. Dieudonné. Sur la séparation des ensembles convexes. *Mathematische Annalen*, **163**, 1966, 1–3.

[61] V.F. Demyanov and N. Malozemov. The theory of nonlinear minimal problems. *Russian Mathematical Surveys*, **26**, 1971, 55–115.

[62] S. Deng. Computational error bounds for convex inequality systems in reflexive Banach spaces. *SIAM J. Optimization*, **7**, 1997, 274–279.

[63] A.L. Dontchev and T. Zolezzi. *Well-Posed Optimization Problems*. Lecture Notes in Mathematics, **1543**, Springer New York, 1993.

[64] A.Y. Dubovitzkii and A.A. Miliutin. Extremum problems in the presence of restrictions. *USSR Computational Mathematics and Mathematical Physics*, **5**, 1965, 1–80.

[65] H.G. Eggleston. *Convexity*. Cambridge Tracts, **47**, Cambridge University Press, 1958.

[66] I. Ekeland. On the variational principle. *Journal of Mathematical Analysis and Applications*, **47**, 1974, 324–353.

[67] I. Ekeland and R. Temam. *Convex Analysis and Variational Problems*. Elsevier Publishing Company, New York, 1976.

[68] J. Eckstein and M. Ferris. Smooth methods of multipliers for complementarity problems. *Mathematical Programming*, **86**, 1999, 65–90.

[69] K. Fan. Minimax theorems. *Proceedings of the National Academy of Sciences*, **39**, 1953, 42–47.

[70] A.V. Fiacco and G.P. McCormick. *Nonlinear Programming: Sequential Unconstrained Minimization Techniques*. Classics in Applied Mathematics,

[71] W. Fenchel. On conjugate functions. *Canadian J. of Mathematics*, **1**, 1949, 73–77.

[72] W. Fenchel. *Convex cones, sets and functions*. Mimeographed Notes. Princeton University, 1951.

[73] M. Frank and P. Wolfe. An algorithm for quadratic programming. *Naval Research Logistics Quaterly*, **3**, 1956, 95–110.

[74] D. Gabay. Applications of the method of multipliers to variational inequalities. In *Augmented Lagrangian Methods: Applications to the solution of boundary value problems*, Editors M. Fortain and R. Glowinski, North Holland, Amsterdam 1983.

[75] D. Gale and V. Klee. Continuous convex sets. *Mathematica Scandia*, **7**, 1959, 379–391.

[76] D. Gale, H.W. Kuhn, and A.W. Tucker. Linear programming and the theory of games. In *Activity Analysis of Production and Allocation*, edited by T.C. Koopmans, Wiley New York 1951.

[77] M. Goberna and A. Lopez. Optimization value function in semi-infinite programming. *Journal of Optimization Theory and Applications*, **59**, 1998, 261–279.

[78] O. Guler, A.J. Hoffman, and U. Rothblum. Approximations to systems of linear inequalities. *SIAM Journal of Matrix Analysis and Applications*, **16**, 1995, 688–696.

[79] A.J. Hoffman. On approximate solutions of systems of linear inequalities. *J. Research National Bureau of Standards*, **49**, 1952, 263–265.

[80] L. Hormander. Sur la fonction d'appui des ensembles convexes. *Arkiv fur Matematik*, **3**, 1954, 181–186.

[81] D. Klatte and G. Thiere. Error bounds for solutions of linear equations and inequalities. *ZOR-Mathematical Methods of Operations Research*, **41**, 1995, 191–214.

[82] D. Klatte and W. Li. Asymptotic constraint qualifications and global error bounds for convex inequalities. *Mathematical Programming*, **84**, Serie A, 1999, 137–160.

[83] H.W. Kuhn. Nonlinear programming: a historical view. In *Nonlinear Programming*, edited by R.W. Cottle and C.E. Lemke, SIAM-AMS Proceedings, American Mathematical Society, Providence, 1976.

[84] B. Kummer. Stability and weak duality in convex programming without regularity. Preprint. Humboldt Univiversity, Berlin 1978.

[85] P.J. Laurent. *Approximation and Optimization*. Herman Editions, Paris, 1972.

[86] A.S. Lewis and J.-S. Pang. Error bounds for convex inequality systems. Proceedings of the fifth International Symposium of Generalized Convexity, Luminy, 1996.

[87] A.S. Lewis. Convex Analysis on the Hermitian matrices. *SIAM J. Optimization*, **6**, 1996, 164–177.

[88] W. Li. The sharp Lipschitz constant for feasible and optimal solutions of a perturbed linear program. **187**, *Linear Algebra and Applications*, 1993, 15–40.

[89] W. Li and I. Singer. Global error bounds for convex multifunctions and applications. *Mathematics of Operations Research*, **23**, 1998, 443–462.

[90] P.L. Lions. Two remarks on the convergence of convex functions and monotone operators. *Nonlinear Analysis*, **2**, 1978, 553–562.

[91] J.-L. Lions and G. Stampacchia. Variational Inequalities. *Communication in Pure and Applied Mathematics*, **20**, 1967, 493–519.

[92] X.D. Luo and Z.Q. Luo. Extension of Hoffman's error bound to polynomial systems. *SIAM J. Optimization*, **4**, 1994, 382–392.

[93] Z.Q. Luo and J.-S. Pang. Error bounds for analytic systems and their applications. *Mathematical Programming*, **67**, 1994, 1–25.

[94] Z.Q. Luo and P. Tseng, "Perturbation analysis of a condition number for linear systems. *SIAM J. Matrix Analysis and Applications*, **15**, 1994, 636–660.

[95] Z.Q. Luo. On the solution set continuity of the convex quadratic feasibility problem. Department of Electrical and Computer Engineering, McMaster University, Hamilton, Ontario, Canada, September 1995.

[96] Z.Q. Luo and S. Zhang. On the extension of Frank–Wolfe theorem. *Comput. Optim. Appl.*, **13**, 1999, 87–110.

[97] O.L. Mangasarian. Condition number for linear inequalities and equalities. *Methods of Operations Research*, **43**, 1981, 3–15.

[98] O.L. Mangasarian. A condition number for convex differentiable inequalities. *Mathematics of Operations Research*, **10**, 1985, 175–189.

[99] O.L. Mangassarian. Error bounds for nondifferentiable convex inequalities under a strong Slater constraint qualification. *Mathematical Programming* **83**, 1998, 187–194.

[100] D. McFadden. Convex analysis. In *Production Economics: A dual approach to theory and applications*, Volume 1, eds. M. Fuss and D. McFadden. 1979.

[101] L. McLinden. Dual operations on saddle functions. *Transactions of the American Mathematical Society*, **179**, 1973, 363–381.

[102] L. McLinden. An extension of Fenchel's duality theorem to saddle functions and dual minimax problems. *Pacific Journal of Mathematics*, **50**, 1974, 135–150.

[103] J.J. Moreau. *Fonctionelles Convexes*. Séminaire sur les équations aux dérivées partielles. Collège de France, Paris, 1966.

[104] G. Minty. On the maximal domain of a monotone function. *The Michigan Mathematical Journal*, **8**, 1961, 135–137.

[105] U. Mosco. Dual variational inequalities. *J. of Mathematical Analysis and Applications*, **40**, 1972, 202–206.

[106] J.-S. Pang. Error bounds in mathematical programming. *Mathematical Programming*, **79**, Serie B, 1997, 299–332.

[107] J.-P. Penot. Well posedness and nonsmooth analysis. *Pliska Stus. Math. Bulgar.*, **12**, 1998, 141–190.

[108] J.-P. Penot. Non coercive problems and asymptotic conditions. Preprint 2001, Université de Pau, France.

[109] A.F. Perold. A generalization of Frank–Wolfe theorem. *Mathematical Programming*, **18**, 1980, 215–227.

[110] J. Renegar. Incorporating condition measures in the complexity theory of linear programming. *SIAM Journal on Optimization*, **5**, 1995, 506–524.

[111] S.M. Robinson. Bounds for error in the solution set of a perturbed linear system. *Linear Algebra and Applications*, **6**, 1973, 69–81.

[112] S.M. Robinson. An application of error bounds for convex programming in a linear space. *SIAM Journal of Control and Optimization*, **13**, 1975, 271–273.

[113] S.M. Robinson. Regularity and Stability for convex multivalued functions. *Mathematics of Operations Research*, **1**, 1976, 130–143.

[114] S.M. Robinson. Generalized equations and their solutions: parts I, II. *Mathematical Programming Studies*, **10**, **19** 1977; 1982, 128–141; 200–221.

[115] S.M. Robinson. Composition duality and maximal monotonicity. *Mathematical Programming*, *85*, Serie A, 1999, 1–13.

[116] R.T. Rockafellar. Minimax theorems and conjugate saddle functions. *Math. Scand*, **14**, 1964, 151–173.

[117] R.T. Rockafellar. Level sets and continuity of conjugate convex functions. *Transactions of the American Mathematical Society*, **123**, (1966), 46–63.

[118] R.T. Rockafellar. On the maximality of sums of nonlinear monotone operators. *Transactions of the American Mathematical Society*, **149**, (1970), 75–88.

[119] R.T. Rockafellar. *Convex Analysis*. Princeton University Press, Princeton, New Jersey, 1970.

[120] R.T. Rockafellar. Ordinary convex programs without a duality gap. *Journal of Optimization Theory and Applications*, **7**, 1971, 143–148.

[121] R.T. Rockafellar. *Conjugate Duality and Optimization*. Conference Board of Mathematical Sciences Series, Volume 16, SIAM Publications, Philadelphia, 1974.

[122] R.T. Rockafellar. Monotone operators and augmented Lagrangians in nonlinear programming. In O.L. Mangasarian et al., eds, *Nonlinear Programming 3*, Academic Press, New York, 1978, 1–25.

[123] R.T. Rockafellar and R.J.B Wets. *Variational Analysis*. Springer-Verlag, New York, 1998.

[124] A. Seeger. Convex analysis of spectrally defined matrix functions. *SIAM J. Optimization*, **7**, 1997, 679–696.

[125] M. Slater. Lagrange multipliers revisited: a contribution to nonlinear programming. Cowles Commission Discussion Paper, Math 403, 1950.

[126] E. Steinitz. Bedingt konvergente Reihen und konvexe Systeme, I, II, III, *Journal der Reine und Angewandte Mathematik*, **143,144,146**, 1913–1916, 128–175; 1–40; 1–52.

[127] J.J. Stoker. Unbounded convex sets. *American Journal of Mathematics*, **62**, 1940, 165–179.

[128] M. Valadier. Integration de convexes fermés, notamment d'épigraphes. Inf-convolution continue. *Revue d'Informatique et de Recherche Operationelle*, 1970, 57–73.

[129] F.A. Valentine. *Convex Sets*. McGraw Hill, 1964.

[130] D. Walkups and R.J.B. Wets. A Lipschitzian characterization of convex polyhedra. *Proceedings of the American Mathematical Society*, **23**, 1969, 167–173.

[131] L. Vandenberghe and S. Boyd. Semidefinite Programming. *SIAM Review*, **38**, 1996, 49–95.

[132] J. von Neumann. Zur theorie der Gesellschaftsspiele. *Mathematishe Annalen*, **100**, 1928, 295–320.

[133] J. von Neumann and O. Morgenstern. *The Theory of Games and Economic Behavior*. Princeton Univeristy Press, Princeton, N.J., 1948.

[134] H. Wolkowicz, R. Saigal, and L. Vandenberghe (editors). *Handbook of semidefinite programming*. Kluwer Academic Press Publishers, Boston–Dordrecht–London, 2000.

[135] C. Zalinescu. Recession cones and asymptotically compact sets. *Journal of Optimization Theory and Applications*, **77**, 1993, 209–220.

Index of Notation

$\mathbb{B}$: closed unit ball
$\mathbb{R}$: the real numbers
$\overline{\mathbb{R}}$: the extended real numbers
$\mathbb{R}^n$: Euclidean n-dimensional space
$\|x\|$: the Euclidean norm of x
$(x)_+$: positive part of x
$\mathbb{R}^n_+$: the nonnegative orthant
$\mathbb{R}^n_{++}$: the positive orthant
$\mathbb{N}$: the natural numbers
aff C: affine hull of set C
ri C: relative interior of set C
int C: interior of set C
cl C: closure of set C
bd C: boundary of set C
conv C: convex hull of set C
conv f: convex hull of function f
Δ_n: unit simplex in $\mathbb{R}^n$
dom f: domain of function f
ext C: extreme points of set C
extray: set of extreme rays
C_∞: asymptotic cone
f_∞: asymptotic function
$\mathcal{C}_f$: constancy space
$\mathcal{K}_f$: cone of aymptotic directions
L_f: lineality space

$U(x)$: neighborhood of x
$\operatorname{dist}(x, C)$: distance from C
pos C: positive hull
epi f: epigraph of function f
$\operatorname{lev}(f, \lambda)$: lower level set
σ_C: support function of C
$N_C(x)$: normal cone
$T_C(x)$: tangent cone
$\partial f(x)$: subdifferential set
ker A: kernel of map A
rge A: range of map A
rank A: rank of A
gph S: graph of S
$\nabla f(x)$: gradient of f
$\nabla^2 f(x)$: Hessian of f
$f'(\cdot; d)$: directional derivative
f^*, f^{**}: conjugate, biconjugate
K^*: polar cone
$M^\perp$: orthogonal complement
γ_C: gauge function
δ_C: indicator function
$f \square g$: infimal convolution
S_n: $n \times n$ symmetric matrices
tr A: trace of matrix A
$\lambda(A)$: eigenvalue of matrix A

$\mathrm{diag}(x)$: diagonal matrix
P_C: projection map onto C
$\mathrm{Prox}(f, \lambda)$: proximal map
lsc: lower semicontinuous
als: asymptotically lower stable
awb: asymptotically well behaved
psd: positive semidefinite

Index